Attracting Birds in the Texas Hill Country

MYRNA AND
DAVID K. LANGFORD
BOOKS ON
WORKING LANDS

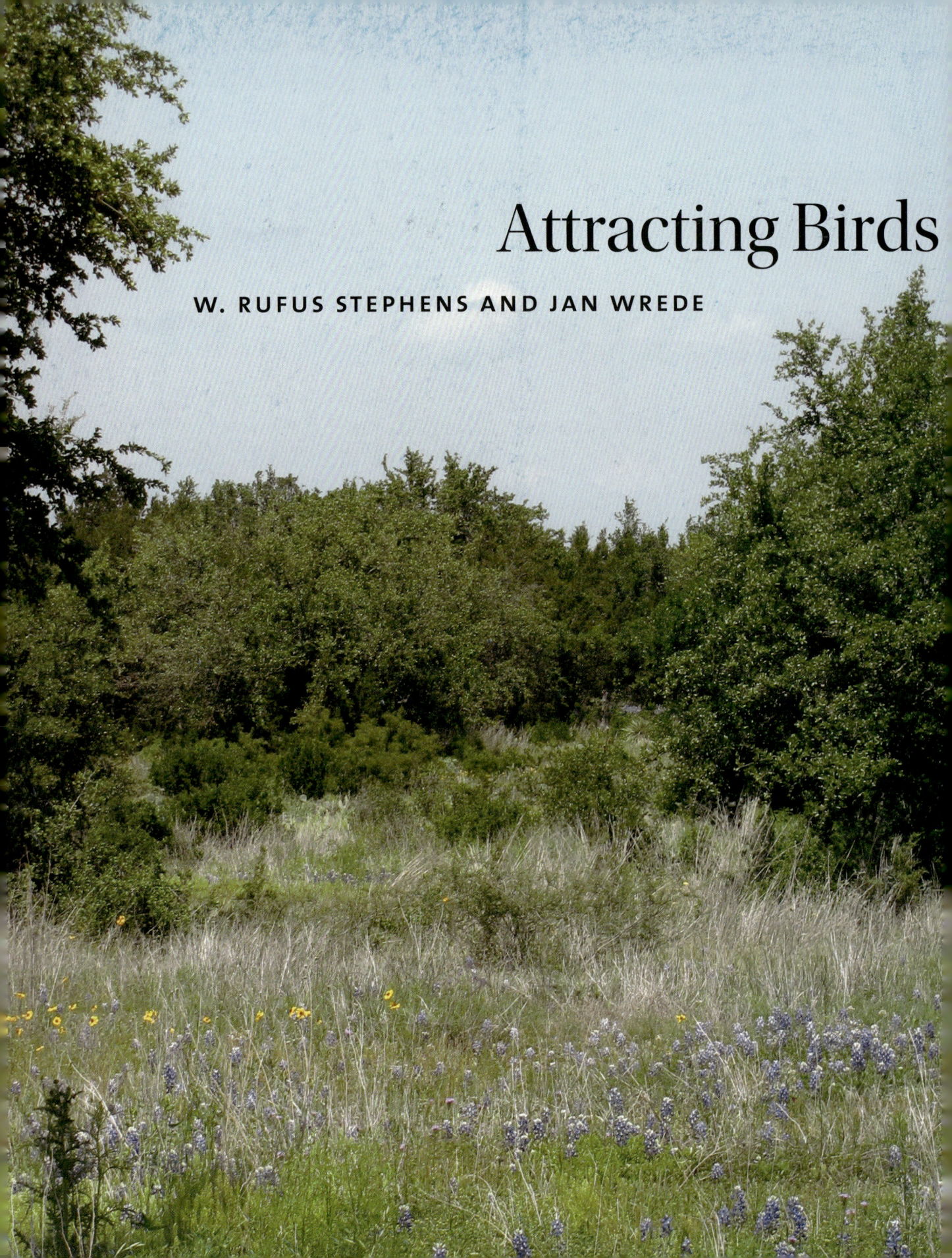

in the Texas Hill Country

A GUIDE TO LAND STEWARDSHIP

TEXAS A&M UNIVERSITY PRESS COLLEGE STATION

Copyright © 2016 by W. Rufus Stephens and Jan Wrede
All rights reserved
First edition

This paper meets the requirements of
ANSI/NISO Z39.48-1992 (Permanence of Paper).
Binding materials have been chosen for durability.

Manufactured in China by Everbest Printing Co., through FCI Print Group

Library of Congress Cataloging-in-Publication Data
Names: Stephens, W. Rufus, author. | Wrede, Jan, author.
Title: Attracting birds in the Texas Hill Country : a guide to land stewardship / W. Rufus Stephens and Jan Wrede.
Other titles: Myrna and David K. Langford books on working lands.
Description: First edition. | College Station : Texas A&M University Press, [2016] | Series: Myrna and David K. Langford books on working lands | Includes bibliographical references and index.
Identifiers: LCCN 2016033461| ISBN 9781623494407 (flexbound (with flps) : alk. paper) | ISBN 9781623494414 (ebook)
Subjects: LCSH: Wildlife habitat improvement—Texas—Texas Hill Country. | Wildlife habitat improvement—Texas—Texas Hill Country—Guidebooks. | Birds—Texas—Texas Hill Country—Identification. | Birds—Texas—Texas Hill Country—Guidebooks. | Texas Hill Country—Guidebooks.
Classification: LCC SK356.W54 S74 2016 | DDC 639.9/209764—dc23
LC record available at https://lccn.loc.gov/2016033461

For Jan's children, Sarah McFarlen and Joel McFarlen, and Rufus' son, W. Rufus "Ruf" Stephens Jr.; with special dedication to the memory and spirit of Ruf, lover of wild places and wild things. To the naturalist and outdoorsman that was, and the conservation professional that might have been.

Contents

Acknowledgments / *ix*

Introduction / *1*

Map of Texas Hill Country / *8*

1. **WOODED SLOPES AND SAVANNAHS** / *9*
2. **GRASSLANDS** / *73*
3. **RIVERS AND CREEKS** / *127*
4. **CANYONS, SPRINGS, AND SEEPS** / *175*
5. **CONSTRUCTED TANKS, PONDS, AND LAKES** / *230*
6. **BACKYARDS** / *284*
7. **PREDATORS AND OTHER "DANGEROUS" ANIMALS** / *355*
8. **DEER MANAGEMENT** / *393*
9. **CEDAR MANAGEMENT** / *410*

Appendix 1: Deer Browse Evaluation / *430*

Appendix 2: Master Bird Chart / *432*

Appendix 3: Priority Woody Plants / *448*

Appendix 4: Master Plant List / *450*

Appendix 5: Purple Martin Monitoring Data Sheet / *458*

Glossary / *459*

References / *463*

Index / *479*

Acknowledgments

We thank our loving spouses, steadfast friends, and generous colleagues at the Texas Parks and Wildlife, Cibolo Nature Center, Native Plant Society of Texas, Natural Resource Conservation Service, Agrilife, Texas Nature Conservancy, and Texas Forest Service, who encouraged and helped us with this book over and over again. Their assistance and advice were invaluable.

Anne Adams and Patsy Inglet supported the writing of this book in far too many ways to mention. Anne's work on our plant lists and Patsy's help with the references were especially important. Whenever we asked for something they always said yes with a smile. The talented Christine Kolbe created all of our diagrams and Ben Eldredge provided the graphic for NO symbols.

We are indebted to many photographers, especially Lora Render and Jeff Forman, who took on the bulk of our bird photography challenges. Our other generous photographers are Grady Allen, Tom Collins, Bart Drees, Ben Eldredge, Jonah Evans, Chase Fountain, Patsy Inglet, Tyra Cox Kane, Bill Kennon, Barbara Kilpper, Ann Mallard, Tom Mast, Kory Perlichek, Tom Rust, Dale Schmidt, and Scott Swearingen.

Our early readers Anne Adams, Patsy Inglet, Robin Stauber, and Coco Brennan made valuable suggestions as we developed our ideas and presentation. Peggy Sankey did early readings and was an invaluable proofreader. Dusty and Norma Bruns, Tyra Cox Kane, Bill Kennon, Barbara Kilpper, Dick Park, and Sue Tracy gave us their personal land stewardship stories. Thank you all.

We are grateful to David and Myrna Langford for supporting publication of this book, to Shannon Davies, TAMU Press director and our advocate and mentor through the process, to Patricia Clabaugh, our project editor, and to Cynthia Lindlof, copyeditor extraordinaire. Thank you, thank you thank you, thank you.

Finally we wish to acknowledge our workshop participants, who asked for more information on bird habitat conservation in the Texas Hill Country. Their questions made us understand the need for this resource book.

Attracting Birds in the Texas Hill Country

Introduction

We wrote this book after years of presenting three-day workshops for landowners, land managers, tax assessors, and others interested in Texas' Wildlife Tax Valuation (WTV). Our focus was on land stewardship for birds because they are charismatic species that attract many landowners to wildlife stewardship. Birds also represent the diversity of Hill Country habitats and can be helped by habitat improvements on small properties as well as large. In addition, for the most part when birds flourish, other wildlife does too.

Following all of these workshops, participants asked for more information and reference books to help them carry out the sustainable wildlife stewardship practices advocated and to qualify for the Wildlife Tax Valuation. We realized that no practical handbook was out there, and we had to write it ourselves. The book took ten years to complete and grew into a comprehensive guidebook. We had time to learn from plenty of good advice and to develop a layout with easy access to information on specific habitats, bird species, and common questions. Attracting Birds in the Texas Hill Country will help readers understand bird habitat requirements, use this information to create a Wildlife Tax Valuation plan, and become more effective land stewards. We hope it will be useful for both rank amateurs and skilled professionals.

WILDLIFE TAX VALUATION

The Texas conservation tax incentive is not a wildlife exemption. It is a special agricultural tax appraisal based on wildlife management. This unique Texas law began in 1995 when Texas voters approved a constitutional amendment known as Proposition 11, which expanded the standard agriculture tax valuation to include wildlife management. The Wildlife Tax Valuation has become especially important in the Hill Country where cattle and goat ranching is decreasing and wildlife management is becoming widespread due to changes in land ownership.

In general, it is most common to qualify for a WTV after land has been in a traditional agricultural valuation for at least 5 years. In the Hill Country, there are minimum acreage requirements that range from 12.5 to 50 acres (depending on county), for a stand-alone property and on the lower end of that range for land that is endangered species habitat or part of a wildlife management cooperative.

WTV is a conservation law that uses 3 principles to protect agricultural land and wildlife habitat.

1. Revenue neutral: To prevent removing essential financial support from counties and schools, the only land eligible for WTV is that which currently have a 1-D-1 agricultural tax valuation.
2. Active management: Landowners and wildlife managers must work actively to meet the legal requirements.
3. Any native wildlife species: All native wild species (including birds) are important to the intricate and dynamic balance of nature. Thus, tax appraisal based on wildlife management is not limited to a select group of animals.

Texas Parks and Wildlife Department (TPWD) has established the wildlife management practices and activities that can be used to qualify for WTV. And wildlife management for birds is especially well suited for the many smaller properties (less than 1000 acres) common in the Hill Country. To satisfy the rules, a landowner must use at least three of the following seven management practices annually: erosion control, habitat improvement, supplemental food, supplemental water, supplemental shelter, predator control and census.

To truly benefit birds and qualify for a WTV, effective land stewards select practices that provide the greatest advantage to bird species that can live in their property's natural habitats. These decisions are based on habitat assessment and habitat management is done in the context of long-term planning. Our book fits perfectly into the WTV process because it provides the "nuts and bolts" on what can be done to help birds and qualify for a WTV in specific situations. As a comprehensive guide to habitat assessment, identification of birds and the habitats they use, plus stewardship practices that will benefit these birds, Attracting Bird in the Texas Hill Country is the ideal "how to manual" for writing an effective WTV plan.

For more information on the rules, guidelines and application forms, go to the TPWD webpage titled Agricultural Tax Appraisal Based on Wildlife Management.

Literature research for the book has been laborious and as thorough as possible given the broad range of topics and limited field research available in many areas. Much more funding is needed for projects to gather information that will enhance conservation of birds not yet endangered. Throughout our work we were alarmed at the number of bird species whose populations were reported in decline—either within the range of this book or in related areas. And we have become more and more determined to help land managers, landowners, and other decision makers do what is necessary to provide healthy habitat. Each bird species has its own special requirements for cover, food, water, nesting, and rearing young. All of these need to be understood and protected. We urge you to act now to conserve a valuable environmental asset before it slips away.

PREPARE YOURSELF TO BE YOUR OWN BEST ADVISER

As land stewards we work within a context of time and place. We need to be present and aware, well informed, and imaginative. Your property is more than just the land. It is the moisture and rock that lie beneath the surface. It is the soil and all the creatures, both plant and animal, living in the soil. It is every bit of vegetation, both living and in decay, including grasses, weeds, wildflowers, lichens, mosses, vines, shrubs, and trees. It is every insect, reptile, amphibian, mammal, and bird living there. Your land comes with its history and its neighbors. You are the present steward and have responsibility for its future.

To be a good steward of your land, you must observe and study to know what you have. You must pay attention not just today but every day. To be a good land steward for birds, you must know that all land stewardship is conducted by habitat because a bird's habitat is where it lives—where it finds essential food, water, shelter, and space to survive. To be a good land steward, you must be able to read the habitats on your land by their plant communities, which are the basis of habitat and control which birds may be present. The condition of your plant communities determines whether you have these birds or not.

Land stewardship for birds, like ranching and farming, includes many elements out of your control: climate, rainfall, base rock, seasons, and past land use. Many other elements may or may not be in your control. What comes down the creek and what washes off higher land depends on weather and on decisions made by your upstream neighbors. White-tailed deer and big exotic animals move over large areas, and unless you have a large property (as much as 500 acres or more), your influence on them is limited to when they are moving through your land. Migratory birds travel over even greater distances. Again, your impact occurs only when they are present.

You can control what you do—or perhaps not entirely. If you are part of a couple or a family that owns your property, all of you influence land stewardship decisions. Communication and compromise are essential.

GETTING STARTED

Here are some important background suggestions for you to think about when using this book. These are starting places enormously helpful in managing your land to attract the bird species that live in healthy habitats. These suggestions are the first steps to success.

Surveys

To learn what you have, map the habitats on your property and survey and list the most common plants in each habitat. Pay special attention to the different kinds of trees, shrubs, vines, and grasses. Make a list of all the wildlife known to live in each of the habitats. In addition, as you set your land management goals and use this book to reach your goals, monitoring is an extremely important tool to reveal your successes and failures.

Goals and Objectives

To develop your skills as a good land steward, you must know what you have but also what you want and how to achieve it. Thus, it is necessary to set land management goals and objectives for your property. Together, they tell you where you are going with your land management and the specific steps to get you there.

Goals are the big picture of what you want your property to look like. You can think of your land stewardship goals as the outcomes or products of all the management work you plan on doing. For example, if you have a wooded slope invaded by cedar, you might want to have a diverse and abundant mix of native trees, shrubs, and grasses in a mosaic to provide for a healthy population of native birds. If you have an old field dominated by mesquite and KR bluestem, you might want it to be a pocket prairie with native grasses and forbs to attract grassland birds.

Land management objectives are the steps necessary to get you to your goals. Useful land management objectives are specific and measurable: for example, divide the work into annual projects and (1) control regrowth cedar on 53 acres over the next five years and (2) install and maintain one guzzler on the northwest corner of the property in 2015.

Monitor to Determine Success

Monitoring is essential to clarify what you have accomplished and to know whether or not you have achieved a desired result. Monitoring might be measuring recovery of vegetation or the change in numbers of a plant or animal. Record your monitoring with photographs or information on data sheets. A well-designed data sheet makes it easy to compare your land before and after making changes. Following are examples of what to monitor:

1. Numbers and species of plants growing inside and outside a deer exclosure
2. Eggs, nestlings, and fledglings produced in a set of nest boxes
3. Browsing pressure on woody plants in a pasture
4. Birds present before, 1 year, 2 years, 5 years, and 10 years after cedar removal from an old field

CHAPTER ORGANIZATION

We discuss birds, bird habitat, and related conservation or stewardship issues in six central chapters organized by common Hill Country habitat types: (1) wooded slopes and savannahs; (2) grasslands; (3) rivers and creeks; (4) canyons, springs, and seeps; (5) constructed tanks, ponds, and lakes; and (6) backyards.

Each chapter begins with a list of priority, favorite, and under-the-radar birds featured in bird summaries at the end of the chapter. Priority birds may be endangered, threatened, rare, or reported to be declining. Many agencies and organizations compile lists of birds needing conservation attention. If we found a bird on any list, we considered it a priority bird because with so little research done on these birds, we chose to be conservative.

Chapters 7, 8, and 9 cover special topics that are so common and complex that they require extra attention: (7) predators and other "dangerous" animals; (8) deer management; and (9) cedar management.

Structure in Chapters 1–6

- Introduction states background information for understanding the special issues in that chapter.
- Description covers the habitat types in that chapter.
- Problems common to each habitat type are the main part of most chapters. This section includes a Problems Summary followed by an in-depth discussion of each problem with suggestions on what you can do to solve it.
- Bird summaries are placed at the end of each chapter to feature birds selected as representatives of Texas Hill Country habitats. They illustrate the birds to be expected in healthy habitats covered in the chapter and include key land stewardship information for the featured bird.

Remember Boxes
These boxes throughout the book bring attention to points that are so important that they require repetition.

Easy Access to Information
This is a handbook with many access points for answering your questions, for example:

> *How do I attract Painted Buntings to my property?*
> Look at the Painted Bunting bird summary in chapter 1, and read about its habitat, cover, nest, and food requirements.

> *What does Black-capped Vireo habitat look like?*
> Go to chapter 1 where the Black-capped Vireo is featured and look at the large photograph of its habitat. Each bird photo is placed with a habitat where it can be found.

> *Can we make an old, abandoned field into good winter bird habitat?*
> Start with problem 7 in chapter 2.

> *What is wrong with our creek, and how do we make it better for birds?*
> Look at table 3.5, "Summary of habitat problems at rivers and creeks," and read the causes and "What You Can Do" sections for each problem in chapter 3.

Use of Bird Summaries
These summaries give quick access to key information on how to manage land for that bird and include photographs of the bird. The description is placed in a particular chapter because that bird is expected to live in habitats covered there. For example, the Golden-cheeked Warbler and the Golden-fronted Woodpecker are described in chapter 1 because wooded slopes and savannahs are their natural habitats; American Widgeon and Green Kingfisher are in chapter 3 because they live on water and in riparian habitat. However, a bird may also use habitats covered in other chapters, which are listed in its summary.

The bird summary tells the bird's status, such as priority, favorite, under the radar, or outlaw; seasons of the year the bird is present in the Hill Country; and useful facts about the bird. It also includes the following information:

- **Identification:** a brief description of the bird's physical characteristics
- **Habitats:** a list of habitat types where the bird can be found
- **Cover:** plant structure used by this bird
- **Breeding Territory:** the area one nesting pair is likely to use, which tells you the number of breeding pairs you can expect on your property
- **Nest:** where the bird nests and what the nest might look like
- **Food:** key foods used by the bird during the season it is present in the Hill Country
- **Stewardship:** goals for its habitat and land management practices used to support this bird

In this book, the Texas Hill Country includes the Balcones Canyonland, Edwards Plateau Woodland, and Llano Uplift in all or part of 28 counties. See the map for an outline of the greater Hill Country. However, the major problems impacting bird habitat—fire suppression, overgrazing, and deer overpopulation—exist in many other areas. Thus, the land management advice given here applies to many other counties.

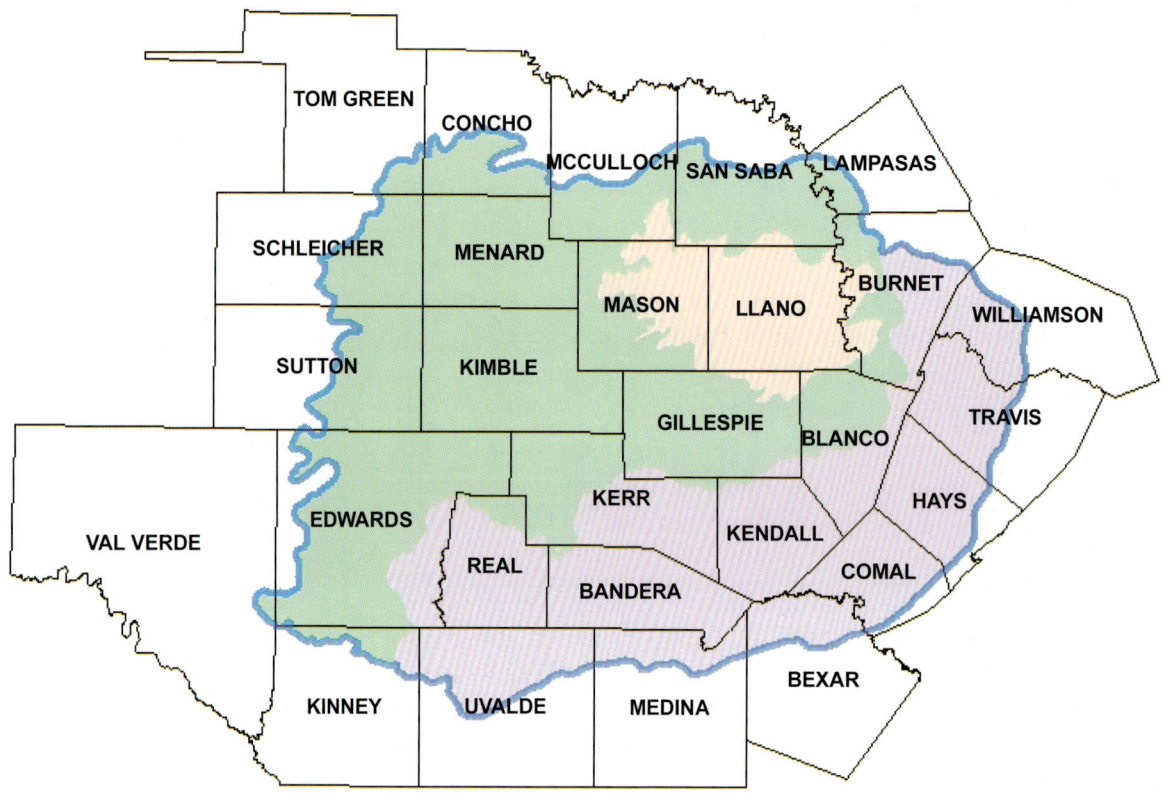

Greater Texas Hill Country

- Area Covered by this Book
- Balcones Canyonlands
- Edwards Plateau Woodland
- Llano Uplift

This book is a guide to habitat management for birds in the Greater Texas Hill Country. As shown on the map, this area includes all or part of 28 counties.

Adapted from Level III and IV Ecoregions of the Conterminous United States, 2009 from Omernik, J. M., US Environmental Protection Agency

1 Wooded Slopes and Savannahs

Birds are indicators of the environment. If they are in trouble, we know we'll soon be in trouble. — Roger Tory Peterson

INTRODUCTION

This chapter covers woodlands other than those along streams. Here we discuss wooded habitat on high ground or upland sites, the wooded or partially wooded plant communities most widespread in the Texas Hill Country. Different types of these communities are influenced by slope, aspect, soil type and depth, and underlying geology. We have classified wooded habitat of the Texas Hill Country into categories:

- Mixed wooded slope
- Moist mixed wooded slope
- Dry mixed wooded slope
- Post oak savannah
- Shin oak savannah
- Live oak savannah

Healthy mixed wooded slopes with closed canopy are Golden-cheeked Warbler habitat. Savannahs are open land dotted with trees or clumps of trees. However, cedar (Ashe juniper) has become so widespread in the Hill Country that it often masks some types of woodlands and savannahs. Mixed wooded slope habitat may be the most extensive woodland of the Texas Hill Country. It covers many hillsides and is a mix of deciduous and evergreen trees with a mostly closed canopy. All wooded upland sites have shallow soil at best.

Post oak savannah, shin oak savannah, and live oak savannah are usually on flat to rolling land with vegetative ground cover ranging from medium to short grasses mixed with forbs. Post oak savannah has stands of tall, full-sized trees. Shin oak savannah and live oak savannah are dotted with mottes having a mixture of trees, shrubs, and vines. Tree species vary from place to place depending on geologic features such as slope and aspect. The trees in shin oak and live oak savannah mottes are stunted when they grow on very shallow soil.

FEATURED BIRDS OF WOODED SLOPES AND SAVANNAHS

PRIORITY
American Kestrel
Black-capped Vireo
Cactus Wren
Canyon Towhee
Carolina Chickadee
Golden-cheeked Warbler
Ladder-backed Woodpecker
Loggerhead Shrike
Northern Bobwhite
Painted Bunting
Rufous-crowned Sparrow

UNDER-THE-RADAR
Ash-throated Flycatcher
Blue-gray Gnatcatcher
Hutton's Vireo
Nashville Warbler
Scott's Oriole
White-crowned Sparrow

FAVORITE
Greater Roadrunner
Red-tailed Hawk
Western Scrub-Jay

OUTLAW
Brown-headed Cowbird

Note:
Priority = on at least one of the lists provided by TPWD or Oak and Prairies Joint Venture naming species in need of conservation assistance or shown by USGS-sponsored Breeding Bird Surveys in the Edwards Plateau to be a species whose population is in decline
Under-the-radar = a regular bird in Texas Hill Country but often unnoticed
Favorite = popular with just about anyone who knows the bird
Outlaw = undesirable because of its behavior and impact on other bird species

SLOPE AND ASPECT

Slope is the slant or angle of the side of a hill, which may be gradual or steep. A slope can be thought of as a solar collector, with the amount of heat gain depending on the angle of slope. A slope of 100% is a vertical cliff, and flat ground has 0% or no slope. In general, a slope with an intermediate angle of about 25%–50% collects more heat than either extremely steep or flat places. Most of Central Texas is covered with hills having intermediate slopes. Aspect is the direction a hillside faces and affects a slope's potential heat gain. Hillsides with a western or southern aspect receive more sun than those with eastern or northern aspects. Thus, a south slope gains more heat than a north slope, and a west slope gains more heat than an east slope. Together, aspect and slope determine the prevailing microclimate temperature of a specific location.

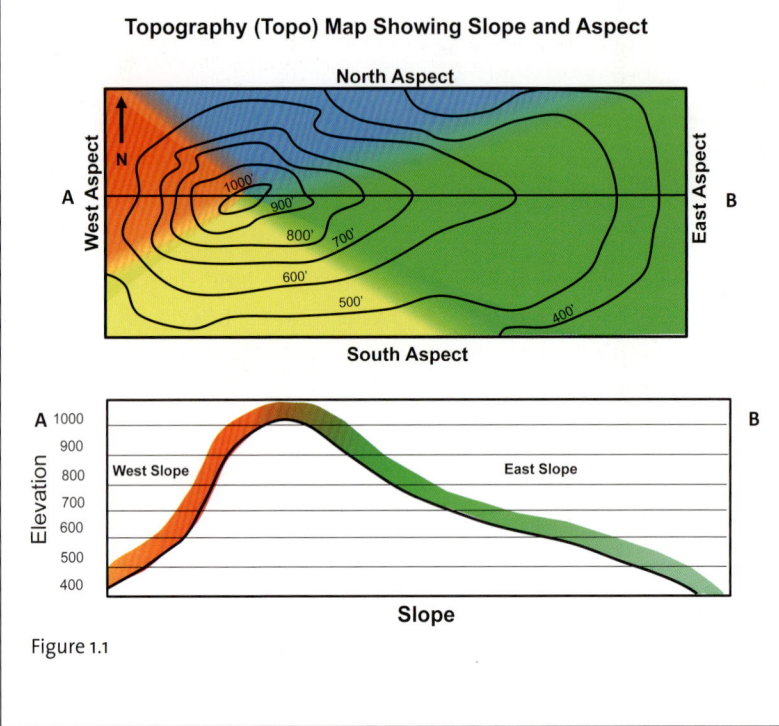

Figure 1.1

Which is most common, mixed wooded slope, post oak savannah, shin oak savannah, or live oak savannah?

Mixed wooded slope plant communities might be the most common in many places, but all Hill Country wooded plant communities are common in certain locations. On the south edge of the Hill Country along the Balcones Escarpment, mixed wooded slopes dominate the

Mixed wooded slope

Golden-cheeked Warbler
(Photo by Lora L. Render)

landscape. On flatter, more rolling portions of the central Edwards Plateau, mixed wooded slopes give way to the live oak savannah, where it may be the most common plant community. On hilltops with very shallow soil, shin oak savannah will prevail. Post oak savannahs are the least common of all Texas Hill Country woodlands. More than any of the other wooded habitat types, post oak savannah locations are determined by soil. Post oaks grow in pockets or bands of neutral to slightly acidic soil.

DESCRIPTION: MIXED WOODED SLOPES

Mixed wooded slopes are common and are dominated by deciduous and evergreen trees with a nearly closed canopy. A healthy mixed wooded

> **PLANTS USUALLY FOUND ON BOTH MOIST AND DRY MIXED WOODED SLOPES**
>
> TREES
> Cedar/Ashe juniper
> Lacey oak
> Plateau live oak
> Spanish oak
> Texas redbud
> Western soapberry
>
> SHRUBS AND VINES
> Agarita
> Aromatic sumac
> Evergreen sumac
> Flameleaf sumac
> Pink mimosa
> Poison ivy
> Texas lantana
> Texas persimmon
> Twist-leaf yucca
> Western soapberry
>
> GRASSES
> Cane bluestem/silver bluestem
> Fall switchgrass
> Green sprangletop
> Hall's panicum
> Plains bristlegrass
> Texas cupgrass
>
> FORBS
> Columbine
> Crow poison
> Texas milkweed
> White heath aster
> Wild petunia
> Yarrow

slope habitat has short to medium ground cover, a well-developed understory, and a diverse canopy with trees of various ages. But do not expect all of any mixed wooded slope in the Hill Country to look the same. At different heights of a hill you will see different plant species and tree ages. This is normal for any healthy site, but due to historical land management practices, it may be difficult to find a mixed wooded slope in healthy condition. Also, on mixed wooded slopes, the slope and aspect are important in determining prevailing temperature and moisture, which influence which plant and bird species are present.

MOIST MIXED WOODED SLOPES

These slopes face north, east, or northeast. They receive morning sunlight and are shaded during hot afternoon hours in spring, summer, and fall. In winter, north slopes are shaded almost all day and remain very cool with a low evaporation rate. Thus, in all seasons, moist mixed wooded slopes are much wetter than slopes with a southern, western, or southwestern aspect and are known for their rich woody plant communities and diverse bird populations, including some Hill Country favorites: the Black-crested Titmouse, Carolina Chickadee, and Summer Tanager.

DRY MIXED WOODED SLOPES

These slopes face south, west, or southwest. In the summer, they are especially hot and dry because they are exposed to long hours of afternoon sun and have a high evaporation rate. In the winter, they also get more direct sunlight and have a higher evaporation rate than east-, northeast-, or north-facing slopes. In the Hill Country, summer and winter are typically seasons of low rainfall. Low rainfall and a high evaporation rate produce very dry conditions. Thus, dry mixed wooded slopes are limited to plants that are heat and drought tolerant. And the birds living there, such as Cactus Wren, are adapted to using the xeric plants found on these slopes.

Moist mixed wooded slope

Carolina Chickadee
(Photo by Lora L. Render)

PLANTS USUALLY LIMITED TO MOIST MIXED WOODED SLOPES

TREES
Bigtooth maple
Blanco crabapple
Carolina buckthorn
Escarpment black cherry
Gum bumelia
Hackberries
Walnuts

SHRUBS AND VINES
Creek plum
Elbow bush
Grape
Gum bumelia
Leatherflower
Lindheimer's morning glory
Mexican buckeye
Milkweed vines
Passionflower vines
Possumhaw
Poverty bush
Rusty blackhaw
Sycamore-leaf snowbell
Virginia creeper

GRASSES
Broadleaf woodoats
Little bluestem
Purpletop
Switchgrass
Texas wintergrass

FORBS
Bundleflower
Cedar sedge
Rock daisy
Wood-sorrel

SEASONAL OCCURRENCE OF SOME BIRDS ON WELL-MANAGED MOIST MIXED WOODED SLOPES

WINTER
Brown Creeper (east)
Cedar Waxwing
Golden-crowned Kinglet
Orange-crowned Warbler
Pine Siskin
Ruby-crowned Kinglet
Spotted Towhee
White-crowned sparrow
Yellow-rumped Warbler

SPRING AND FALL (MIGRANTS)
American Redstart
Black-throated Green Warbler
Blue-headed Vireo
Clay-colored Sparrow
MacGillivray's Warbler (west)
Mourning Warbler (east)
Nashville Warbler
Ruby-throated Hummingbird
Swainson's Thrush
Tennessee Warbler
Wilson's Warbler
Yellow Warbler

SPRING AND SUMMER (BREEDING BIRDS)
Ash-throated Flycatcher
Black-chinned Hummingbird
Blue-gray Gnatcatcher
Bullock's Oriole
Eastern Wood-Pewee (east)
Golden-cheeked Warbler
Summer Tanager
White-eyed Vireo

ALL YEAR (RESIDENTS)
Bewick's Wren
Black-crested Titmouse
Canyon Towhee
Carolina Chickadee
Carolina Wren
Chipping Sparrow
Eastern Phoebe
Golden-fronted Woodpecker
Hutton's Vireo
Ladder-backed Woodpecker
Northern Cardinal
Western Scrub-Jay

east = eastern edge of Hill Country
west = western edge of Hill Country

Dry mixed wooded slope with thorn shrubs

Cactus Wren (Photo by Lora L. Render)

PLANTS USUALLY LIMITED TO DRY MIXED WOODED SLOPES

TREES
American smoke tree
Mesquite
Mexican pinyon
Post oak
Shin oak
Texas madrone

SHRUBS AND VINES
Black dalea
Catclaw acacia
Cenizo
Coyotillo
Desert sumac
Green condalia
Guayacan
Huisache
Mountain mahogany
Texas almond
Texas mountain laurel
Texas sotol
Tickle-tongue
Whitebrush

GRASSES
Curly mesquite
Hairy grama
Lindheimer rosettegrass
Red grama
Seep muhly
Sideoats grama

FORBS
Antelope horns
Damianita
Prairie paintbrush

SEASONAL OCCURRENCE OF BIRDS ON WELL-MANAGED DRY MIXED WOODED SLOPES

WINTER
American Goldfinch
Lincoln's Sparrow
Orange-crowned Warbler

SPRING AND FALL (MIGRANTS)
Clay-colored Sparrow
Swainson's Thrush
White-crowned Sparrow
Wilson's Warbler

SPRING AND SUMMER (BREEDING BIRDS)
Black-chinned Hummingbird
Blue-gray Gnatcatcher
Golden-cheeked Warbler
Summer Tanager

ALL YEAR (RESIDENTS)
Bewick's Wren
Black-crested Titmouse
Canyon Towhee
Canyon Wren
Carolina Chickadee
Chipping Sparrow
Golden-fronted Woodpecker
Hutton's Vireo
Ladder-backed Woodpecker
Western Scrub-Jay

DESCRIPTION: SAVANNAHS

In the Texas Hill Country where cedar has been controlled, the savannah often looks like a park with some live oak mottes and grass as short as a mowed lawn. This scenery is so common that we tend to think of it as normal and desirable, but it is not. Healthy savannahs in the Hill Country have trees and shrubs scattered or growing in clusters with prairie or a mixture of midsized to tall grasses and forbs in between. Savannahs might be thought of as grasslands with scattered trees or as woodlands with so few trees that the tree canopy is open and more grass grows beneath and surrounding the trees. Savannahs have plenty of light at ground level where the grasses and forbs grow. We classify Hill Country savannahs into three types: post oak savannah, shin oak savannah, and live oak savannah.

POST OAK SAVANNAH

Post oak savannahs are essentially grasslands under open woods. They have trees scattered through a healthy mixture of grasses and forbs and are ideal habitat for birds such as the Ladder-backed Woodpecker that specialize in living on large trees. Although still common, the Ladder-backed Woodpecker is on the 2002 US Fish and Wildlife Birds of Conservation Concern list for the Edwards Plateau due to habitat loss.

Post oak savannah

PLANTS FOUND ON WELL-MANAGED POST OAK SAVANNAH

TREES
Blackjack oak
Cedar elm
Gum bumelia
Hackberries
Post oak

SHRUBS AND VINES
Agarita
Balsam gourd
  (Lindheimer's globeberry)
Coral honeysuckle
Greenbrier
Gum bumelia
Texas persimmon

GRASSES
Big bluestem
Dropseed
Hairy grama
Little bluestem
Yellow indiangrass

FORBS
Bush sunflower
Engelmann daisy
Sensitive briar
Straggler daisy
Texas star
Wild petunia

Some post oak savannahs are more wooded, and others are more open. Nearly solid stands of post oak savannah occur on flat to rolling landscape with fairly deep, neutral, or slightly acidic (pH less than 7) soil. These are often referred to as Redland ecological sites, characterized by soil with neutral pH 7. In some areas you may see many more blackjack oaks than post oaks. Mixed stands of these oaks tend to grow in pockets determined by soil composition, often including granite, sand, and gravel.

Ladder-backed Woodpecker
(Photo by Lora L. Render)

SEASONAL OCCURRENCE OF SOME BIRDS ON WELL-MANAGED POST OAK SAVANNAH

WINTER	SPRING AND/OR FALL (MIGRANTS)	SPRING AND SUMMER (BREEDING BIRDS)	ALL YEAR (RESIDENTS)
American Goldfinch	American Redstart	Ash-throated Flycatcher	American Kestrel
Lincoln's Sparrow	Blue-headed Vireo	Bell's Vireo	Bewick's Wren
Northern Flicker	Clay-colored Sparrow	Black-chinned Hummingbird	Black-crested Titmouse
Orange-crowned Warbler	Nashville Warbler	Blue-gray Gnatcatcher	Carolina Chickadee
Ruby-crowned Kinglet	Ruby-throated Hummingbird	Bullock's Oriole	Carolina Wren
Spotted Towhee	Tennessee Warbler	Common Nighthawk	Cassin's Sparrow
Vesper Sparrow	Wilson's Warbler	Golden-cheeked Warbler	Chipping Sparrow
White-crowned Sparrow		Orchard Oriole	Eastern Bluebird
White-throated Sparrow		Painted Bunting	Eastern Meadowlark
		Red-eyed Vireo	Field Sparrow
		Scissor-tailed Flycatcher	Golden-fronted Woodpecker
		Western Kingbird	House Finch
			Ladder-baked Woodpecker
			Lark Sparrow
			Lesser Goldfinch
			Loggerhead Shrike
			Mourning Dove
			Rio Grande Turkey
			Vermilion Flycatcher
			Western Scrub-Jay

SHIN OAK SAVANNAH

Shin oak savannahs tend to be on rocky, shallow soil and often on high, flat hilltops. They may be atop isolated hills or on the high rolling country between two major basins. Shin oak savannah, with its low shrubs and stunted trees, is typical Greater Roadrunner habitat. Even live oaks are short on these sites because of the very shallow soil.

Shin oak mottes are dominated by shin oaks but are actually an assemblage of many woody plant species. Shin oaks produce prolific root sprouts. Typically, a healthy shin oak motte will look like a dense rounded thicket with one or two stunted but mature shin oak trees at the center. Going out to the edges are stair steps of smaller shin oak sprouts mixed with a host of vines and other woody shrubs.

Because grasses and forbs do not grow well in the deep shade of their dense growth, these mottes are almost exclusively woody plants. The mottes may grow scattered or in a continuous spread. When separated, wide grassy areas lie between the scattered mottes. Mottes growing in an extensive and continuous stand will look like a big thicket of woody plants with a little grass growing between them. Historically, fire has played an important ecological role in the maintenance of shin oak savannahs.

Shin oak savannah with short grasses

Greater Roadrunner
(Photo by Lora L. Render)

PLANTS FOUND IN WELL-MANAGED SHIN OAK SAVANNAH

TREES
Carolina buckthorn
Gum bumelia
Live oak
Shin oak
Texas redbud

SHRUBS AND VINES
Agarita
Elbow bush
Flameleaf sumac
Gum bumelia
Pearl milkweed
Plateau milkvine
Texas persimmon
Twist-leaf yucca

GRASSES
Curly mesquite
Hairy grama
Little bluestem
Sideoats grama
Slim tridens
Threeawn

FORBS
Bundleflower
Cedar sedge
Dayflower
Gayfeather
Hairyfruit chervil
Milkpea
Penstemon
Texas bush-clover
Zexmenia (wedelia)

SEASONAL OCCURRENCE OF SOME BIRDS ON WELL-MANAGED SHIN OAK SAVANNAH

WINTER
American Goldfinch
Lincoln's Sparrow
Northern Flicker
Ruby-crowned Kinglet
Spotted Towhee
Vesper Sparrow
White-crowned Sparrow

SPRING AND FALL (MIGRANTS)
American Redstart
Blue-headed Vireo
Clay-colored sparrow
Swainson's Thrush
Wilson's Warbler

SPRING AND SUMMER (BREEDING BIRDS)
Ash-throated Flycatcher
Bell's Vireo
Black-capped Vireo
Blue-gray Gnatcatcher
Common Nighthawk
Golden-cheeked Warbler
Orchard Oriole
Painted Bunting
Scott's Oriole
Yellow-billed Cuckoo

ALL YEAR (RESIDENTS)
Bewick's Wren
Black-crested Titmouse
Black-throated Sparrow
Cactus Wren
Canyon Towhee
Cassin's Sparrow
Chipping Sparrow
Eastern Meadowlark
Field Sparrow
Grasshopper Sparrow
Greater Roadrunner
Killdeer
Ladder-backed Woodpecker
Lark Sparrow
Lesser Goldfinch
Loggerhead Shrike
Mourning Dove
Northern Bobwhite
Rio Grande Turkey
Rufous-crowned Sparrow
Vermilion Flycatcher

LIVE OAK SAVANNAH

Live oak savannah is the typical landscape many picture when thinking of the Texas Hill Country. It is a nearly even mix of open grassland blanketed by spring wildflowers among clumps of live oaks. Live oak savannah is sometimes called evergreen savannah or evergreen woods because the dominant plants here appear to be evergreen. Live oaks are in fact deciduous and shed their leaves once a year in March, with new leaves emerging immediately.

Live oak mottes grow on flat to rolling landscape with moderate soil depth. Generally, the soil of live oak savannah is 0.5–2.0 feet deep. That is less than under post oak savannah (2.0 feet or deeper) and deeper than under shin oak savannah (0.5 feet or less). When healthy, live oak mottes include other woody species. Tall live oak trees form the upper canopy over a well-defined understory with some grasses thriving on the ground

Live oak savannah

Painted Bunting (Photo by Jeff Forman)

below. This common Hill Country habitat type is home to many birds, including the popular Black-chinned Hummingbird and incredibly colorful Painted Bunting.

PLANTS FOUND IN WELL-MANAGED LIVE OAK SAVANNAH

TREES	SHRUBS AND VINES	GRASSES	FORBS
Carolina buckthorn	Agarita	Big bluestem	Black samson
Cedar (Ashe juniper)	Catclaw acacia	Buffalograss	Bundleflower
Cedar elm	Elbow bush	Cane/silver bluestem	Bush sunflower
Eve's necklace	Evergreen sumac	Curly mesquite	Cedar sedge
Hackberries	Flameleaf sumac	Green sprangletop	Echinacea (purple coneflower)
Honey mesquite	Green condalia	Hairy grama	Engelmann daisy
Plateau live oak	Greenbrier	Little bluestem	Gayfeather
Texas redbud	Gum bumelia	Plains lovegrass	Greenthread
	Mustang grape	Sideoats grama	Maximilian sunflower
	Nolina	Southwestern bristlegrass	Mealy blue sage
	Pricklypear	Tall dropseed	Penstemon
	Rusty blackhaw	Tall grama	Prairie clover
	Silktassel	Texas wintergrass	Snoutbean
	Texas kidneywood	Threeawn	Zexmenia (wedelia)
	Texas persimmon	Vine mesquite	
	White honeysuckle	Wildrye	
		Yellow indiangrass	

SEASONAL OCCURRENCE OF SOME BIRDS ON WELL-MANAGED LIVE OAK SAVANNAH

WINTER	SPRING AND FALL (MIGRANTS)	SPRING AND SUMMER (BREEDING BIRDS)	ALL YEAR (RESIDENTS)
American Goldfinch	American Redstart	Ash-throated Flycatcher	American Kestrel
Lincoln's Sparrow	Blue-headed Vireo	Bell's Vireo	Bewick's Wren
Northern Flicker	Clay-colored Sparrow	Black-capped Vireo	Black-crested Titmouse
Orange-crowned Warbler	Mourning Warbler (east)	Black-chinned Hummingbird	Black-throated Sparrow
Ruby-crowned Kinglet	Nashville Warbler	Blue Grosbeak	Cactus Wren
Spotted Towhee	Ruby-throated Hummingbird	Common Nighthawk	Canyon Towhee
White-crowned Sparrow	Swainson's Thrush	Common Poorwill	Carolina Chickadee
	Tennessee Warbler	Golden-cheeked Warbler	Carolina Wren
	Wilson's Warbler	Orchard Oriole	Cassin's Sparrow
	Yellow Warbler	Painted Bunting	Chipping Sparrow
		Scissor-tailed Flycatcher	Eastern Meadowlark
		Scott's Oriole	Eastern Screech-Owl
		Yellow-billed Cuckoo	Field Sparrow
		Yellow-breasted Chat	Golden-fronted Woodpecker
			Greater Roadrunner
			Killdeer
			Ladder-backed Woodpecker
			Lark Sparrow
			Lesser Goldfinch
			Loggerhead Shrike
			Northern Bobwhite
			Rio Grande Turkey
			Rufous-crowned Sparrow
			Vermilion Flycatcher
			Western Scrub-Jay

east = eastern edge of Hill Country

WOODED SLOPE AND SAVANNAHS HABITAT MANAGEMENT GOALS

Birds need food (mast, fruit, seeds, nectar), water, nest sites, and cover from predators and bad weather. In managing this habitat for birds and other wildlife, the goal is to ensure the presence of mature seed-producing trees, shrubs, forbs, and grasses that are essential to plant diversity and healthy wildlife habitat. Wooded slope and savannah habitat is common, diverse, and home to many Hill Country birds. These include some favorites such as the Loggerhead Shrike, Greater Roadrunner, Painted Bunting, and Summer Tanager, which landowners can attract and protect through stewardship practices that restore and maintain healthy woodland and savannah habitat.

> Remember: In all Hill Country savannah and wooded slope habitats, land with an overgrazed or "parklike" appearance is unnatural and damaged. It lacks essential food and cover for birds and other wildlife.

PROBLEMS OF WOODED SLOPES AND SAVANNAHS

Problem 1: Loss of Plant Diversity from Lack of Natural Disturbances to Savannah Grasses and Forbs

Maintaining a healthy mix of forb and grass ground cover in savannah, wooded slope, and grassland habitats is important for the rodents and birds that eat these plants. A healthy plant community is also essential to birds of prey such as the Red-tailed Hawk that feed on these smaller critters. Maintenance of a healthy grassland and savannah habitat requires ecological disturbance. Historically, this natural disturbance was fire or the hooves and mouths of migratory bison. Fires converted dry plants to ash and sped up the return of nutrients to the soil. Bison ate the fresh grasses and forbs, their waste fertilized the next generation of plants, and their hooves cut and aerated the soil.

With no regular disturbance, ground cover habitat becomes too dense for many of the creatures that live on or near the ground, including birds such as sparrows, finches, and quail. In addition, without these birds or the small mammals and reptiles that also live in healthy natural ground cover, the habitat can no longer support raptors such as the Red-tailed Hawk, Swainson's Hawk, Red-shouldered Hawk, and American Kestrel.

What You Can Do: Imitate Natural Disturbances

Before the mid-1800s, bison periodically migrated through this area, and Native Americans lived off these animals. The bison herds put heavy grazing pressure on the plants but only intermittently. These are the conditions under which the native plants and habitats of our region developed for thousands of years. Today, land management practices such as prescribed burning and rotational grazing can be used to mimic natural disturbance and maintain or restore bird habitat. They are particularly effective when done over large areas and can be carried out through co-

LOGGERHEAD SHRIKE STORY

The 1967–2007 breeding bird survey for the Texas Edwards Plateau showed the Loggerhead Shrike population to be on a slightly upward trend. However, this species is declining throughout most of the rest of its range and consequently is in need of special attention, even here. During the last 20 years, the Loggerhead Shrike has lost 79% of its population in North America. And although the Loggerhead Shrike has recently been extirpated from many parts of its northern breeding range, Texas is still home to wintering birds from Canada and the Dakotas.

Researchers speculate that the problems causing the bird's decline may be in its migration and wintering grounds. Consequently, it is critical that Texas landowners manage habitat for this threatened and popular bird. Loggerhead Shrikes require three (or possibly four) habitat components:

1. Short to mid-height mixed grass/forb ground cover where they can forage for small mammals and insects
2. Low-level "hunting" perches sprinkled throughout their forage habitat
3. Shrubs or small trees for nest sites

Space for a breeding "colony" may be another critical environmental factor necessary for this bird's survival. Loggerhead Shrike breeding territory is about 20 acres/pair, and there is some evidence that breeding pairs prefer territories near other breeding pairs.

You may wonder if the shrikes that you see nesting year after year in a particular small tree are the same individuals. Research shows that Loggerhead Shrikes have low site fidelity by individual pairs but high site reuse by the species. Thus, your tree is definitely an important nest site for this species, but the birds you see nesting there year after year are probably not the same individuals. Protect that tree and provide others in similar and nearby savannah habitat.

Live oak savannah with hunting perches
Loggerhead Shrike (Photo by Lora L. Render)

Other somewhat aggressive birds such as Northern Mockingbirds and Brown Thrashers have been observed to compete successfully with shrikes for nest sites and hunting perches. Thus, it is important to manage for plenty of healthy woodland and savannah habitat having abundant understory of shrubs and small trees for all the birds that need them.

Deformed feet and beaks have been found on some Loggerhead Shrikes. These may be due to exposure to pesticides or herbicides that have become concentrated through the food chain and are dangerous to this predator. Thus, whenever possible, avoid using pesticides and herbicides in Loggerhead Shrike habitat.

Prescribed burn

operative land management, which involves neighbors working together to accomplish a common conservation goal.

Use of both prescribed burning and rotational grazing most closely replicates historical disturbance. Each plays a part in reducing grass and litter buildup and creating ground-level openings for forbs. But for various reasons, it may not be possible to do both. When only one can be used, prescribed burning may be preferred because it also controls another problem: cedar invasion. Rotational grazing does not control cedar invasion, but it has the advantage of being cheaper and safer than fire. For more information, see problems 1 and 2 in chapter 2.

1. Use prescribed burns on savannahs.

Post oak savannah has a high canopy, and the large trees usually grow and spread throughout a mixture of grasses and forbs. Because of this high canopy and ground cover, prescribed fire can move through and restore a more natural mix of low plants with little effect on the savannah's mature trees.

Shin oak and live oak savannahs have mottes with shrubby understory; a mixed grass and forb ground cover grows between the dense mottes. In these savannahs (as in the post oak savannah), a prescribed burn will improve the health of ground cover growing between the mottes. Fire also burns the trees and shrubs on the edge of the mottes, thus increasing regrowth of new woody plants. But for the most part, fire does not impact the big core trees of larger mottes. In addition, some small mottes of a shin oak savannah may be completely burned; however, they return in a few years, and the new mottes are prime nesting habitat for Black-capped Vireos.

2. Use rotational grazing on savannahs.

When well managed, cattle can be good for the prairie plant community of a savannah. Think of them as walking compost machines. They eat, quickly break down nutrients in grasses, and return them to the soil in manure. Thus, cattle cycle plant nutrients within the habitat where they live and remove only what they need to live and grow. When livestock are removed from the habitat and sent to market, the nutrients in their bodies are removed too. This is different from the situation of a wild animal that is born, lives, and dies in the habitat. How much nutrient value livestock remove may not be significant, but it is a change from the long-ago, natural processes that occurred before the Hill Country became ranchland.

Do not burn or graze a mixed wooded slope. A mixed wooded slope typically has steep topography with limited ground cover or duff, and often the only natural materials to burn are the trees and the canopy they create. When the trees burn, you lose the habitat! Not good. On the other hand, there is some work being conducted at Balcones Canyonlands National Wildlife Refuge that indicates prescribed burning in carefully planned situations can be beneficial in some woodlands. These burns are designed to maintain the signature canopy of these woodlands at approximately 70% or more. Before you consider a prescribed burn in mixed wooded slopes, we suggest you seek advice from experts at Texas Parks and Wildlife Department (TPWD), Natural Resources Conservation Service (NRCS), or US Fish and Wildlife Service (USFWS); a private burn consultant; or your local prescribed burn association. Also, since a mixed wooded slope typically is steep and has little grass, it is easily damaged by grazing livestock. Consequently, livestock should be excluded from wooded slope habitat. This is usually not a problem on property where cattle are present because they do not readily use such hazardous terrain.

Problem 2: Overgrazed Habitat with Bare Ground, Very Short Grass, and KR Bluestem

Throughout the Hill Country, overgrazing is a very serious problem. Almost all savannahs and mixed wooded slopes have been overgrazed, many for generations. Severe overgrazing is especially common on savannahs with too many goats or cattle for the land to support. If they

Live oak savannah with small-mammal habitat

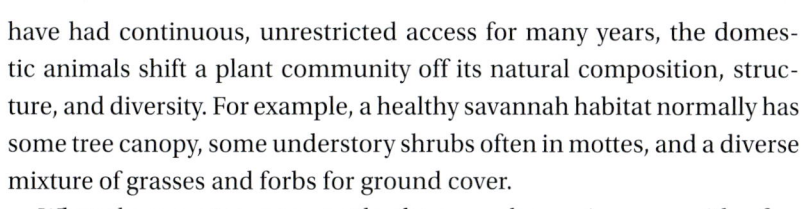

Red-tailed Hawk
(Photo by Lora L. Render)

American Kestrel
(Photo by Lora L. Render)

have had continuous, unrestricted access for many years, the domestic animals shift a plant community off its natural composition, structure, and diversity. For example, a healthy savannah habitat normally has some tree canopy, some understory shrubs often in mottes, and a diverse mixture of grasses and forbs for ground cover.

When there are too many cattle, the ground cover is sparse with a few unpalatable forbs, some resistant short grasses, and bare ground. Too many cattle or goats can change a diverse motte of small trees and shrubs into a thin collection of unpalatable species such as Texas pricklypear, mesquite, cedar, whitebrush, tasajillo, Texas persimmon, and agarita. This change in structure and diversity limits the birds and other wildlife that live in the habitat. Birds such as Black-chinned Hummingbird, Ash-throated Flycatcher, White-crowned Sparrow, and Northern Bobwhite depend on a healthy mix of native grasses and forbs growing among wooded mottes and cannot live in severely overgrazed savannahs. And the presence of cattle herds creates prime habitat for the Brown-headed Cowbird. This "outlaw" bird parasitizes many different species of birds breeding in the Hill Country.

Because mixed wooded slopes have little natural ground cover and plenty of woody understory, damage from domestic animals here is a loss of understory. Both cattle and goats have this impact, but impact from goats is much worse. If you are able to look through a wooded area at tabletop height (crouched, not standing) and see an open space under the trees, you are looking at a browse line that shows damage made by livestock or deer. Once this kind of change happens, it takes the land a long time to recover. For more information, see problem 1 in chapter 2.

What You Can Do: Restore Overgrazed Savannah Pastures and Mixed Wooded Slopes

1. Check for damage.

Different types of livestock favor different plants. Overuse by cattle is usually seen first when desirable grasses like little bluestem are grazed down to only a few inches high. As overgrazing continues, little bluestem and other desirable grasses decrease and less desirable grasses like the threeawns increase and eventually dominate the pasture. Goats are browsers, and overuse in their pastures shows first with some hedging on woody plants; continued overbrowsing produces a distinct browse line throughout the pasture.

2. Rest overgrazed pasture.

Remove grazing animals long enough for grass and forbs to recover naturally. Time necessary for adequate recovery of their root systems will vary depending on the amount of rainfall. Regeneration is especially slow for large trees, because they take so long to mature. Be patient. Consult with experts in your area for information on your particular site.

3. Switch to rotational grazing.

Whether you have cattle, sheep, or goats, rotational grazing permits regrowth of plant species preferred by the livestock and allows recovery of wooded slope and savannah plant communities. Annual plants go to seed and rapidly reestablish themselves. Perennials also go to seed and respond quickly. Woody plants require more time because they grow

KR BLUESTEM

KR bluestem, a drought-tolerant exotic species, now dominates many overgrazed pastures. First introduced on the King Ranch, this invasive Eurasian grass is poor bird habitat and has replaced native grasses such as little bluestem, the gramas, and yellow indiangrass that are important to birds and other wildlife. KR bluestem is easiest to recognize in dry seasons, (usually late summer, fall, and winter) when it is a mass of prostrate, pale grass 2–18 inches high. KR grows in dense turf mats that allow little light to reach the ground and exclude other grass species.

Field of King Ranch bluestem grass

much more slowly. For information on rotational grazing, see the diagram showing multiple pasture rotational grazing in chapter 2. For more information on the best grazing system and timing for your site, consult with the NRCS office in your area.

4. Look for recovery in fall and winter.

In pastures being rested, check cleared areas next to old live oak mottes. When you see conspicuous upright grasses fresh green in spring and rust-colored in fall, you are looking at recovery. Native grasses rebound because some bunchgrasses will have survived under "ball" cedar where grazers could not reach. Others will have survived in the dense cedar thickets. Deep shade under dense cedar prevents their normal robust growth, and the bunchgrasses grow fast when exposed to bright

Shin oak savannah in recovery from overgrazing

Overgrazed next to healthy live oak savannah

sunlight after cedar clearing. Two or three years after clearing, tall, diverse stands of native grasses can be seen right next to solid patches of KR bluestem. The image is striking: pale KR and swaths of rust-colored bunchgrasses meandering next to the live oak mottes. For more information, see problem 1 in chapter 2.

Live oak savannah recovering from overgrazing

Problem 3: Overbrowsed Habitat with No Regeneration, Little Diversity, and No Understory

Healthy wooded slope and savannah habitat must have trees and shrubs of different ages. For a continuous supply or regeneration of mature trees and shrubs, there must be seeds sprouting into seedlings, seedlings growing into healthy saplings, and some saplings surviving to become mature trees. In the Hill Country, too many browsing animals interrupt this chain of events for preferred plants. Typically, goats, exotic deer, or white-tailed deer eat the seedlings, so they never grow into saplings. Too many browsers interfere with natural reproduction and prevent regeneration. As these woody plants decline, the habitat loses natural food, structure, and shelter essential to its bird populations. When this type of damage occurs, bird populations also decline, and many species, including the Canyon Towhee, Black-capped Vireo, Clay-colored Sparrow, American Redstart, and Rufous-crowned Sparrow, are affected. To determine if you have a deer browse problem, see appendix 1.

What You Can Do: Assure Woody Plant Regeneration

Removing goats and exotics and controlling white-tailed deer overpopulation are especially important to the Black-capped Vireo, which is a Hill Country songbird listed federally as a threatened species. This rare bird thrives on conditions found in the mid-regeneration stages for woody plants.

Northern Bobwhite
(Photo by Jeff Forman)

Live oak savannah with grasses and forbs

1. Reduce goat stocking rate levels or remove entirely.

Because of their preference for woody plants, goats present a special management problem for mixed wooded slopes and savannahs. They should be excluded or allowed on steep wooded slopes only occasionally and for a limited amount of time. Also, do *not* use goats to control cedar. They reduce essential plant diversity by consuming more palatable woody plants before eating the cedar. Wherever goats are overstocked for a long time, palatable shrubs, forbs, and tree branches within their reach are lost, which has a significant impact on birds such as the wintering White-crowned Sparrow and Spotted Towhee. These birds need

White-crowned Sparrow
(Photo by Lora L. Render)

Shin oak savannah after prescribed burn

the cover of small trees and shrubs near the ground where they feed on seeds and insects. When restoring damage by goats, remove the animals to rest the land and allow plant regrowth.

2. Reduce deer overpopulation.

Deer also browse on woody plants, so when you remove the goats, you also must control deer overpopulation. Consult your local NRCS range specialist for information on sustainable stocking levels for your property. Be sure to state that your land stewardship goals include improving wooded slope and savannah habitat for birds. Almost everywhere in the

HIGH FENCING

A high fence is too often considered a panacea for controlling deer, and it is better to think of a high fence as a big responsibility because it comes with many more problems than solutions:

- High fencing to control white-tailed deer access must be "overbuilt" and is expensive. This is especially true on streams where flooding is inevitable and on steep, rocky terrain where fence building is most difficult.
- High fences are never impenetrable. They must be maintained frequently.
- High fencing prevents immigration but also prevents emigration, and deer density within the fence often increases very quickly to an unsustainable level. Preventing severe overbrowsing within a high fence requires consistent and intense management. If you do not continuously remove deer, woody plants within your high fence will be damaged dramatically.
- Within a high fence, a deer herd that used to roam over a wide area becomes a captive herd. Now it is your responsibility to feed them 24/7, and it becomes more important than ever to manage your plant communities so the deer do not ruin your greatest asset—your plants.
- If you keep exotic animals *and* white-tailed deer within your high fence, your problems and responsibilities double or triple. It is not simply a matter of dropping a herd of your favorite exotics in a large pasture and letting them take care of themselves. Most Hill Country land already has an overpopulation of white-tailed deer, so by adding exotics, you are probably adding an impossible number of mouths to feed, and you now have to control the numbers of all these animals to protect your plants.
- All animals have special impacts, and managing a mixture of exotics and white-tailed deer requires knowledge, awareness, and active control of the balance between all the animals and their food plants.
- Supplemental feeding is not the answer. Too often supplemental feeding is seen as an easy way to help animals on overused land, but instead it escalates a spiral of degradation that steadily gets worse, not better.
- If you have a high fence, do not let your land end up more damaged than adjacent land. It can be very hard to remove favorites or ones that have been there a long time, but usually, the most sensible answer is to remove the animals and simply let the plants recover.

Remember: Your greatest wealth is the plant community on your land, *not* the animals.

Hill Country, an overpopulation of white-tailed deer is having a negative impact on their preferred plants. To allow regeneration and restore plant diversity, reduce deer density with long-term and aggressive annual deer hunting. Often it is necessary to cooperate with neighbors hunting on their property so you can have a meaningful, long-term reduction of deer density. Once their numbers are reduced, continue an annual deer harvest to maintain their population at a moderate, sustainable level. For information on controlling deer, see chapter 8.

Small deer exclosure made with cedar logs

3. Add exclosures to protect woody plant production.

In the Hill Country, exclosures may be necessary to keep the deer away from woody plants. Even small exclosures increase woody understory by protecting seed-bearing plants from being eaten by deer. They allow damaged plants to return to healthy reproductive status and maximum seed production. Exclosures also provide bird habitat, including food and nest sites. Land managers also use them to monitor the effects of overbrowsing (or overgrazing). All exclosures, like all fences, must be maintained.

Large exclosures (quarter acre or more) provide a big block of habitat for nesting birds, woody plant recovery, and insects that live on the native plants. The space inside a large corral can be excellent space for plant recovery, and you can make an exclosure out of any fencing material. Large exclosures are least expensive to build when clearing cedar. If a corral exclosure is constructed with an elongated pile of cedar slash, the perimeter is also a brush pile giving birds protection from predators and inclement weather.

Pile cedar in a circle to create the corral. Use whole trees with branches intact so the pile will be as dense as possible. Stack them up to make a tall, wide perimeter band that effectively excludes deer. Also, to maintain adequate height, plan to add more cedar slash in the following years as the pile compresses. The challenge here is to have enough cedar within

Large cedar exclosure made with cleared cedar

Axis deer

easy hauling distance to replenish the cedar fence as it decomposes. Plan ahead.

4. Remove exotic browsing animals.

If exotics such as axis deer, aoudad, and fallow deer are on your property, they are eating the same plants that white-tailed deer and goats eat. Thus, if your land is overbrowsed, these exotics must also be reduced or removed.

5. Add brush piles for a "quick fix."

When missing understory, brush piles can immediately attract wrens, sparrows, and other birds. Brush piles do not permanently replace shrubs as cover but can provide good protection and some feeding areas as woody plants grow large enough to be useful.

Problem 4: Overgrown with Cedar or Indiscriminate Cedar Clearing

Both extremes are common in the Hill Country because landowners tend to either ignore cedar encroachment or overreact. And in both extremes, birds lose essential habitat. Where the land is covered with dense cedar thicket, you lose birds such as Cedar Waxwing, Nashville Warbler, Orange-crowned Warbler, and Black-crested Titmouse, which depend on insects and berries common to the diverse mixture of shrubs and small trees that occurs naturally.

Where all the cedar has been cleared, you lose birds such as Western Scrub-Jay, Golden-cheeked Warbler, and Greater Roadrunner, which depend on healthy mixed wooded slope habitat with some cedar for food, nesting and shelter from predators, and dangerously cold north winds. Where cedar and nearly all other underbrush is cleared, a surprising number of birds are affected. For example, wintering Orange-crowned Warblers and small migrants like Wilson's and Mourning Warblers feed near the ground and need a diverse mix of understory plants to find insects and berries or fruit. Breeding birds are affected too: Yellow-breasted Chat and Summer Tanager need understory plants where they forage for food. In the Hill Country, cedar is also important habitat for mammals, reptiles, and insects. For example, white-tailed deer hide among cedar trees with low branches, and cedar is host plant to the bright, olive-green juniper hairstreak butterflies.

What You Can Do: Use Sustainable Cedar Management and Protect Snags

In the case of "reckless" cedar clearing, convert to cedar management using preferred clearing methods done by a well-informed and coopera-

Canyon Towhee
(Photo by Lora L. Render)

Dry mixed wooded slope with plant diversity

tive operator. For areas already extensively cleared, look at where young cedar trees are regenerating and decide which ones to keep and which might be culled.

1. Use sustainable cedar management.

See chapter 9 to find cedar clearing guidelines and cedar clearing instructions for specific habitat types, as well as how to make a cedar management plan. Pay particular attention to cedar management on mixed

Shin oak savannah with scattered low thickets (Photo by Jeff Forman)

Black-capped Vireo
(Photo by Lora L. Render)

wooded slopes and live oak savannahs, where cedar is a particularly important part of the plant community.

2. Keep or add snags for cavity-nesting birds.

A standing, partly or completely dead tree, often missing the top or small branches, is called a snag. In the past you might have removed such trees because you thought they were ugly or were worried about the diseases they could spread. As a land steward for birds you are wise to keep most snags as a vital part of the woodland ecosystem. A snag may not work in your front yard, but it is usually okay to leave snags in other less prominent places. Hill Country breeding birds that regularly

Nashville Warbler
(Photo by Lora L. Render)

Moist savannah with diverse understory

use snags include Golden-fronted Woodpecker, Ladder-backed Woodpecker, Eastern Bluebird, Black-crested Titmouse, Eastern Screech-Owl, Black-bellied Whistling-Duck, Carolina Chickadee, and wrens.

Snags do not spread disease, but some insects attracted to the snags may do physical damage to other trees or carry fungal spores from one tree to another. As predators on these insect pests, birds limit the problems. Thus, by keeping the snags needed for the survival of these birds,

Western Scrub-Jay
(Photo by Lora L. Render)

Dry mixed wooded slope with cedar

you help moderate disease and maintain the sustainability of your woodland. This relationship is a good example of ecosystem stability and the good health that comes from diversity.

Another reason to keep a snag is that when the dead tree falls to the ground, its decomposing wood, called detritus, reduces soil erosion and dramatically influences soil morphogenesis. Detritus is rich in nutrients, holds water, and is a seedbed for certain plant species. Snags are habitat for microbes, invertebrates, and vertebrates, and they are an important element in the cycling of carbon through the ecosystem.

Because snags are nest sites and food sources for many birds and

Snag left standing for bird habitat

other wildlife in the Texas Hill Country, in woodlands in which standing dead trees are rare, we recommend creating snags to improve bird habitat. (Before undertaking snag development, conduct an initial survey of appropriate trees for snags. If, for example, you have only a few hackberry trees, it may be best to keep this important food source. However, if you have many, this may be a viable species for snag development.) Snags may be made when removing trees for another purpose or undertaken separately.

Because of the insects living on dead wood, live oak tree snags are good feeding areas for woodpeckers and the small gleaning birds like wrens, chickadees, and titmice that pick up insects on the bark. There is a continuum of nest cavity use from less to more with live, blackjack, post, and Spanish oaks. However, because of the spongy nature of the dead wood containing a fungus, trees killed by oak wilt are not the best trees for nest cavities. Cedar trees do not make good snags because they are resistant to pests. Sycamore, pecan, cedar elm, and hackberry make good snags because they decompose rapidly and quickly become an easy place for birds to excavate a nest cavity. As they decompose, these trees become home to many insects that birds eat.

To convert live trees to snags, girdle them and simply let them stand. Details on ideal snag production for the Texas Hill Country are not well researched. When considering snag production, be sure to consult your local wildlife biologist or forester to get specific recommendations for your property.

> Remember: Big dead trees, including those lost to oak wilt or drought, are precious habitat for insects, insect-eating birds, and cavity-nesting birds.

3. Add nest boxes.

Mixed wooded slope and savannah birds that normally nest in tree cavities are often limited by the absence of natural nest sites. In the Hill Country, many of these birds, such Ash-throated Flycatcher, Carolina Chickadee, Black-crested Titmouse, Carolina Wren, and Bewick's Wren, readily use a typical bluebird box. Natural nest sites may also be a limiting factor for the Eastern Screech-Owl. You can add a special box for this species anywhere in the Hill Country. It may also be helpful to add a special box for Wood Ducks in the southeastern counties and for American Kestrels in the northern counties. For more information on building, mounting, and monitoring these boxes, see chapter 6.

Problem 5: Rapid Loss of Topsoil

Erosion and loss of topsoil are pervasive throughout the Texas Hill Country, negatively impacting the natural woodland and savannah plant communities and their birds. Birds from the Blue-gray Gnatcatcher, which lives on shrubs and small trees, to the Rufous-crowned Sparrow, which nests and feeds among healthy ground cover, are lost along with the soil and plants. There are many causes of rapid topsoil loss. Any one or more of them may be causing erosion and soil loss on your property.

CURRENT OVERGRAZING

Severe overgrazing is common on small properties, especially where horses or goats are kept and the ground becomes so eroded and bare that it appears to be "growing rocks." More moderate but still serious overgrazing is present on a broad scale throughout the Hill Country. In many

Ash-throated Flycatcher
(Photo by Lora L. Render)

Dry live oak savannah with nest box

cases, pastures have been overgrazed for generations. The ground cover has been reduced to nothing more than a thin layer of very short grass too sparse to be an effective sponge for rain. Thus, runoff and sheeting are rampant and carry soil away after moderate to heavy rainfall. Because so much of the Hill Country is in this condition, many visitors and residents consider this normal or "the way it should look." But it is not normal.

Dry live oak savannah with understory

Blue-gray Gnatcatcher
(Photo by Jeff Forman)

What You Can Do: Develop Protective Vegetative Ground Cover
Developing protective vegetative ground cover can be accomplished by excluding livestock or switching to a rotational grazing system.

PAST OVERGRAZING ON SAVANNAHS
Overgrazed savannah that may have been rested for years can still have sparse vegetation and remain relatively ineffective at soaking up and

Overgrazed goat lot "growing" rocks

holding rainwater. The thin vegetation allows sheeting after even moderate rainfall and contributes to continuing downslope erosion and soil loss.

What You Can Do: Exclude Livestock

After excluding livestock, when possible, cover and protect the ground with no more than 2 inches of mulch.

DENSE CEDAR THICKETS

Dense cedar will shade out healthy understory and ground cover. The problem is compounded by heavy deer browse that prevents healthy ground cover renewal. Where there is no ground cover, any rain that penetrates the cedar canopy quickly washes soil particles downhill. Distinct lines of plant litter or duff meandering parallel to the slope indicate that this type of erosion is taking place.

What You Can Do: Depends on Slope and Age of Cedar

Wooded slopes with a mix of oak and mature cedar are Golden-cheek Warbler habitat. Here, avoid disturbing breeding birds during spring and early summer. Thin rather than clear cedar so you preserve 70%–80% of the canopy. Provide space for seedlings of deciduous trees such as red mulberry, cedar elm, black cherry, and Spanish oak. On flat ground or gently rolling savannahs, you can be more aggressive when clearing cedar. Let in sunlight to stimulate grass and forb growth. At the same time, control deer overpopulation or they will eat the plants you want to become ground cover.

TOO MUCH CEDAR CLEARING

Excessive cedar clearing on steep wooded slopes produces erosion and soil loss after moderate to heavy rainfall. A big storm dropping intense rain on a slope after it has been cleared of all its cedar and before any new ground cover has had time to grow causes severe erosion and irreparable soil loss.

What You Can Do: Cedar Management
Convert to cedar management that uses preferred clearing methods. Be sure to use a well-informed and cooperative operator. For more information on managing cedar, see chapter 9.

SPARSE VEGETATION ON SOUTH- AND WEST-FACING SLOPES

These hot, dry slopes do not have much soil in the first place, and they are susceptible to loss of what little soil they have. An indicator of severe erosion problems here are pedicles, or hummocks with plants growing on their tops. Pedicles are usually spread throughout a flaky limestone matrix.

What You Can Do: Pray for Rain
Recovery in this dry climate is very slow. Be patient. The land needs long-term rest from grazing. In years of better-than-average rainfall, recovery will accelerate. If you have lost most ground cover and a lot of soil, maintain a layer of surface compost that will protect what soil you have and will in time decompose and add new soil.

GRASS MOWED VERY SHORT

If grass is mowed so short that it cannot capture rainwater and slow runoff, the soil will be washed away after moderate to heavy rainfall. To reduce runoff and prevent soil loss, allow the grass to grow tall and develop a healthy root system.

What You Can Do: Stop or Reduce Mowing
Avoid mowing in spring and early summer:

- The soil needs protection from erosion during these months.
- Forbs provide critical habitat cover for many birds during these seasons.

> **Remember: It takes many, many years of plant renewal for continuous, long-term soil retention and replacement. Considering the Hill Country's dry climate, soil conservation scientists estimate that it takes hundreds of years (many generations) to add 1 inch of topsoil in the Hill Country.**

- Wildfire danger is least during these relatively wet and green months of the year.
- It is breeding season, and ground-nesting birds such as the Rufous-crowned Sparrow are on their nests.

Reduce mowing in fall and winter. If you must mow to reduce fire hazard, mow only in late summer, fall, or early winter. By waiting until after native forbs go to seed, you allow them to produce their seeds, which are food for birds and assurance of more plants in the future. Limit mowing to areas immediately around structures that need protection. The Texas A&M Forest Service recommends a 200-foot buffer in all directions around rural homes. Also, when mowing is unavoidable, set the mower at the highest setting to protect enough ground cover for absorbing rain and preventing erosion.

Problem 6: Too Dry

Almost any place in the Hill Country with a moderate to steep slope on a southern, western, or southwestern aspect will be naturally dry. These sites will not support lush vegetation no matter what you do. However, in spite of their dry conditions these locations may be home to some special birds, such as the Hutton's Vireo, which is adapted to an arid climate.

What You Can Do: Protect Habitat and Prevent Soil Loss

Relatively low annual rainfall is normal in the Texas Hill Country, and rain often comes in intense downpours. Thus, if you live on a naturally dry western, southern, or southwestern slope, it is very important to prevent unnecessary erosion and soil loss. Good ground cover will make the soil surface resistant to erosion and speed the addition of new soil. However, many years of good stewardship are necessary to build new soil; be patient.

The following practices are especially important in these dry sites:

- Mulch with cedar. Spread it in a thin layer to jump-start the soil-building process.
- Place cedar slash on the ground parallel to the slope. This helps hold the little soil that is present until grass cover is established and can do the job.
- Use rotational grazing and reduce stocking to moderate or light levels.

Prevent runoff, replace lost surface water, or increase groundwater recharge by the following methods. A small difference in groundwater makes a big difference in the plants and birds present.

Rufus-crowned Sparrow
(Photo Lora L. Render)

Dry shin oak savannah with ground cover

- Improve grass cover to provide maximum absorption and retention of rain.
- Thin dense stands of trees such as cedar to limit interception of light rain before it reaches the ground.
- Add water guzzlers to replace water that birds and wildlife previously found in springs and seeps now dry due to a lower water table.

Soil moisture determines the type of woodlands present, and a small difference in soil moisture makes a dramatic impact on what you see.

Hutton's Vireo
(Photo by Lora L. Render)

Dry shin oak savannah with dense cover

Just look at the live oaks growing on top of a low hill and at the bottom of the same hill. Both places get the same amount of rain, but the hilltop trees are stunted, and in the valley below they are normal-sized trees. In addition, the hilltop will have only a few woody plant species specialized for survival in arid conditions. The valley below will have a diverse plant community with the same species as on the hilltop plus many others.

On Hill Country hilltops there are probably live oak, shin oak, netleaf hackberry, evergreen sumac, agarita, cedar, Texas mountain laurel, and

Dry upland live oak savannah with yucca

Scott's Oriole
(Photo by Lora L. Render)

pricklypear. In the valley, these species will be present along with cedar elm, Spanish oak, black walnut, Texas redbud, Carolina buckthorn, and rusty blackhaw. This results in different bird habitat in the two places. For example, the Pine Siskin and Ruby-crowned Kinglet prefer relatively low areas with diverse vegetation where they can find plenty of seeds and insects to eat. And the Scott's Oriole lives in higher, drier habitat, where it finds yucca plants and strips threads from the leaves for nest-building material.

SUMMARY OF HABITAT PROBLEMS IN WOODED SLOPES AND SAVANNAHS

HABITAT PROBLEMS	STEWARDSHIP SOLUTIONS	AFFECTED BIRDS
Problem 1: Loss of plant diversity from lack of natural disturbances to savannah grasses and forbs	Prescribed burns on savannah land Rotational grazing Cooperative land management	American Kestrel Black-capped Vireo Loggerhead Shrike Painted Bunting Summer Tanager Swainson's Hawk
Problem 2: Overgrazed habitat with bare ground, very short grass, and KR bluestem	Rest overgrazed land Restore with rotational grazing	Ash-throated Flycatcher Black-chinned Hummingbird Northern Bobwhite Spotted Towhee White-crowned Sparrow
Problem 3: Overbrowsed habitat with no regeneration, little diversity, and no understory mottes	Reduce deer population Add deer exclosures Do not use goats to control cedar Remove exotic browsers Add brush piles Add nest boxes	American Redstart Ash-throated Flycatcher Black-capped Vireo Black-crested Titmouse Canyon Towhee Carolina Chickadee Eastern Bluebird Eastern Screech-Owl Golden-fronted Woodpecker Ladder-backed Woodpecker Rufous-crowned Sparrow Wrens
Problem 4: Overgrown with cedar	Start long-term, regular cedar control Use cleared cedar for mulch, fence posts, deer exclosures, trail building, erosion control	Black-crested Titmouse Cedar Waxwing Nashville Warbler Northern Flicker
Problem 4: Indiscriminate cedar clearing	Cut small regrowth cedar Keep understory Maintain ground cover Keep snags	Golden-cheeked Warbler Greater Roadrunner Western Scrub-Jay
Problem 5: Rapid loss of topsoil	Develop ground cover Exclude livestock Thin cedar thickets Use preferred cedar-clearing methods Stop mowing	Blue-gray Gnatcatcher MacGillivray's Warbler Meadowlarks Rufous-crowned Sparrow White-crowned Sparrow
Problem 6: Too dry	Prevent soil loss Improve grass cover Thin dense cedar thickets Mulch with cedar Place cedar slash on ground over slopes Use rotational grazing Add water guzzlers	Blue-gray Gnatcatcher Hutton's Vireo Scott's Oriole Western Scrub-Jay Wrens

Bird Summaries: Featured Birds

AMERICAN KESTREL
PRIORITY BIRD
ALL YEAR

Small falcon often seen on wires or fence lines; commonly hunts by hovering and then dropping to the ground on its prey. Some nest in the Hill Country, but many more winter here where they can find food. Breeding bird surveys since 1966 show a decline of American Kestrel populations throughout most of its range.

Identification: About 10.5 inches head to tail; male has slate-blue head and wings with rust-red back and tail; female similar but red-brown instead of blue wings and slightly larger; both have strong black-and-white facial pattern, long narrow tail, and pointed wing tips.

Habitats: Old field, plateau prairie, river valley prairie, pocket prairie, post oak savannah, shin oak savannah, live oak savannah, river, creek, canyon, seep, springs, pond, tank, lake, and backyard.

Cover: High perches or wooded fencerows in open grassland areas or urban and suburban roadsides.

Breeding Territory: Little information, but apparently varies widely; usually 27–430 acres/pair, but some studies show breeding territory at only 10 acres/pair; variability may be related to nest site or food availability.

Nest: No nest material; in woodpecker holes, natural cavities, niches in cliffs or buildings, and specially designed nest boxes.

Food: Large insects, mice, bats, birds, and small reptiles.

Stewardship: Encourage diverse grasslands or savannah with some shrubby cover and a few large trees for perches.
- Avoid broad-spectrum pesticides, and control red imported fire ants with an ant-specific insecticide to promote insect diversity and maintain food supply.
- Use prescribed burns to minimize woody plant encroachment and maintain diverse plant community.
- Control regrowth cedar with appropriate mechanical or chemical methods to maintain shrubs and woody plant diversity.

American Kestrel

- Use rotational grazing at moderate levels to maintain diverse plant community and medium height.
- Use periodic mowing in limited areas to promote forbs (and therefore small mammals). Mow only from July 15 until first frost.
- Retain snags with cavities (including drought-killed trees) to enhance nest sites and provide important feeding sites.
- Install nest boxes when available nest sites are lacking to increase nesting success.
- Add brush piles where brushy habitat is lacking to provide important cover for small mammals.

Other Potentially Useful Management:
- Control white-tailed deer and other browsing animals to encourage woody plant recruitment, healthy shrubs, and forbs for prey.
- Use deer exclosures to encourage woody plant recruitment.

ASH-THROATED FLYCATCHER
UNDER-THE-RADAR BIRD
SPRING AND SUMMER

One of the larger flycatchers; may be more common on fairly dry sites. In the wild, are dependent on woodpeckers to create cavities used for nesting but will use many different kinds of human-made structures, including nest boxes, pipes, fence posts, and ledges under eaves or porches.

Identification: About 8 inches head to tail; gray-brown above; throat and breast pale gray; washed-out yellow low on belly; rufous tail. Very similar to great crested flycatcher, which has more distinct yellow.
Habitats: Mixed wooded slope, post oak savannah, shin oak savannah, live oak savannah, river, creek, and canyon.
Cover: Shrubs and open woodland.
Breeding Territory: May not be clearly defined and can vary widely; size may depend on food and nest site availability; medium-sized territories about 12.5–37 acres/pair.
Nest: In a variety of cavities at least 1 foot above ground; usually created by woodpeckers; will readily use nest boxes.
Food: Mainly arthropods, especially wasps and native bees, with occasional seeds and fruits.

Stewardship: Encourage diverse shrubby savannahs with healthy forb and insect diversity and sufficient cavities for nesting.
- Avoid broad-spectrum pesticides, and control red imported fire ants with an ant-specific insecticide to promote insect diversity and maintain food supply.
- Retain snags for natural cavities (including drought-killed trees) to enhance nest sites and provide important feeding sites.
- Manage for healthy woodpecker populations to create cavities.
- Install nest boxes at a rate of at least one per 12.5 acres where natural cavities are limited.
- Use prescribed burns in savannah to increase forbs, create snags, and maintain woody plant communities.

Ash-throated Flycatcher

- Control regrowth cedar in shrubby savannahs and healthy diverse woodlands, using appropriate mechanical or chemical methods to maintain woody plant diversity.
- Control white-tailed deer and other browsing animals to encourage woody plant recruitment, healthy shrubs, and forbs.

Other Potentially Useful Management:
- Use light to moderate rotational grazing to encourage a balance of grasses and forbs.
- Use deer exclosures to encourage woody plant recruitment.

BLACK-CAPPED VIREO

PRIORITY BIRD

SPRING AND SUMMER

Low shin oak savannah thickets provide ideal nesting habitat for this Hill Country endangered species. May stay hidden but during breeding season males often call from top of shrubs and are easy to spot.

Identification: About 4.5 inches head to tail; black head with distinct white mask is diagnostic; olive above, light below; yellowish flanks.
Habitats: Shin oak savannah and live oak savannah.
Cover: Scattered dense, shrubby mottes.
Breeding Territory: About 7–10 acres/pair.
Nest: Baglike nest 3–8 feet above the ground and attached to ends of thin branches.
Food: Mostly insects, often collected from small branches.

Stewardship: Encourage deciduous shrubby mid-successional-growth savannah interspersed with diverse herbaceous vegetation.
- Conduct periodic (every five to seven years) prescribed burns to maintain low shrubs with good plant diversity, which is essential for nesting habitat.
- Do manual cutting when a prescribed burn is not possible to encourage shrubby regrowth.
- Control regrowth cedar with appropriate mechanical or chemical methods to maintain shrubs and woody plant diversity.
- Conduct any brush control only between September and February to avoid disturbing nesting activities.

Black-capped Vireo

- Trap and remove cowbirds to reduce nest parasitism and increase nesting success.
- Avoid broad-spectrum pesticides, and control red imported fire ants with an ant-specific insecticide to promote insect diversity and maintain food supply.
- Control white-tailed deer and other browsing animals to encourage woody plant recruitment and healthy shrubs.

Other Potentially Useful Management:
- Use rotational grazing at light to moderate levels to maintain a diverse plant community.
- Keep cats indoors, and control stray and feral cats to reduce predation.
- Use deer exclosures to encourage woody plant recruitment.

BLUE-GRAY GNATCATCHER
UNDER-THE-RADAR BIRD
SPRING AND SUMMER

Often overlooked because of its indistinct gray color; never still; its quick, nervous movement is distinctive and obvious.

Identification: About 4.5 inches head to tail; active, blue-gray bird; long, slender black tail with white outer tail feathers.

Habitats: Mixed wooded slope, post oak savannah, shin oak savannah, live oak savannah, river, creek, canyon, seep, and springs.

Cover: Shrubs and small trees of edges and open woodlands.

Breeding Territory: About 1.5–20 acres/pair.

Nest: Tidy nest 3–25 feet high in shrubs or small trees.

Food: Small insects gleaned from small branches.

Stewardship: Encourage diverse mix of deciduous woody plants with healthy understory or shrubby habitat and good insect diversity.

- Use prescribed burns to control cedar and encourage deciduous shrubs and trees with a relatively open habitat.
- Avoid broad-spectrum herbicides and insecticides, and control red imported fire ants with an ant-specific insecticide to increase forbs and promote insect diversity.
- Manage regrowth cedar with appropriate mechanical or chemical methods to maintain woodlands and savannah.

Blue-gray Gnatcatcher

- Control white-tailed deer and other browsing animals to encourage woody plant diversity, recruitment, and forbs.
- Trap and remove cowbirds to reduce nest parasitism and increase reproduction.

Other Potentially Useful Management:
- Use light to moderate rotational grazing to encourage forbs and insect diversity.
- Use deer exclosures to encourage woody plant recruitment.

CACTUS WREN
PRIORITY BIRD

ALL YEAR

Common and conspicuous desert bird found in the western half of the Hill Country; fierce defender of nest against predators; the male's loud distinctive song often associated with desert habitats.

Identification: About 7.0–8.5 inches head to tail; big body and long tail; large dark crown; broad white eyebrow; breast with dense black spots; known for its chattering call sung from a high perch.
Habitats: Shin oak savannah, live oak savannah, canyon, and backyard.
Cover: Thorn shrub on arid savannahs.
Breeding Territory: About 2.4–11.5 acres/pair.
Nest: Wide, round clump; with entrance tube to main body; often 2.5–9.0 feet above the ground in thorny shrubs.
Food: Mostly insects but also some fruit and an occasional small reptile or amphibian foraged on ground and low shrubs.

Stewardship: Encourage deciduous, shrubby, mid-successional-growth savannah interspersed with diverse herbaceous vegetation.
- Control regrowth cedar with appropriate mechanical or chemical methods to maintain shrubs and woody plant diversity.
- Avoid broad-spectrum pesticides, and control red imported fire ants with an ant-specific pesticide to promote insect diversity and maintain food supply.
- Use rotational grazing at light to moderate levels to maintain diverse plant community.

Cactus Wren

Other Potentially Useful Management:
- Use prescribed burns to maintain diverse plant communities and adequate bare ground, which *may* be beneficial.
- Use deer exclosures to encourage diverse woody plant recruitment.

CANYON TOWHEE

PRIORITY BIRD

ALL YEAR

Lives in southwestern United States and Mexico; will eat insects off parked cars and hide under the cars when disturbed; flies short distances into shrubs for safety and food and to roost; makes long-term pair bond; partners usually remain within yards of each other and sing a "squeal duet" several times per day. Breeding bird surveys since 1966 show an alarming decline of Canyon Towhee populations in the Texas Hill Country.

Identification: About 9 inches head to tail; grayish bird with long tail; streaks on upper breast in form of necklace; rusty crown; might be mistaken for a large sparrow.
Habitats: Mixed wooded slope, live oak savannah, shin oak savannah, river, creek, and canyon.
Cover: Low, grassy, dense thickets of cedars, oaks, and often thorny shrubs.
Breeding Territory: Thought to be about 4.0–7.5 acres or more/pair.
Nest: Deep cup built in thick trees or shrubs 3–12 feet above the ground.
Food: Seeds and some fruits and insects collected on the ground.

Stewardship: Encourage diverse shrubs and small trees with well-developed understory interspersed with a diverse herbaceous composition.
- Use prescribed burns to control cedar and encourage deciduous shrubs and trees with relatively open habitat and balance of grasses, forbs, and bare ground.
- Manage regrowth cedar with appropriate mechanical or chemical methods to maintain savannahs and shrubs.

Canyon Towhee

- Control white-tailed deer and other browsing animals to encourage woody plant diversity, recruitment, and forbs.
- Use light to moderate rotational grazing to encourage a balance of grasses, forbs, and bare ground.
- Install guzzlers or accessible water troughs to provide consistent water supply.

Other Potentially Useful Management:
- Keep cats indoors, and control stray and feral cats to reduce predation.
- Avoid broad-spectrum insecticides, and control red imported fire ants with an ant-specific insecticide to promote insect diversity.
- Use deer exclosures to encourage woody plant recruitment.

CAROLINA CHICKADEE
PRIORITY BIRD
ALL YEAR

Bright, friendly, cheerful little bird. Gleans its food from tree bark and foliage. In winter, moves around with mixed feeding flocks that often include titmice and kinglets.

Identification: About 4.25 inches head to tail; black cap, black bib, gray wings and back with whitish underside.
Habitats: Mixed wooded slope, post oak savannah, shin oak savannah, live oak savannah, river, and creek.
Cover: Deciduous woodland with understory shrubs and small trees with many branches.
Breeding Territory: Approximately 4–6 acres/pair.
Nest: Made of moss and bark strips and lined with soft material, in natural cavity 2–25 feet above the ground or nest box.
Food: In winter half insects and spiders and half seeds and fruit; summer, 80%–90% insects and spiders. Most insects collected on small branches less than 2 inches in diameter. Seeds and fruit include sunflower, ragweed, redbud, mulberry, and Virginia creeper.

Stewardship: Encourage diverse woody plant communities adjacent to herbaceous vegetation for seed production and sufficient snags for nesting.
- Keep snags with cavities (including drought-killed trees) to enhance nest sites and provide important feeding sites.
- Where natural cavities are limited, increase nesting by adding nest boxes.
- Encourage forbs in order to increase food (seeds) production.
- Use prescribed burns to increase forbs, create snags, and maintain diverse woody plant communities.
- Manage regrowth cedar with appropriate mechanical or chemical methods to maintain savannah and relatively open woodland.
- Control white-tailed deer and other browsing animals to encourage woody plant recruitment, healthy understory, and a diversity of forbs.
- Use light to moderate rotational grazing to encourage balance of forbs, grasses, and woody plants.
- Control red imported fire ants with ant-specific insecticides, and limit use of broad-spectrum pesticides to increase forbs and encourage insect diversity.

Other Potentially Useful Management:
- Use deer exclosures to encourage woody plant recruitment.
- Maintain bird-friendly, natural or artificial water source during dry winters to supply reliable and safe water.
- Keep cats indoors, and control stray and feral cats to reduce predation.

Carolina Chickadee

GOLDEN-CHEEKED WARBLER

PRIORITY BIRD

SPRING AND SUMMER

An endangered species that nests only in the Texas Hill Country; breeding pairs favor moist mixed wooded slopes; habitat loss, primarily through suburban development, in its limited breeding range continues to be a threat.

Identification: About 5 inches head to tail; black-and-white songbird with vivid yellow cheek; female resembles much more common Black-throated Green Warbler, but male is more vivid with black back and black eye line.
Habitats: Mixed wooded slope, post oak savannah, live oak savannah, river, creek, and canyon.
Cover: Large woodland blocks with dense canopy of mix deciduous and evergreen (cedar) woody plants.
Nest: Cup, lined with strips of bark from mature cedar trees (6–20 feet tall).
Breeding Territory: About 40 acres/pair.
Food: Insects and spiders often collected on twigs and leaves of deciduous and evergreen trees such as Spanish oak and cedar.

Stewardship: Encourage large blocks (>80 acres) of woodlands with woody plant diversity and a balance of mature cedar and deciduous trees and a relatively closed canopy.
- Limit removal of cedar in woodlands to regrowth and pole-sized cedar using mechanical methods, and retain mature trees to maintain balance of deciduous and evergreen (cedar) woody plants and relatively closed canopy.
- Control white-tailed deer and other browsing animals to encourage diverse woody plant recruitment.

Golden-cheeked Warbler

- Avoid broad-spectrum insecticides, and control red imported fire ants with an ant-specific insecticide to promote insect diversity.
- Trap and remove cowbirds, and maintain large woodland blocks to reduce nest parasitism and increase reproduction.

Other Potentially Useful Management:
- Use deer exclosures to encourage woody plant recruitment.
- Control feral and stray cat populations to reduce predation.

GREATER ROADRUNNER

FAVORITE BIRD

ALL YEAR

Huge brown bird that runs with bursts of speed and clatters from overhead perches. Warms itself in the sun by lifting feathers to expose skin. Native Americans believed the roadrunner to possess special powers.

Identification: Males about 21 inches and females about 20.5 inches head to tail; ragged crest that lifts up and down; long tail; brown back with light streaks and streaked breast.
Habitats: Old field, plateau prairie, pocket prairie, post oak savannah, shin oak savannah, live oak savannah, river, creek, and canyon.
Cover: Shrubs and small trees up to about 10 feet tall.
Breeding Territory: Need an exceptionally large area of 60–250 acres/pair.
Nest: Stick nest; 3–15 feet high in dense cover.
Food: Mostly hunts insects, lizards, snakes, small mammals, and birds on the ground in openings between mottes.

Stewardship: Promote diverse dense shrubs interspersed with a mixed herbaceous plant community.
- Use prescribed burns to control cedar and maintain a relatively open shrubby habitat with diverse grassland community.
- Manage regrowth cedar with appropriate mechanical or chemical methods to maintain savannah and shrubland.
- Control white-tailed deer and other browsing animals to improve forb diversity and maintain a healthy shrub community.

Greater Roadrunner

- Construct low brush piles to increase shrub-level habitat for prey.
- Avoid broad-spectrum insecticides to promote insect diversity.
- Apply ant-specific insecticides to control red imported fire ants and increase insect diversity.

Other Potentially Useful Management:
- Use light to moderate rotational grazing to encourage balance of forbs, grasses, and woody plants.
- Mow no more than 30% of total herbaceous ground cover periodically during fall or winter dormant seasons to promote forbs and short vegetation.
- Use deer exclosures to encourage woody plant recruitment.

HUTTON'S VIREO

UNDER-THE-RADAR BIRD

ALL YEAR

Very difficult to distinguish from a kinglet either by looks or behavior and probably regularly overlooked in mixed winter flocks as "just another kinglet." Examine winter flocks of kinglets, for these birds may travel together; vocal differences may be best way to tell them apart.

Identification: About 4.5 inches head to tail; grayish body with dusky underparts; light wing bars; dark eye with white eye ring; black bill; learn its monotonous, whiny song to distinguish from kinglets.

Habitats: Mixed wooded slope, shin oak savannah, live oak savannah, river, creek, and canyon.

Cover: Prefers fairly dense live oak or mixed wooded habitat with well-developed mid- to upper canopy.

Breeding Territory: About 1.5–7.0 acres/pair.

Nest: Baglike, 7–35 feet high and at the tip of a branch.

Food: Up to 98% insects and spiders collected in the mid- to upper canopy.

Hutton's Vireo

Stewardship: Encourage woody plant diversity and maintain relatively closed canopy.
- Manage regrowth cedar with appropriate mechanical or chemical methods to maintain diverse woodland.
- Control white-tailed deer and other browsing animals, and use rotational grazing to encourage woody plant recruitment.
- Avoid broad-spectrum pesticides, and control red imported fire ants with an ant-specific pesticide to promote insect diversity and maintain food supply.
- Use light to moderate rotational grazing at moderate levels to encourage a balance of grasses and forbs.

Other Potentially Useful Management:
- Use deer exclosures to encourage woody plant recruitment.

LADDER-BACKED WOODPECKER

PRIORITY BIRD

ALL YEAR

Active, relatively small woodpecker adapted to dry climate of southwestern United States and parts of Mexico; flutters, twists, turns, and hops sideways as it hunts insects; on the 2002 US Fish and Wildlife Birds of Conservation Concern list for the Edwards Plateau.

Identification: About 6.5–7.0 inches head to tail with females slightly smaller; thin black and white stripes on back and wings; spotted sides; distinct black triangle on face; male has obvious red crown; immature has small red crown, and adult female has black crown.

Habitats: Mixed wooded slope, post oak savannah, shin oak savannah, live oak savannah, river, creek, and backyard.

Cover: Large trees, mesquite, brush and dense cactus, and woodland edges.

Breeding Territory: About 16.5 acres/pair, but breeding pair may require a larger area where food is limited, such as on dry uplands.

Nest: Nest height 2–30 feet above the ground in willows, oaks, hackberry, and utility poles, posts, or dead trees.

Food: Mostly insects but also some seeds and cactus fruit; drills for wood-boring beetles; gleans caterpillars, ants, and other insects on trees, shrubs, cactus, and the ground.

Stewardship: Encourage diverse shrub and woody plant communities interspersed with herbaceous habitat and abundant insects.

- Keep snags with cavities (including drought-killed trees) to enhance nest sites.
- Use prescribed burns to maintain woody plant communities and create snags.
- Control red imported fire ants with ant-specific insecticides, limit use of broad-spectrum pesticides, and increase forbs to encourage insect diversity.

Ladder-backed Woodpecker

- Control regrowth cedar with appropriate mechanical or chemical methods to maintain savannah and relatively open woodland.
- Control white-tailed deer and other browsing animals to encourage woody plant recruitment.
- Increase nesting by installing nest boxes where natural cavities are limited.

Other Potentially Useful Management:
- Restore overgrazed pastures to promote healthy habitat.
- Use rotational grazing to maintain woodland diversity.
- Use deer exclosures to encourage woody plant recruitment.

LOGGERHEAD SHRIKE

PRIORITY BIRD

ALL YEAR

Can be mistaken for a Northern Mockingbird. Also known as blockheaded shrike or "butcher bird" because of its behavior of impaling prey on a sharp object like a thorn or barbed-wire fence, which is a mating display indicating how good a provider the male is, acts as a food supply or "pantry" for rapidly growing young, and marks a pair's territory boundary. Has been extirpated from much of its northern breeding grounds.

Identification: About 8.25 inches head to tail; distinctive black mask and large head; swooping flight and white wing patches most noticeable in flight distinguish it from Northern Mockingbird.

Habitats: Old field, plateau prairie, river valley prairie, post oak savannah, shin oak savannah, live oak savannah, river, and creek.

Cover: Isolated trees and thorny shrubs mixed with short to midlevel grasses and forbs.

Breeding Territory: Very territorial, ranging from 2 to 20 acres/pair.

Nest: Often in thorny tree or shrub 2.5–7.5 feet above the ground; large open cup made of twigs with soft lining such as animal fur and string. (Brown Thrasher nests are very similar in appearance but usually placed lower in the nest tree or shrub and lined with small rootlets instead of fur.)

Food: Mainly insects such as grasshoppers and beetles but also spiders and small animals, including rodents, snakes, frogs, lizards, and small birds.

Stewardship: Encourage native grasses and forbs with a diversity of insects and scattered thorny shrubs and trees.

- Use prescribed burns to increase forbs, maintain grasses and a diverse woody plant community, and possibly help prevent spread of fire ant colonies.
- Add brush piles to provide important cover for small mammals where brushy habitat is lacking.
- Control regrowth cedar with appropriate mechanical or chemical methods to maintain shrubs and woody plant diversity.
- Mow no more than 30% of herbaceous area in fall or winter (dormant season) to promote forbs.
- Control white-tailed deer and other browsing animals to encourage woody plant recruitment, healthy shrubs, and forbs.
- Keep thorny trees such as tickle-tongue and gum bumelia to allow breeding behavior and protect nest sites.
- Maintain brushy fence lines to protect hunting perches and nesting.
- Curtail use of pesticides, and take special care when using ant-specific insecticides to control imported fire ants and aid breeding success.
- Keep cats indoors, and control stray and feral cats to reduce predation.

Other Potentially Useful Management:
- Use light to moderate rotational grazing to encourage a balance of grasses and forbs.
- Use deer exclosures to encourage woody plant recruitment.

NASHVILLE WARBLER

UNDER-THE-RADAR BIRD

SPRING AND FALL MIGRANT

One of many warblers that migrate through Texas Hill Country feeding on insects, often found in healthy understory of shrubby habitat or woody regrowth.

Identification: About 4.75 inches head to tail; males bright yellow breast and throat, with gray head, greenish back; coloration the same in females but more muted; both sexes have distinct white eye ring and thin, sharp bill.

Habitats: Pocket prairie, mixed wooded slope, post oak savannah, shin oak savannah, live oak savannah, river, creek, canyon, seep, springs, pond, tank, lake, and backyard.

Cover: Open deciduous woodlands and savannah with well-developed shrub layer or understory.

Breeding Territory: N/A

Nest: N/A

Food: Almost exclusively insects collected from the ends of branches.

Stewardship: Encourage diverse open woodlands with healthy understory, shrubs, and forbs and abundance of insects.
- Avoid broad-spectrum pesticides to increase forbs and promote insect diversity.
- Apply ant-specific insecticides to control red imported fire ants and encourage insect diversity.
- Use prescribed burns to maintain healthy deciduous plant community and encourage forbs.

Nashville Warbler

- Control white-tailed deer and other browsing animals to encourage woody plant recruitment, healthy understory, and a diversity of forbs.
- Control regrowth cedar with appropriate mechanical or chemical methods to maintain shrubs and woody plant diversity.

Other Potentially Useful Management:
- Use light to moderate rotational grazing to encourage a balance of grasses and forbs.
- Mow no more than 30% of herbaceous area in fall or winter dormant season to promote forbs.
- Use deer exclosures to encourage woody plant recruitment.

NORTHERN BOBWHITE

PRIORITY BIRD

ALL YEAR

One of the most popular upland game birds in North America; has dramatically declined throughout most of its range, including the Texas Hill Country. Presence easily detected by its signature *bob-white* call. Breeding bird surveys since 1966 show a distinct population decline of this species in the Hill Country. Very large blocks of quality habitat may be necessary for a sustainable population.

Identification: About 9.75 inches head to tail; compact, medium-sized bird with white throat and extensive white eyebrow (buffy in female) separating rusty cap from cheek; rusty sides streaked with white and lightly speckled breast.

Habitats: Old field, plateau prairie, river valley prairie, pocket prairie, shin oak savannah, live oak savannah, river, and creek.

Cover: Native grasses and forbs with shrubs for escape.

Breeding Territory: About 17.5 acres/pair.

Nest: Depression on the ground lined with native bunchgrasses.

Food: Mainly forb seeds, but insects important to chicks in late spring and early summer.

Stewardship: Maintain native bunchgrasses for nest cover and good balance of forbs and bare ground. Manage for a mosaic of low brush within 50 yards of each large clump for shelter.

- Control white-tailed deer and exotic browsing animals to improve forb diversity and maintain a healthy shrub community.
- Use prescribed burns to control cedar and maintain a relatively open shrubby habitat with healthy bunchgrasses.
- Manage regrowth cedar with mechanical or chemical methods to maintain savannah and shrubland.
- Construct tepee-style brush piles to increase shrub-level habitat.
- Mow periodically on no more than 30% of herbaceous area only during fall or winter dormant seasons to promote forbs.
- Use light to moderate rotational grazing to encourage balance of forbs, grasses, and woody plants.
- Install water devices such as guzzlers where surface water is lacking to provide consistent water.
- Use light disking to increase forb production.
- Avoid broad-spectrum herbicides and insecticides to increase forbs and promote insect diversity.

Other Potentially Useful Management:
- Control feral hogs to minimize potential nest predation.
- Control red imported fire ants with an ant-specific insecticide to increase insect diversity.
- Organize a quail habitat association with neighbors to create large blocks of habitat for sustainable populations.
- Use deer exclosures to encourage woody plant recruitment.

Northern Bobwhite

PAINTED BUNTING

PRIORITY BIRD
SPRING AND SUMMER

Incredibly striking multicolored bird; probably the one that every landowner would like to attract. Nests in shrubs/small trees and feeds on grass seed, so is perfectly suited to life in a savannah habitat. Breeding bird surveys since 1966 show an alarming decline of Painted Bunting populations in the Texas Hill Country.

Identification: About 5.25 inches head to tail; males have blue head, green back, and red breast; females, with a mostly green body and lighter green breast, stay concealed.

Habitats: Old field, river valley prairie, pocket prairie, post oak savannah, shin oak savannah, live oak savannah, river, creek, seep, springs, and backyard.

Cover: Brushy thickets interspersed with grasses with good access to ground.

Breeding Territory: About 5 acres/pair and may be smaller in ideal habitat.

Nest: Deep, well-built cup; average height about 3.5 feet above the ground.

Food: Seeds, in particular bristlegrass often picked up off the ground, and some insects, especially during the breeding season.

Painted Bunting

Stewardship: Encourage shrubby mottes with diverse herbaceous vegetation and open savannahs with healthy understory.
- Use prescribed burns to maintain shrubs and good plant diversity, including bare ground.
- Control regrowth cedar with appropriate mechanical or chemical methods to maintain shrubs and woody plant diversity.
- Trap and remove cowbirds to reduce nest parasitism and increase nesting success.
- Avoid broad-spectrum pesticides, and control red imported fire ants with an ant-specific insecticide to promote insect diversity and maintain food supply.
- Control white-tailed deer and other browsing animals to encourage woody plant recruitment and healthy shrubs.

Other Potentially Useful Management:
- Use rotational grazing at light to moderate levels to maintain a diverse plant community.
- Use deer exclosures to encourage woody plant recruitment.

RED-TAILED HAWK

FAVORITE BIRD

ALL YEAR

One of the most recognizable and widespread hawks in North America; adapted to a wide variety of habitats from semi-open forests to deserts; common throughout Texas; some individuals specialize in hunting bats as they exit their caves.

Identification: Males about 20 inches and females 22 inches head to tail; most distinguishing feature is the rusty red tail.

Habitats: Old field, plateau prairie, river valley prairie, pocket prairie, mixed wooded slope, post oak savannah, shin oak savannah, live oak savannah, river, creek, canyon, seep, springs, pond, tank, lake, and backyard.

Cover: Scattered trees in a variety of settings, including grassland, woodland edges, wet and dry wooded slopes, agricultural and urban areas where there are high nest sites and perches.

Nest: Made of sticks; up to 30 inches in diameter and several inches thick; usually located in tall trees and occasionally on cliffs, with a good view of the surrounding area.

Breeding Territory: Around 550 acres/pair; more or less depending on food and hunting perch availability; fiercely territorial; breeding territories often have distinct boundaries such as roads and rivers.

Food: Nearly all small mammals; some birds and reptiles; occasionally carrion; usually hunts from a high perch.

Stewardship: Encourage healthy grassland and savannah habitat with sufficient brush for small mammals and scattered tall trees for nesting and perches.
- Add brush piles where brushy habitat is lacking to provide important cover for small mammals.
- Use prescribed burns in savannah to increase forbs and maintain diverse woody plant communities.

Red-tailed Hawk

- Control regrowth cedar with appropriate mechanical or chemical methods to maintain shrubs and woody plant diversity.
- Mow periodically in limited areas to promote forbs (and therefore small mammals). Mow only after July 15 until first frost.
- Use rotational grazing at moderate levels to encourage a balance of grasses and forbs.

Other Potentially Useful Management:
- Control white-tailed deer and other browsing animals to encourage woody plant recruitment, healthy shrubs, and forbs.
- Use deer exclosures to encourage woody plant recruitment.

RUFOUS-CROWNED SPARROW

PRIORITY BIRD

ALL YEAR

An attractive and inconspicuous ground bird; one of the LBJs (little brown jobs) often overlooked as it moves through shrubs and on the ground; may draw attention to itself when it sings on top of a rock to defend its breeding territory. Breeding bird surveys since 1966 show an alarming decline of Rufous-crowned Sparrow populations in the Texas Hill Country.

Identification: About 5.5 inches head to tail; male and female similar; rust crown; white eye ring; gray-brown above with reddish streaks; gray below; long, rounded tail.

Habitats: Plateau prairie, shin oak savannah, live oak savannah, canyon, seep, and springs.

Cover: Mixed shrubs and grasses on rocky terrain.

Breeding Territory: About 3.5 acres/pair.

Nest: Usually on the ground but occasionally up to 1.5 feet above the ground.

Food: Insects and seeds often picked up off the ground or in grassy ground cover.

Stewardship: Encourage diverse shrub and open woodland with balanced mix of grasses, forbs, and bare ground.
- Use prescribed burns to control cedar and maintain shrubby and relatively open habitat.
- Avoid broad-spectrum herbicides to increase forbs.
- Avoid broad-spectrum pesticides, and control red imported fire ants with an ant-specific insecticide to promote insect diversity and maintain food supply.
- Manage regrowth cedar with appropriate mechanical or chemical methods to maintain savannah and shrubland.
- Control white-tailed deer and other browsing animals to encourage forb and woody plant diversity.

Rufus-crowned Sparrow

- Use light to moderate rotational grazing to encourage balance of forbs and grasses.
- Mow in fall or winter dormant season on no more than 30% of herbaceous area to promote forbs.
- Install guzzlers or accessible water troughs to provide consistent water supply.
- Keep cats indoors, and control stray and feral cats to reduce predation.

Other Potentially Useful Management:
- Control feral hogs to minimize damage to nesting habitat and possible nest predation.
- Use deer exclosures to encourage woody plant recruitment.

SCOTT'S ORIOLE

UNDER-THE-RADAR BIRD

SPRING AND SUMMER

Gorgeous, yellow-and-black desert bird is not common and seen most frequently on western portions of Edwards Plateau.

Identification: Size varies by sex from 7.25–8.25 inches with males larger; males have brilliant yellow and black colors, extensive black on head, back, and breast and bright yellow underparts, white wing bars; females have dusty yellow-green underparts, olive-gray back, and faint wing bars.

Habitats: Post oak savannah, shin oak savannah, live oak savannah, river, creek, canyon, seep, springs, pond, and tank.

Cover: Dry woodland shrubs, trees, and grasses.

Breeding Territory: Unknown; size may be limited by quality of cover and space in its habitat rather than food availability.

Nest: Prime nest sites in high, dry hills with yucca; sturdy, woven cup made with grass and yucca leaf fibers; attached to a yucca, clump of mistletoe, or tree branch about 5–10 feet above the ground.

Food: Primarily insects but also fruits and nectar; will visit hummingbird feeders for sugar water.

Stewardship: Maintain patchwork of grass and shrubby habitat less than 10 feet high, and encourage insects and nectar-producing plants such as yuccas.

- Use prescribed burns to control cedar and maintain shrubby and relatively open habitat and to encourage nectar producing forbs.

Scott's Oriole

- Avoid broad-spectrum herbicides and insecticides. and control red imported fire ants with an ant-specific insecticide to increase forbs and promote insect diversity.
- Manage regrowth cedar with appropriate mechanical or chemical methods to maintain savannah and shrubland.
- Control white-tailed deer and other browsing animals to encourage forb and woody plant diversity.

Other Potentially Useful Management:
- Use light to moderate rotational grazing to encourage a balance of forbs and grasses.
- Use deer exclosures to encourage woody plant recruitment.

WESTERN SCRUB-JAY

FAVORITE BIRD

ALL YEAR

Common in the dry hills; the bird most associated with cedar thickets; big, active, and noisy but not especially aggressive toward other birds at feeders; will steal food from other birds where food is not so abundant. Habitat loss from excessive cedar clearing and competition from Blue Jays are potential concerns for management of this bird.

Identification: About 11.5 inches head to tail; powder-blue jay; pale throat; long tail; no crest.

Habitats: Mixed wooded slope, post oak savannah, shin oak savannah, live oak savannah, river, creek, canyon, seep, springs, and backyard.

Cover: Shrubby oak and cedar woodlands and savannahs.

Nest: Open cup; 6.5–13 feet above the ground; at the ends of branches but well hidden.

Breeding Territory: About 5.5–8.0 acres/pair.

Food: Opportunistic; mainly fruit and nuts and some insects; occasionally young birds, mammals, or amphibians.

Stewardship: Encourage woody plant diversity in savannahs and woodlands with healthy understory and a balance of cedar and other evergreens.

- Use prescribed burns in savannahs to maintain diverse, shrubby, and relatively open habitat.
- Avoid broad-spectrum herbicides and insecticides, and control red imported fire ants with an ant-specific insecticide to increase forbs and promote insect diversity.
- Keep snags with cavities (including drought-killed trees) to provide feeding sites.

Western Scrub-Jay

- Limit control of cedar in woodlands to regrowth/pole-sized cedar, using appropriate mechanical methods while retaining mature trees to maintain balance of deciduous and evergreen (cedar) woody plants.
- Control white-tailed deer and other browsing animals to encourage a diversity of woody plants, especially those that produce fruits or nuts.
- Install guzzlers or accessible water troughs to provide consistent water supply.

Other Potentially Useful Management:
- Use light to moderate rotational grazing to encourage forbs and promote insect diversity.
- Use deer exclosures to encourage woody plant recruitment.

WHITE-CROWNED SPARROW
UNDER-THE-RADAR BIRD
WINTER

Large sparrow with long tail and small beak; commonly seen in winter flocks; easy to identify by zebra-striped head. Frequently on ground under feeders; often "chicken scratches" with both feet.

Identification: About 6.25 inches head to tail; males somewhat larger than females; adult has obvious black-and-white stripes on head, plain gray chest, throat, face, and nape. Immature has dull-colored head stripes, plain gray chest, throat, face, and nape plus thin white wing bars.

Habitats: Old field, mixed wooded slope, post oak savannah, shin oak savannah, live oak savannah, canyon, seep, springs, and backyard.

Cover: Low shrubs and small trees; feeds within about 3.5 feet of cover.

Breeding Territory: N/A

Nest: N/A

Food: Mainly forb and grass seeds; also some fruit, berries, and insects.

Stewardship: Encourage a diverse healthy shrub community including a mix of grasses, forbs, and bare ground.
- Control white-tailed deer and other browsing animals to improve forb diversity and maintain a healthy shrub community.
- Use prescribed burns to control cedar and maintain shrubby and relatively open habitat.
- Manage regrowth cedar with appropriate mechanical or chemical methods to maintain savannah and shrubland.

White-crowned Sparrow

- Construct tepee-style brush piles to increase shrub-level habitat.
- Use light to moderate rotational grazing to encourage a balance of forbs and grasses.
- Install guzzlers or accessible water troughs to provide consistent water supply.
- Avoid broad-spectrum herbicides to increase forbs.

Other Potentially Useful Management:
- Mow in fall or winter dormant season on no more than 30% of herbaceous area to promote forbs.
- Use deer exclosures to encourage woody plant recruitment.
- Avoid broad-spectrum insecticides, and control red imported fire ants with an ant-specific insecticide to promote insect diversity.

2 Grasslands

A thing is right when it tends to preserve the integrity, stability and beauty of the biotic community. It is wrong when it tends otherwise. —Aldo Leopold

INTRODUCTION

This chapter discusses the nearly level places where grasses and forbs (including wildflowers and weeds) prevail. These habitats are home to many birds, from great birds of prey to tiny wrens and sparrows such as the inconspicuous Cassin's Sparrow. When discussing land stewardship for these birds, whose lives depend on a plant community of grasses and forbs, it is difficult to find the best word to describe where they live.

Is it rangeland, grassland, prairie, or plains?

Rangeland is often thought of as an open, grassy place where cattle and horses can graze. Rangelands are distinguished from pastureland because they consist of primarily native vegetation rather than plants established by humans. Rangelands are also managed principally with extensive practices such as managed livestock grazing and prescribed fire rather than more intensive agricultural practices of seeding, irrigating, and fertilizing. Some range ecologists refer to most of the land west of IH-35 as rangeland. However, rangeland includes grasslands, woodlands, brushland, and other plant communities that are not suited to farming but are often used for grazing livestock.

Grassland is another description of the low, open habitat addressed in this chapter, but it incorrectly implies that we are talking about a place with only grasses. Prairies or plains are open landscapes with a mixture of grass and forb species. These words most often engender a picture of the Great Plains, the vast expanse occupying the center of our country. The Texas Hill Country has much smaller prairie habitats, so the difference is a matter of scale.

Throughout this chapter, the words "rangeland," "prairie," and "grassland" refer to the mostly level places where grasses and forbs dominate the landscape. All these terms may be used interchangeably. There are four types of grasslands in the Texas Hill Country, all of which sustain populations of resident, winter, migrant, and breeding birds:

Plateau prairie
River valley prairie
Pocket prairie
Old field

FEATURED GRASSLAND BIRDS

PRIORITY
Black-throated Sparrow
Cassin's Sparrow
Eastern Meadowlark
Field Sparrow
Grasshopper Sparrow
Lark Sparrow
Le Conte's Sparrow
Lesser Goldfinch
Mourning Dove
Rio Grande Turkey
Scissor-tailed Flycatcher

UNDER-THE-RADAR
American Pipit
Savannah Sparrow
Vesper Sparrow

FAVORITE
Eastern Bluebird

OUTLAW
Brown-headed Cowbird
House Sparrow

Note:
Priority = on at least one of the lists provided by TPWD or Oak and Prairies Joint Venture naming species in need of conservation assistance or shown by USGS-sponsored Breeding Bird Surveys in the Edwards Plateau to be a species whose population is in decline
Under-the-radar = a regular bird in Texas Hill Country but often unnoticed
Favorite = popular with just about anyone who knows the bird
Outlaw = undesirable because of its behavior and impact on other bird species

Spring wildflower field

Remember: For a bird, good habitat means a place to nest, the right kind of food, accessible water, and protection from bad weather and predators.

DESCRIPTION: GRASSLANDS

Because the differences between types of grasslands are not clear-cut but involve gradations, these divisions are imperfect. And unlike woodlands and savannahs, the different grassland habitat types usually are composed of similar plant species.

PLATEAU PRAIRIE

Plateau prairie is dry plain of the western half of the Hill Country covered primarily with a mix of forbs and short to medium grasses with occasional wooded mottes. Due to low rainfall, and in contrast to the eastern half of the Hill Country, these dry plains remain high and flat, uncut by water channels. Plateau prairie habitat is also found on wide hilltops through much of the eastern Hill Country.

    The short grasses of a plateau prairie are important habitat for a number of birds that make their nests on the ground: Cassin's Sparrow, Grass-

ECOLOGICAL SUCCESSION AND GOOD BIRD HABITAT

Ecological succession is the process of natural, gradual change over time in an ecological community. This process is divided into primary succession, ecological change beginning on a bare lifeless surface, and secondary succession, ecological change in a place where some life already exists. Fire (both natural and human-made) initiates secondary succession. The ecological changes that occur in old fields—abandoned pasture and fallow cropland—are also considered secondary succession. This chapter explains how land stewards can guide secondary succession to an ecologically balanced community that is good bird habitat.

Cassin's Sparrow
(Photo by Lora L. Render)

Dry grassland with scattered trees and shrubs

Black-throated Sparrow
(Photo by Lora L. Render)

Dry plateau prairie with scattered trees and shrubs

hopper Sparrow, Killdeer, Field Sparrow, Rufous-crowned Sparrow, Lark Sparrow, and both Eastern and Western Meadowlarks. Although not a ground nester, the beautiful Black-throated Sparrow is also known throughout much of the dry western Hill Country, but its population is decreasing across much of its range. Thus, it is important to improve its habitat by keeping the bird's needs in mind.

RIVER VALLEY PRAIRIE
River valley prairies provide important habitat for birds and other wildlife in the Balcones Canyonlands of the eastern portion of the Hill Country. Relatively high rainfall has sculpted this landscape with deep

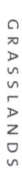

Grasshopper Sparrow
(Photo by Lora L. Render)

River valley prairie

canyons, rivers, and creeks. These popular streams have cool, running water that drains the hills and deposits alluvial soil. Through the millennia, water has carved channels and deposited sediments along their waterways. This is the process by which the valleys are first cut and then made flat again.

The alluvial soil deposited along rivers is relatively deep and supports river valley prairie habitat, which is home to many grassland birds, including the big, beautiful Red-shouldered Hawk and the much smaller and less well-known Grasshopper Sparrow. This tiny species gets priority status and deserves special attention because it once was common but has declined drastically during the last 50 years.

Lesser Goldfinch
(Photo by Lora L. Render)

Pocket prairie with forbs

POCKET PRAIRIE

Also called wet meadows, pocket prairies vary depending on how much water comes their way and how much water can be held by soil and vegetation. These are microhabitats found throughout Central Texas wherever water collects long enough to produce water-loving plants. They may be in a seasonal drainage, near a spring, or along a seep. Lesser Goldfinches use these wet meadows during years when rainfall is sufficient to produce healthy grass and forb cover with plenty of seeds and insects.

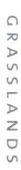

Mourning Dove
(Photo by Thomas Collins)

Old field with abundant forbs

OLD FIELD

Old fields are abandoned fields or fallow agricultural crops. When in early-succession stages with plenty of exposed soil and annual forbs or weeds (recent signs of cultivation), they are easy to recognize. Later stages may be more difficult. Depending on how long ago it was abandoned, an old field in a later-succession stage may be dominated by just a few grasses or a combination of some grass and invasive woody plants such as cedar, poverty bush, or mesquite. The Mourning Dove is a popular game bird of old fields and will flourish when these areas are disturbed periodically to maintain more forbs and exposed ground and curb woody plant invasion.

Eastern Meadowlark
(Photo by Lora L. Render)

Old field with grasses and forbs

Because the natural ecological succession of old fields is often "hijacked" by invasive species, they need active management to make them good bird habitat. When the succession process is halted by the domination of either native or exotic invasive plants, the whole system becomes "stuck" in a biologically impaired community that lacks food or nest sites and is useless habitat for most grassland birds. When old field succession is progressing without excessive interference from exotic or invasive species, the early-succession predominance of forbs, annual grasses, and bare ground provides plenty of insects and seeds, as well as the space where they can be hunted by grassland birds such as Mourning Dove, Field Sparrow, Lark Sparrow, and Cassin's Sparrow.

SEASONAL OCCURRENCE OF BIRDS ON WELL-MANAGED GRASSLANDS

WINTER
American Pipit
American Robin
Harris's Sparrow
Lark Bunting
Le Conte's Sparrow
Northern Harrier
Orange-crowned Warbler
Savannah Sparrow
Song Sparrow
Vesper Sparrow
White-crowned Sparrow

SPRING AND/OR FALL (MIGRANTS)
Clay-colored Sparrow
Mississippi Kite
Swainson's Hawk
Upland Sandpiper
Yellow-headed Blackbird

SPRING AND SUMMER
(BREEDING BIRDS)
Bell's Vireo
Common Nighthawk
Dickcissel
Painted Bunting
Scissor-tailed Flycatcher

ALL YEAR (RESIDENTS)
Bewick's Wren
Black-throated Sparrow
Cassin's Sparrow
Eastern Bluebird
Eastern Meadowlark
Field Sparrow
Grasshopper Sparrow
Killdeer
Lark Sparrow
Lesser Goldfinch
Loggerhead Shrike
Mourning Dove
Northern Bobwhite
Red-shouldered Hawk
Rio Grande Turkey
Rufous-crowned Sparrow
Vermilion Flycatcher

Early- to mid-succession stages with plenty of annual forbs but few bunchgrasses or woody plants have abundant food for doves, sparrows, meadowlarks, and quail, but they do not provide good ground-nesting habitat for Western Meadowlark, Eastern Meadowlark, or Northern Bobwhite. Later, the normal mid-succession stage habitat of an old field develops as native clump grasses become more common.

In late succession, large, dense stands of the taller bunchgrasses provide abundant well-hidden nest locations needed by meadowlarks and bobwhite. When conducting large-scale grassland management, the best combination for these birds is plenty of late-succession stage habitat with good nest cover and a smaller area in early to mid-succession stages for food.

GRASSLANDS HABITAT MANAGEMENT GOALS

The land stewardship goal for all Texas Hill Country grasslands is the same. Grassland birds need high-quality prairie with a diverse mix of native grasses and forbs that supplies appropriate food, water, and cover for the species specializing in each of the four types of habitat. To understand how land management affects grasslands and birds, it is useful to think about the ecological processes that operated long ago. Migratory grazing animals such as bison and fires set by Native Americans or lightning strikes were intense disturbances to the grassland habitat and consequently determined the historic distribution of plant communities and the variety of plants in these communities.

After European settlers arrived, the sedentary farm and ranch practices of plowing and livestock grazing had their own major impact on the composition of native prairie grasslands. On every Hill Country property, historic or current livestock grazing, mowing, lack of wildfires, fragmentation, and brush management play a major role in influencing plant composition and ecology. Today, tools like proper livestock grazing, prescribed burning, and managing land cooperatively can mirror the historic natural processes and dramatically improve bird habitat.

Grazing and prescribed burning, especially in combination, can be used to improve a plant community's composition and health by recycling nutrients in a form readily available to plants and provide space for germination. Preferred management methods, in descending order, are grazing and prescribed burns in combination, prescribed burns alone, and grazing alone. Mowing is a poor substitute for other options. However, when neither grazing nor prescribed burns are possible, mowing could be somewhat useful to improve forbs or give a slight benefit to warm-season grasses.

Mowing Basics

When done incorrectly, mowing damages plant community composition, bird habitat, and nesting birds.

- Frequent mowing suppresses warm-season bunchgrasses and many forbs needed by birds.
- Mowing in spring and early summer destroys the nests of ground-nesting birds.
- Mowing is inefficient at removing accumulated litter that prevents some grassland birds from getting enough forbs and insects to eat and from gaining access to bare ground for feeding.

The following recommendations are based on the current understanding and ongoing research of how mowing impacts grassland systems. All instructions apply to the predominant mixed grass habitat of the Texas Hill Country grasslands and savannahs. Timing and frequency are critical when you use mowing for bird habitat management. Depend-

> **INSECTS**
>
> Insects are critical to nearly all birds. We do not often think about insect production except in a negative way because many insects, such as mosquitoes, are pests. But think about insects as critical bird food. For example, an outbreak of caterpillars in the spring or grasshoppers in the summer means there is a lot of food available for your favorite birds.
>
> - Insects are a high-protein food eaten by most birds during some part of their life cycle.
> - Migrant birds need to store large amounts of energy before migration and will eat large numbers of insects.
> - Nearly all birds feed insects to their young during their maximum growth period from hatching through fledging.
> - For insectivorous birds such as the flycatchers, insects are a major portion of their diet.
> - Many priority species in all habitats have insects as a major component of their diet.
> - Forb abundance and diversity are important to insect production because many insects live on forbs.

ing on the bird species in question, somewhat different timings and frequencies apply. Be very careful. More is not better!

MOWING FOR NATIVE GRASSES

To promote warm-season grasses, mow only once during the dormant season (late fall or winter) and limit area mowed in any one year to one-third or less of a grassland or savannah habitat. Mowing for native grasses can partially recycle the standing vegetative growth from past years, maintain healthy native warm-season grasses, and slightly improve native forbs in overall grass community composition.

MOWING FOR FORBS

When mowing for forbs, the increase in forbs comes at the expense of native warm-season grass, so the area mowed must be limited. You can create a mosaic where forbs and native warm-season grasses dominate in difference areas. To accomplish this, start mowing in midsummer—after many forbs have gone to seed and most ground-nesting birds have finished nesting for the season; mow one-fourth or less of a grassland or savannah habitat; mow three to four times from midsummer to the first

Remember: More forbs = more insects = more birds.

RIO GRANDE TURKEY

The Rio Grande Turkey is a charismatic and historic wildlife icon. Its Hill Country populations can benefit from multiple management tools because this huge bird requires a large area and has varied habitat needs during its life cycle. Unlike for many nongame bird species, a great deal of research has been done on the wild turkey. Landowners can use information on the Rio Grande Turkey's habitat requirements to help with stewardship decisions:

- Rio Grande Turkey habitat requires a roost.
- Rio Grande Turkeys form large winter flocks that roost together at night, usually near seasonal or perennial streams.
- Big trees at least 40 feet tall with many horizontal roosting branches that supply 50%–70% canopy may be used as a winter turkey roost for decades.
- Preferred roost trees are surrounded by 10–15 acres of mixed woods on savannah or grassland that is at least 10%–15% mixed wooded habitat.
- Understory within about 6 feet of the best roosts has low to moderate woody plant density.

TPWD prohibits propagation, sale, purchase, transport, or release of turkeys for the purpose of establishing a free-ranging wild turkey population. This is important for several reasons:

Pocket prairie with tall bunchgrasses
Rio Grande Turkey (Photo by Lora L. Render)

- Pen-raised birds become tame and have poor survivability because they lack natural wariness.
- Disease can be spread into wild turkey populations.
- Wild birds are genetically adapted to survive in the region where they evolved and may not survive in different habitats.
- Interbreeding may introduce detrimental traits that impair survival.

frost (beginning of the dormant season); and do not mow again until the midpoint of the next summer. Mowing for forbs will suppress growth of warm-season grasses, open up more space for forbs to germinate, and allow more light for forbs to grow. Increased forb diversity provides more seeds for seed-eating birds and creates good brood habitat by attracting insects, which most birds feed to their nestlings.

MOWING FOR GRASSES AND FORBS

You cannot mow for both grasses and forbs in the same space, which is a major problem with mowing. When managing for forbs and native grasses, mow no more than one-third of a grassland or savannah habitat. *Never* mow in spring or early summer while birds are nesting and forbs are going to seed. When managing for native warm-season grasses, mow only *once* during the dormant season.

PROBLEMS OF GRASSLANDS

Overgrazed rangeland, brush invasion, and dominance of exotic species have caused grassland birds of the Texas Hill Country to decline throughout recent history. First, when settlers arrived and brought cattle to this vulnerable, dry land, overgrazing of sensitive grasses depleted the natural grassy ground cover on a grand scale. Then, cedar, mesquite, KR bluestem, and native forbs unpalatable to cattle invaded and changed huge areas of the Hill Country beyond suitability for many grassland birds.

Problem 1: Inferior Bird Habitat Resulting from Overgrazed Grasslands

Throughout the Hill Country, overuse by domestic animals is the most prevalent land management abuse. Overgrazing by cattle damages plants by removing so much of the upper, green portion that their root systems are stunted and they become weak. Consequently, livestock removal or reduction produces the most significant improvement in grassland community health. Overgrazed rangeland is easily recognized by the presence of very short ground cover, bare ground, and an abundance of grasses resilient to heavy grazing, such as KR bluestem and the three-awns.

Damage from overgrazing in all Hill Country grasslands is rapid and pervasive, and recovery is lengthy. Grassland habitat in the western portion of the Hill Country is more fragile than in the eastern counties. Because there is less rain in the west, overgrazing damages grasses more quickly and they recover more slowly.

What You Can Do: Restore Overgrazed Rangeland

Prairie birds such as the Rio Grande Turkey, Eastern and Western Meadowlarks, Dickcissel, Northern Shrike, flycatchers, bluebirds, and sparrows, including the Vesper Sparrow, benefit when overgrazed rangeland is rested. Most of these birds thrive on a complex mix of native prairie forbs and grasses that supply the seeds and insects they need. They also require a little bare ground within the natural ground cover plus a scattering of trees and shrubs. Long-term, severe overgrazing over a

Vesper Sparrow
(Photo by Lora L. Render)

Prairie with medium grasses and forbs

large area may result in an invasion of brush, with few palatable plants remaining to produce seed. If so, a number of measures should be undertaken, and recovery may take a very long time.

1. Control deer density.

Throughout the Hill Country, it is necessary to control deer density so their favorite native forbs and woody plants have a chance to mature and reproduce. For more information, see chapter 8.

2. Use mulch and slash to reduce soil loss.

Spread cedar mulch less than 2.5 inches deep and/or place slash in contact with the ground to prevent erosion and promote recovery. On

Slash on gentle slope

even slight slopes with short grass and bare ground, heavy downpours cause rainwater sheeting or overland flow that leaves behind curvy lines of litter. Sheeting erodes soil particles and should be prevented. You might want to experiment with different mulch depths to see which allow new plants to sprout quickly. The only slash that will stop or slow sheeting is the part in direct contact with the ground. Slash left in a jumble with at least some surfaces touching the ground is acceptable because it provides a protected area for seeds to collect and new woody plants, grasses, and forbs to grow, cover, and hold the soil.

3. Use fire and continued rest to accelerate recovery.

If overgrazing was neither too severe nor long term, you still may have some desirable grass seed, such as little bluestem and yellow indiangrass, in the soil. In this case, a prescribed burn and continued rest will accelerate native grass and forb recovery.

4. Reseed if necessary.

Land that has been severely damaged by heavy grazing for a very long time will require reseeding or planting of native forbs, grasses, and even woody plant species. Select ones that are well adapted to your site and most useful to the birds you wish to attract. For information on native plants for specific habitats on your property, consult your local Native Plant Society of Texas chapter, NRCS, or TPWD. If most of the plants preferred by livestock remain on overgrazed land, it can return to a healthy natural state on its own. In many cases, livestock removal for two to three

> **Remember: Excellent supplemental feed does not stop animals from eating plants that taste good to them.**

years or long-term stocking-level reduction allows restoration of plant diversity and promotes sustainable ground cover.

5. Practice rotational grazing.

Throughout the Hill Country, sustainable livestock management is essential for maintaining healthy rangeland. The natural grassland plants eaten by livestock must remain strong enough to grow vigorous root systems to maintain their health. Since grassland birds also depend on these native grassland plants for food and shelter, sustainable livestock management is critical to their survival. Annual and perennial forb species have different growth patterns.

To protect both annual and perennial forbs, use a suitable rotational grazing system. Annual forbs need a rest period in fall or winter for germination and also in the spring or summer for growth and seed production. Perennial forbs store food and energy in their roots and are more resilient to grazing than are annual forbs. A good rotational system rests pastures during the months when perennial forbs grow in spring, summer, and early fall. When livestock are removed for short periods from March through October, perennial forbs can recover and sustain themselves.

Rotational grazing moves livestock through several pastures in sequence through the grazing period. Some types of rotational grazing are slow enough for grasses to replenish themselves during the growing season and are especially useful for healthy bird habitat. Rotational grazing that controls frequency, timing, and intensity is best.

- Frequency = how often a pasture is grazed
- Timing = months or seasons when it is grazed
- Intensity = how many animals are in the herd

Multiple pasture rotational grazing benefits grassland birds. This type of grazing management uses two or more pastures with one rested throughout the growing season. Thus, grasses are allowed maximum growth during rest periods when no grazing occurs. The "Merrill" system uses four pastures. As cattle are rotated through the pastures, at all times three pastures are stocked with livestock and one pasture has none so its plants are rested from grazing pressure. Short duration grazing (SDG) and high-intensity/low-frequency (HILF) grazing are more intensive. In these systems, pastures have short periods of intense use by many animals followed by relatively long rest periods. For information on the best choice for your specific situation, consult your local AgriLife Extension or NRCS office.

Rotational Grazing

This example of Rotational Grazing shows 2 years of a system that uses 4 pastures for a single herd with 3 pastures grazed and 1 pasture rested each year. Thus, each pasture is grazed during the same season every 5th years.

Year 1

Grazed January - April	Rested All Year
Grazed May - August	Grazed September - December

Year 2

Grazed May - August	Grazed January - April
Grazed September - December	Rested All Year

Figure 2.1

Is there any way to do rotational grazing if you have only one pasture?

On less than 40 acres, unless you do cooperative grazing with a neighbor, the short answer is no. If your land is top quality, you need at least 20 acres for only one cow. Thus, if you want to have two cows and their calves, you need more than 40 acres. An excellent solution is to partner with a neighbor who has enough land to complement your acreage and do cooperative grazing. Another solution is to simply buy hay and grain for your animals and feed them in dry lot for much of the year and then let them out to graze periodically after the grass has recovered from the last grazing period.

If you have 40 acres or more, here are some possibilities to consider:

- You already may have more than one pasture. If one is much smaller than the other, you might have been keeping the gate open between the two and using them as one. If so, try using the smaller one as a second pasture but keep the animals in it for less time than in the other.
- Cross-fencing may not be as difficult as you think. Electric fence is an inexpensive alternative used by many owners for the most intensive multiple pasture systems. Once you have the charger, adding electric fence as well as moving it around can be much easier than traditional fence building. It is necessary to train your animals to the electric fence, but they will adapt.
- If you are leasing land to a neighbor, start using both parcels of land as two pastures and rotate between them. Set up a plan in which the animals are moved at different times each year. In this way, many more plant species are rested and thus have a chance to recover.

For information on the best grazing systems and timing for your land, consult with a local NRCS range specialist. Be sure to make your land stewardship goals clear.

6. Balance light, medium, and heavy grazing.

Think mosaic! Balanced grassland management for birds benefits the maximum number of bird species. And different birds are attracted to the different ground cover heights produced by light, medium, and heavy grazing levels. Thus, balanced grassland management should produce a patchwork of plant cover that supports an array of both breeding and nonbreeding birds. This patchwork includes areas of short, medium, and tall grasses that meet the specific requirements of different grassland bird species. Short vegetation like that created by intensive grazing provides habitat for a relatively small number of grassland birds, but dense ground cover provides for a variety of grassland species. So when managing a small patch of prairie (less than about 50 acres), be sure to maintain plenty of relatively dense vegetative cover, as it will provide for more grassland birds than thinner, shorter grasses.

Light grazing, long rests, or even complete removal of grazers for sev-

DROUGHT

During periods of extended drought, all pastures must be rested until enough rain falls to replenish grass growth. A long drought often requires severe culling or even selling of all livestock until the drought is over and the plant community recovers. Good land managers do not become too attached to their animals. Think of your job as manager of valuable land covered with an impressive plant community. After the livestock are removed, it may be slow to recuperate. Be patient. Let the land dictate when to replenish your herd.

eral years benefits some grassland birds. Le Conte's Sparrows and Sedge Wrens winter in the Hill Country and need tall native bunchgrasses (such as little bluestem) for food and shelter. Sedge Wrens are especially sensitive and cannot live on normally grazed rangeland. Dickcissel, Northern Bobwhite, and Rio Grande Turkey nest in grassland and savannah habitat. In spring and early summer, they need medium to tall native bunchgrasses, where they hide their nests on the ground. Rangeland research shows that rotational grazing at light to moderate levels can increase plant diversity and vigor.

Moderate grazing usually involves rotational grazing and/or limited stocking levels that produce short to medium grasses in a mixed plant community typical of Hill Country prairies. When the natural prai-

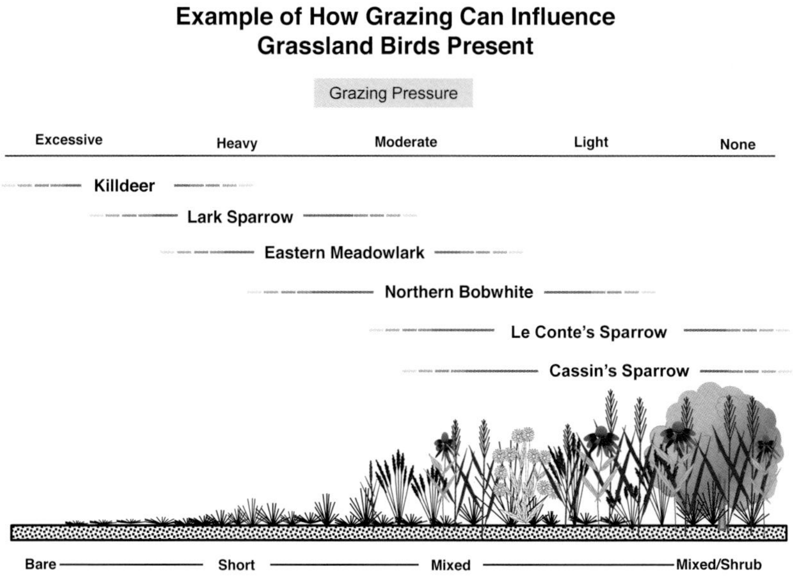

Example of How Grazing Can Influence Grassland Birds Present

Grazing Pressure

Excessive	Heavy	Moderate	Light	None

- Killdeer
- Lark Sparrow
- Eastern Meadowlark
- Northern Bobwhite
- Le Conte's Sparrow
- Cassin's Sparrow

Bare — Short — Mixed — Mixed/Shrub

Figure 2.2 (Adapted from Knopf, "Prairie Legacies—Birds," 137)

rie plant community is maintained in this way, birds, including Rufous-crowned Sparrow, Vesper Sparrow, Eastern and Western Meadowlark, Savannah Sparrow, Sprague's Pipit, American Pipit, and Northern Bobwhite, respond favorably.

Heavy grazing short term, especially as part of a rotational grazing system, can be acceptable because a few birds such as Killdeer and Lark Sparrow prefer short-grass habitat. They nest in very short grass and may even increase soon after heavy grazing. Both species nest on open, bare ground and feed on seeds and insects gathered off the ground.

Long-term, continuous heavy grazing reduces the root systems of preferred grasses and forbs, degrades ground cover, and limits plant diversity. Other factors, such as soil depth and type, weather, and type of livestock, also play a role in the impact of heavy grazing on prairie habitat. But how long the overgrazing has occurred has a great influence. If 10 years or less, preferred plants, such as little bluestem or yellow indiangrass, can recover when the livestock are removed and the land is rested. If 10 years or more, plants preferred by the livestock are lost and replaced with short grasses and/or species such as western ragweed, frostweed, and Texas persimmon that are tolerant to heavy grazing.

Also, during long-term heavy grazing, bare ground increases and both native and exotic invaders such as cedar, mesquite, Texas persimmon, and KR bluestem move in. Long-term overgrazing results in a persistent shift in the plant community until very few if any of the animals' preferred plants survive. Long-term, continuous heavy grazing also decreases the insects that live on native forb and grass species. These insects are essential food for insectivorous birds such as Eastern Bluebirds and Scissor-tailed Flycatchers.

When long-term overgrazing changes the plant community, landowners often switch to another class of livestock that will eat what is present. For example, grazers such as cattle may be replaced by forb-eating sheep. Sheep then are replaced with goats that browse and can live on the limited community of woody plants that remain.

Does the type of animal make any difference when restoring land damaged by livestock?
Yes. The two ranch animals that have had the greatest effect on Hill Country grasslands and savannahs are goats and cattle. Horses and sheep have less effect on these plant communities. Their impacts are very different, so restoration is also different. Goats are browsers. They eat shrubs, trees, forbs, and some grasses. Land that has been overbrowsed by too many goats has few forbs, so tree branches and only the

most resilient, unpalatable shrubs are within reach of the goats. This has a significant impact on birds like Spotted Towhees and Painted Buntings that are dependent on understory for protection and/or nest cover. When restoring areas damaged by goats, remove animals and rest the land to allow plant regrowth. Also, since deer and goats have similar food preferences, you must aggressively control deer numbers or the deer will consume the valuable plant species even though the goats are gone. Cattle are grazers. They mostly eat grasses, some forbs, and few woody plants (unless they are very hungry). Land heavily overgrazed by cattle will have lost or have many fewer midsized, native grasses such as little bluestem, sideoats grama, dropseed, and green sprangletop. Ground-nesting birds such as Northern Bobwhite and Rio Grande Turkey, as well as winter sparrows, that feed on the ground need the protection of these midsized grasses growing in dense clumps. Through the years, cattle have had a great impact on Hill Country grass populations so that grass communities in many areas have shifted to species that are less palatable (the threeawns) or are able to reproduce very close to the ground and withstand heavy grazing (KR bluestem). These grasses do not form the dense clumps needed to protect turkey and quail from predators and bad weather. Restoration after overgrazing requires reduced stocking levels, conversion to a rotational grazing system that allows midsized clump grasses to reproduce, and time.

Horses primarily eat grass. In the past, the Hill Country had many cattle ranches and few horse farms, so horses have had less impact on grass communities. Where land has been damaged by too many horses, restoration is similar to that for land overgrazed by cattle: rest, reduced stocking levels, and rotational grazing. Good horse pasture grass is precious and needs strenuous protection from overuse. To prevent plant loss and avoid erosion, limit grazing to one horse per 20–25 acres of good pasture or one horse per 60 acres of poor pasture. Thus, Hill Country horse farms should keep animals stalled or on dry lots and allow them out to graze on a limited basis—when there is enough rain to produce plenty of fresh grass. Sheep have never been as numerous as cattle or goats in the Hill Country so have had less impact on wildlife habitat. However, if you do have sheep and want to manage for wildlife too, it is important to know that sheep love forbs, and their preferred food puts them in competition with doves, quail, and seed-eating songbirds for the same plants. Forbs are also critical to healthy and diverse populations of insects that many Hill Country birds need. To attract the birds you want, you must not have too many sheep. The main components of a sheep's diet are forbs, but sheep also eat woody plants, and these need to be considered when managing for birds. On land where sheep are pastured, be sure to keep the stocking level low enough to protect the maturation of woody plant saplings. If you see a browse line, you have too many animals.

Problem 2: Poor Ground Cover That Allows Erosion and Soil Loss during Seasonal Storms

Tall bunchgrasses native to our area absorb and hold rainwater even in heavy rainfall. They prevent erosion by absorbing rainwater so that it does not flow across the surface and remove bits of plant litter and soil particles. Unfortunately, water erosion has been a major source of soil loss for many generations. Because the Hill Country has an arid climate, here it takes hundreds of years to replace 1 inch of soil.

Overgrazing increases stream channel erosion. When ground cover is severely damaged, the bare ground and extremely short grass on merely the slightest slope cannot absorb even a small amount of rainwater. As rainwater runs over the ground, sheeting is not inhibited in any meaningful way. Consequently, a much larger volume of water enters the stream channels and contributes to stream channel erosion.

In addition, the root systems of very short grasses are significantly shallower and less robust than the roots of native bunchgrasses. Thus, short grasses cannot effectively hold the soil, so heavy rains erode more and more soil particles that are carried away through the drainage system. Erosion in a drainage channel also increases as runoff intensifies and becomes a more powerful erosive force.

What You Can Do: Maintain Deep Ground Cover That Acts like a Sponge

To maintain deep ground cover, follow balanced range management, avoid overgrazing, and rest overgrazed rangeland until it recovers. The necessary restorative outcome of balanced grassland management is ground cover that holds and absorbs rainwater and builds the soil from the old plant material that dies and remains on the ground. Desirable and stable grassland habitat with deep ground cover can receive a 9-inch deluge and retain small plant material and soil particles without any runoff. Tall grasses also provide essential nest sites for ground-nesting birds such as Dickcissel, Northern Bobwhite, and Rio Grande Turkey.

Remember: A healthy watershed protected by the extensive roots of deep ground cover is essential to erosion control.

Problem 3: Bird Diversity Limited by Invasive, Dense Second-Growth Cedar or Mesquite

Many Hill Country grasslands are now overgrown by cedar or mesquite. Invasion by these woody plants could have occurred generations ago, and what once was prairie, pasture, or cropland may now be a dense

Prairie with plant diversity and hunting perches

Scissor-tailed Flycatcher
(Photo by Lora L. Render)

thicket. Or the invasion could have happened within the last 10–20 years, and the one-time prairie, pasture, or cropland may now have a sprinkling of small to medium-sized second growth cedar or mesquite.

What You Can Do: Brush Management

The sooner invasive species are controlled, the better, and grassland brush management improves bird habitat in many ways. Brush management has these effects:

- It increases or maintains plant diversity and results in a diverse and stable food supply for grassland birds that depend on rich native grassland habitat.
- It increases grass cover, thus improving habitat for ground-nesting birds such as the Northern Bobwhite, meadowlarks, and Rio Grande Turkey.
- It reestablishes openings among diverse herbaceous plants where birds like Eastern Bluebirds, Scissor-tailed Flycatchers, and Northern Bobwhites find insects to feed to their nestlings.

1. Manage mesquite.

Fire kills cedar (*Juniperus ashei;* not redberry juniper, *J. coahuilensis*), but it only top-kills mesquite trees. After being burned, mesquite roots immediately send up new sprouts. Mesquite control can be mechanical, chemical, or both. Mesquite trees do not shade out grasses and forbs, so after removing mesquite, it is not necessary to plant ground cover habitat. Also, with sufficient ground cover, it may be possible to immediately begin other grassland stewardship practices such as a prescribed burn to speed recovery of grassland composition and diversity. Some larger trees may be left as habitat for birds such as Bullock's Oriole and Vermilion Flycatcher. You can also "half-cut" a mesquite tree trunk to lay the tree down and keep it alive. This improves shrub-level bird habitat and is especially beneficial to species such as Northern Bobwhite.

MECHANICAL CONTROL (IN RELATIVELY DEEP SOIL)

- For young mesquite up to 6–8 feet tall and light to moderate density, hand grubbing works well and is cost effective. Use a long-handled tool with a very sharp, flat cutting edge. Be sure to cut plants below the soil surface so you remove the basal bud zone growth center.
- For large mesquite trees above 6–8 feet tall and growing over a large area, machine grubbing works well with an excavator, but it is expensive. The excavator with its backhoe arm is better than a bulldozer or chaining because it can take out specific trees, does not disturb as much soil, and removes the bud zone growth center.

Usually, after being grubbed or uprooted, trees can be left stacked in brush piles for bird habitat. There is no need to burn them unless you have too many to use for habitat improvement.

CHEMICAL CONTROL

- For young mesquite up to 6–8 feet high, spray herbicide on foliage.
- For a small number (about 50 or fewer trees) of large mesquite trees over 6–8 feet tall growing in a relatively small area, it might be easiest and least expensive to cut with a chainsaw and hand-paint stumps with an appropriate herbicide.

MECHANICAL PLUS CHEMICAL CONTROL

For large mesquite trees growing over a large area but fewer than 150 trees per acre, there is a technique using machine sheer and herbicide spray. In some cases it has been only 36% effective, but effectiveness may be increased significantly if done correctly. Spray stumps immediately after cutting, and use a special spray nozzle that makes a fine, broad, cone-shaped mist over the surface of the stump. For more information on chemical control methods for mesquite, call your county AgriLife Extension agent. Most, if not all, of these chemicals require special training and licensing from Texas A&M AgriLife Extension Service.

2. Manage cedar.

For information on clearing cedar, see chapter 9. After cedar is cleared, the open ground will probably look like very little is growing. Recovery is a process that takes time. Be patient. In most cases, a few individual plants are enough to regenerate native grasses and some forbs. After cedar clearing, the ground is no longer shaded, and sunlight now reaching ground-level plants also helps them flourish. As the herbaceous grassland species increase, they will produce more and more seeds for further improvement in the future.

In addition, when removing or thinning cedar, it is important to minimize soil loss. Lay cut cedar slash in windrows across the slope to catch soil and reduce runoff. Mulch the cut cedar, and spread a thin layer 1–3 inches deep to reduce runoff, control erosion, and conserve soil moisture.

Problem 4: Native Forbs of Healthy Grassland Bird Habitat Reduced or Eliminated by Deer Overpopulation

Deer overpopulation damages native plants, so many species are not being maintained at a sustainable level. Birds and other wildlife depend on those plant communities that are damaged by excessive numbers of deer.

What You Can Do: Reduce Deer Population Density

To foster bird-friendly habitat for all grassland birds, maintain long-term deer management. On Fort Hood, where deer density is controlled, biologists studied grassland birds during the winters of 2008–2009 and 2009–2010. They found more than 25 species living in grassland habitat. Savannah Sparrows were by far the most common, followed by Field, Vesper, Le Conte's, Grasshopper, White-crowned, and Song Sparrows. Because these birds need the forbs and shrubs that the deer love to eat, deer overpopulation must be reduced in the restoration of all Hill Country grassland habitats. For more information, see chapter 8.

Problem 5: Quality of Grassland Bird Habitat Limited by Dense Litter

The openings between grass clumps in the prairie are places where many grassland birds hunt insects, search for forb and grass seeds, find shelter, and hide from predators. Over time, these spaces may be covered by old plant pieces, or litter, so that the birds can no longer find food there. Thus, the grassland deteriorates into inferior habitat.

What You Can Do: Conduct a Prescribed Burn

Natural wildfires once occurred more regularly and cleaned the ground. Today, a prescribed burn is the most effective way to eliminate dense litter buildup, control cedar, and improve grassland bird habitat. Reducing litter buildup through prescribed burning benefits birds in many ways:

- Fire makes openings on the soil between bunchgrasses, which become growing sites for forbs that sustain seed-eating birds.
- Fire creates a place for ground-feeding birds such as the American Pipit to forage for both seeds and insects. This "snowbird" nests in the arctic and alpine tundra of the far north and comes here in the winter where it can escape the cold and find food.
- Fire is a tool to increase small-mammal populations needed by raptors. Burning vegetation increases plant diversity, especially forbs that are food for cottontails, jackrabbits, cotton rats, and pygmy mice. These species in turn are food for Loggerhead Shrike and raptors such as the Red-tailed Hawk, Swainson's Hawk, Northern Harrier, and Red-shouldered Hawk.

If you are considering a prescribed burn, consult with your local prescribed burn association, TPWD, NRCS, or a private burn consultant for advice and assistance.

Le Conte's Sparrow
(Photo by Lora L. Render)

River valley prairie with tall grasses and forbs

Problem 6: Native Plants Needed for Food and Shelter Reduced by Invasion of Exotic Plant Species

Native grasslands consistently support more bird species than do grasslands dominated by exotic plants. Exotic plants reduce plant diversity, damage plant community structure, and provide an unstable food supply. KR bluestem is the most widespread problem.

What You Can Do: Remove Invasive Exotic Plants on a Regular Basis

Just as cedar management is never ending, exotic plant control is an ongoing process. Each invasive exotic plant species has its own unique

Remember: Exotic plants are detrimental to birds because birds depend on native plants for food and shelter. When exotics move in, they displace native plants.

American Pipit
(Photo by Lora L. Render)

Prairie with mixed grasses, forbs, and bare ground

problems and control methods. But with good long-term control, the job gets easier in time.

1. Control KR bluestem.

KR bluestem, the most prevalent invasive exotic grass species in the Texas Hill Country, grows in masses that are inferior bird habitat. Its structure does not provide good ground-level nest sites or escape cover. Birds such as Rio Grande Turkey, Northern Bobwhite, and Rufous-crowned Sparrow must have the taller, native, warm-season bunchgrasses for nesting and cover. Help native grasses compete with invasive KR bluestem: mow infrequently, allow only periodic livestock grazing, use light to moderate stocking levels, and do not allow overgrazing. Summer burns, especially very hot, intense fires, may be beneficial in setting back KR enough to provide space for natives. More research is needed to further clarify the benefits of summer burns.

If you have a dense stand of KR bluestem and want native bunchgrasses and perennial forbs for better bird habitat, mowing or burning alone will not work. You must use herbicide to kill it and then plant the desired native grasses and forbs. Be sure to use the herbicide according to directions. Do not use preemergent herbicide to kill KR bluestem seeds because it also kills the native seeds that you are planting. When done right, this herbicide control system may produce 50%–75% native grasses and forbs but may take two to three years to show these results. Follow these steps:

- Spray herbicide in late summer or early fall when KR bluestem is growing actively.
- Disk the area a few weeks later.
- At this time, you may not see satisfactory reduction of KR bluestem plants. If not, apply a second herbicide treatment on actively growing plants.
- Be careful with replacement seed selection. Consult experts to find species native to your location. Be sure to use a reliable source of native seeds.
- Plant native grass seed the next spring.
- For best results, wait until the following fall to plant forbs. Because they are expensive, you may need to plant a limited selection. This is okay. Avoid aggressive forbs like Mexican hat.

It is best to complete the first two steps in late summer and early fall and delay the last three steps until the following spring. Waiting to plant in spring allows the disked soil to increase moisture and potentially freeze and thaw in the winter, which creates much better planting conditions. If you must do the full herbicide treatment in a single season, prepare and plant within two months between March and May. Rain and other weather conditions may cause delays and make it difficult to fit in all the steps.

> *Is herbicide treatment for KR bluestem control worth all the trouble?*

The short answer is yes. KR bluestem is well established in the Hill Country and is so well adapted that it is very hard to remove. Fol-

Remember: The more KR bluestem is mowed or grazed, the more abundant it becomes.

lowing a successful herbicide treatment to control KR bluestem, the new plant community composition will benefit grassland birds by creating improved food production and structural diversity that provides protection from predators and bad weather.

2. Control johnsongrass.

Johnsongrass is an invasive exotic grass from the Mediterranean region that limits the variety of available nest sites and native food for grassland birds. It is hard to control but does not spread as fast as KR bluestem. Johnsongrass grows most quickly in the spring and spreads primarily by underground rhizomes from existing plants. Its leaf blades are long and wide, making them easy targets for early-spring "wick application" of herbicide.

Mow johnsongrass anytime from late fall through the winter, and instead of spraying, swab the herbicide on the fast-growing young leaves with the wick applicator in the early spring. A wick applicator looks like a long hockey stick; it allows you to treat individual plants growing near desirable species, avoids the indiscriminate chemical drift of spraying, and requires less herbicide.

How does johnsongrass impair grassland bird habitat?

Even though many grassland birds eat its seeds, johnsongrass has a negative effect on bird habitat. As an invasive exotic, johnsongrass crowds out native grassland plants, reduces plant species composition, limits insect diversity, and degrades habitat structure. These changes limit the variety of available nest sites and native food for grassland birds. Johnsongrass probably facilitates these changes by suppressing germination and growth of other plants through allelopathy—the release of seed-germination and/or growth-inhibiting substances into the soil.

3. Control Chinese tallow, chinaberry, pyracantha, and ligustrum.

Chinese tallow, chinaberry, pyracantha, and ligustrum are four of the most persistent invasive, exotic woody plants in the Hill Country grasslands. All are difficult to control. Simple cutting does not kill them because they resprout from the roots.

- Use mechanical levers such as a Pullerbear, Extractigator, UpRooter, or Root Jack to extract plants with a trunk diameter less than 2.5 inches.
- For larger plants, cut and paint stumps with an appropriate herbicide such as triclopyr (e.g., Remedy or Brush-B-Gone). Use all pesticides according to the label directions.

EXOTIC PLANTS DAMAGE GRASSLANDS BIRD HABITAT

Exotics reduce plant diversity. Invasive exotic plants take over habitat space and limit diversity. As plant and seed diversity diminishes, grassland can no longer support bird species dependent on the native plants.

Exotics damage plant community structure. Invasive exotic species change the structure of a plant community and limit escape cover and food, thus making the habitat less attractive to some wintering and breeding birds. The common Savannah Sparrow and uncommon Sedge Wren are examples of wintering species that depend on native grassland habitat structure.

Exotics supply an unstable food supply. Where grassland is dominated by a few exotic species, the birds that might be able to survive there have a greater risk of failed food supply. In bad years, the limited plant species produce very few seeds. The birds that live on these seeds cannot survive and are driven away. They may not return.

Exotic plant seeds are spread by birds. American Robins, Northern Mockingbirds, and Cedar Waxwings may survive feeding on exotic plants, but the birds also spread them all over the countryside. As seeds pass through their gut and are deposited on the ground, they are distributed to wherever the birds fly and perch.

4. Control invasive exotic forbs.

Exotic annual bastard cabbage, Malta star-thistle, Brazilian vervain, musk thistle, and bur-clover are some of the exotic invasive forb species that spread readily along roadsides and through old fields and other disturbed areas. Once established, these species are difficult and costly to control. Watch for them so they can be removed as soon as they appear.

- If you find a few plants, pull or grub them out with a hoe before they go to seed.
- Kill larger infestations with a broadleaf herbicide or horticultural vinegar. As in the case of exotic invasive grasses, no chemical control method is completely effective. For chemical control of a specific species, consult with your county AgriLife Extension Service agent.
- The key is seed control. Agricultural weed control experiments suggest that oversowing infested areas with species having biological traits similar to the those of the invasive species can significantly reduce seed productivity. For example, using Indian

Savannah Sparrow
(Photo by Lora L. Render)

Prairie with mixed grasses and forbs

blanket wildflower seed to oversow established seedling colonies of annual bastard cabbage can reduce the exotic invasive plant's seed set by 83%.

Problem 7: Poor Bird Habitat Resulting from Old Field Succession Stuck in an Unproductive Plant Community

Invasive exotic grasses have been widespread in the Hill Country for many years, and they now dominate old field succession. KR bluestem is the most aggressive and well-adapted exotic grass. It hijacks old field succession, changing a relatively productive forb community into a stagnant and inferior habitat. KR forms large, dense stands and excludes most competing species. A stagnant old field plant community dominated by KR bluestem is a deficient food source lacking natural diversity of seeds or insects. Wildlife biologists consider a KR bluestem field to be a poverty-stricken "wasteland" instead of the rich native grassland needed by birds and other wildlife.

For bird stewardship, old fields provide varying degrees of food and cover at each stage of ecological succession. When fire burned through an area or early pioneer farmers plowed their fields, they created good habitat for doves, pipits, quail, and lark sparrows. These major disturbances produced early-succession stages with abundant forb and grass seed eaten by these birds. The downside of this early stage may be lack of adequate cover. The Field Sparrow needs a little more cover from a few shrubs and is a grassland and savannah bird well adapted to the later, intermediate succession stages of a healthy (little or no KR bluestem) old field.

Our emphasis here has been on KR bluestem. Cedar and mesquite are invasive native species that also hijack secondary ecological succession in old fields. For information on controlling cedar or mesquite, refer to chapter 9 and the section in this chapter on dealing with mesquite.

What You Can Do: Manage Restoration of Old Fields
In the Hill Country, many old fields may have been used to grow oats, haygrazer, or milo.

1. Learn about the history of your property, and verify crop use in your suspected old field.

- Ask a neighbor who has been around for a while.
- Review old aerial photos for signs of past crop production, or consult with a farm service organization like NRCS.
- Look for indicators on the ground, such as old fences that might have separated an old field from adjacent grazed pastures. Sometimes you will see only brushy tree lines where the fences once stood.
- If an old field is on a slope, it may have artificial terraces that were installed to control erosion. These terraces are ridges that wind across the field and follow the contour of the land.

2. Know the natural stages in sequence of natural ecological succession for such an old field.

- Stubble or remnants of the annual crop
- Annual forbs (broadleaf weeds)

- Annual grasses
- Mostly native perennial forbs and grasses
- Diverse mix of mostly native forbs and grasses with some woody plants

3. Manage the old field.

Old fields can be managed through the natural changes of ecological succession from the annual crops to a biologically diverse savannah community.

4. Figure out a way to skip or halt the "wasteland"
stage dominated by KR bluestem.

This is not easy. It requires drastic measures, and you cannot expect to completely eliminate such a well-established exotic species. For details on control of invasive exotic grasses, see the KR bluestem section in problem 6 in this chapter.

5. Add diversity for bird habitat in old field restoration.

For additional habitat diversity, you can maintain a small portion of an old field in an early-succession stage by disking it every year or two. This will keep the area in annual forb production. Disking an area essentially creates a food plot within the larger block where the birds have nest and cover habitat.

Problem 8: Natural Bird Habitat Damaged by Excessive Mowing

Mowing too much and too short damages natural bird habitat. This practice is especially common in suburban areas where owners want their property to look like a park. Excessive mowing eliminates habitat needed by many grassland birds. Mowing reduces native grass diversity. On small acreages, mowing is often used to create a parklike look. Mowing is also used for fire prevention by removing plants that could fuel a fire. Excessive mowing is common, and excessive mowing of perennial native grass (little and big bluestem, yellow indiangrass, switchgrass, eastern gamagrass, sideoats grama, and dropseed) gradually shifts the naturally complex composition of a native grass community to one of few species. Such a monoculture does not support many native bird species and attracts exotic birds that can eat almost anything and thus can survive in places with little diversity.

Mowing reduces bird food and cover. Excessive mowing removes food and cover and can dramatically reduce bird species. Too much mowing interrupts the regular reproductive cycle of annual forbs such as bluebonnet, beebalm, and firewheel, which are bird food, as well as native

Field Sparrow (Photo by Jeff Forman)

Old field with native grasses, forbs, and bare ground

bunchgrasses that provide bird food and cover. Because they can grow and reproduce below a mower blade, exotic and turf-forming grasses can survive in spite of regular mowing and then replace many of the essential native plants.

Mowing destroys bird nests. Mowing during the nesting season will physically destroy almost all nests of ground-nesting birds, including Rio Grande Turkey, Northern Bobwhite, Field Sparrow, Rufous-crowned

Lark Sparrow (Photo by Lora L. Render)

Old field with short grasses and bare ground

Sparrow, and Lark Sparrow; however, because meadowlarks nest in a slight depression in the ground, their nests sometimes survive mowing.

Mowing alters habitat structure. Many ground-nesting birds need grasses with significant aboveground structure for concealment. Because excessive mowing reduces perennial bunchgrasses and favors turf-forming grasses, essential structure is lost.

Mowing decreases forbs and insects in grassland communities. Excessive mowing can also decrease forb species. These plants are home to insects; when the plants go, so do the insects. The insects are food for Scissor-tailed Flycatcher, Eastern Bluebird, and virtually all other grassland bird species.

Eastern Bluebird
(Photo by Lora L. Render)

River valley prairie with nest box

What You Can Do: Restore Grassland Habitat Damaged by Mowing

In this case, the land stewardship correction is easy. Stop mowing or change how you mow. Possible changes include mowing with the highest setting, only once a year, no more than one-third of the habitat in any year, or late in the season when all plants are dormant (immediately after the first frost is a good time). With adequate rainfall, these choices will immediately benefit winter, migrant, and breeding prairie birds.

Problem 9: Grasslands Lacking Good Nest Sites for Cavity-Nesting Birds

The Hill Country has only eight species of cavity-nesting birds. Of these, three or four routinely nest at prairie edges. They prefer building their nests in dead snags or in live trees with natural holes or ones created and abandoned by woodpeckers. Where snags and trees with dead sections are absent, cavities suitable for nests are a limiting factor for all these birds. Since dead tree removal is very common, cavity-nesting birds are routinely eliminated from otherwise suitable habitat because they cannot find a good place to build their nests.

Eastern Bluebirds are especially popular cavity-nesting birds of open grasslands and savannahs. They thrive in habitat with a healthy mix of grasses and forbs adjacent to a wooded area with nest sites. Bluebirds commonly perch on a branch, post, or wire while looking for insects such as caterpillars, grasshoppers, and spiders and can spot them at a distance of 60 feet or more. They usually catch their food on the ground, so they prefer an open landscape with some bare ground between clumps of grass.

What You Can Do: Provide Nest Sites for Cavity-Nesting Birds
1. Add bird boxes.

The species that routinely live in or near our grasslands are Eastern Bluebird, Ash-throated Flycatcher, Bewick's Wren, and Carolina Wren. Fortunately, all these birds use the same kind of nest box. (Purple Martins nest in colonies near open areas. Black-crested Titmice and Carolina Chickadees prefer wooded habitat.) For information on bird box design, placement, maintenance, predator guards, and monitoring, see chapter 6.

2. Let dead trees stand and fall apart naturally.

These snags and trees with dead sections are preferred by cavity-nesting birds. Where woody vegetation is taking over grassland habitat, you can conduct a prescribed burn to create snags and open the area again. For information on snags, see chapter 1.

SUMMARY OF HABITAT PROBLEMS IN GRASSLANDS

HABITAT PROBLEMS	STEWARDSHIP SOLUTIONS	AFFECTED BIRDS
Problem 1: Inferior bird habitat resulting from overgrazed grasslands	Restore overgrazed land Use rotational grazing Control deer density Reduce soil loss with slash and/or mulch Use prescribed burns Reseed as needed	Dickcissel Eastern Bluebird Eastern Meadowlark Le Conte's Sparrow Loggerhead Shrike Northern Bobwhite Rio Grande Turkey Scissor-tailed Flycatcher
Problem 2: Poor ground cover that allows erosion and soil loss during seasonal storms	Manage for deep ground cover that absorbs maximum rainfall	Lesser Goldfinch Rio Grande Turkey Rufous-crowned Sparrow Vesper Sparrow
Problem 3: Bird diversity limited by invasive, dense second-growth cedar or mesquite	Manage cedar and mesquite for maximum native plant and wildlife diversity	Eastern Bluebird Eastern Meadowlark Field Sparrow Northern Bobwhite Rio Grande Turkey Scissor-tailed Flycatcher
Problem 4: Native forbs of healthy grassland bird habitat reduced or eliminated by deer overpopulation	Reduce deer density to a sustainable level so deer are not damaging native plant species	Field Sparrow Grasshopper Sparrow Le Conte's Sparrow Northern Bobwhite Rio Grande Turkey Savannah Sparrow Song Sparrow Vesper Sparrow White-crowned Sparrow
Problem 5: Quality of grassland bird habitat limited by dense litter	Remove litter with prescribed burn or mow and bale	Black-throated Sparrow Cassin's Sparrow Eastern Meadowlark Grasshopper Sparrow Lark Sparrow Loggerhead Shrike Mourning Dove Northern Harrier Red-shouldered Hawk Swainson's Hawk
Problem 6: Native plants needed for food and shelter reduced by invasion of exotic plant species	Control invasive exotic species	Eastern Meadowlark Lesser Goldfinch Northern Bobwhite Rio Grande Turkey Rufous-crowned Sparrow
Problem 7: Poor bird habitat resulting from old field succession stuck in an unproductive plant community	Eliminate stagnant communities dominated by invasive species	Cassin's Sparrow Eastern Meadowlark Field Sparrow Lark Sparrow Mourning Dove Northern Bobwhite
Problem 8: Natural bird habitat damaged by excessive mowing	Limit mowing Mow no more than 30% each year Mow at highest setting Mow when plants are dormant	Eastern Meadowlark Lark Sparrow Northern Bobwhite Rio Grande Turkey Rufous-crowned Sparrow Scissor-tailed Flycatcher
Problem 9: Grasslands lacking good nest sites for cavity-nesting birds	Add, monitor, and maintain boxes for cavity-nesting birds Leave standing dead trees as snags	American Kestrel Eastern Bluebird

Bird Summaries: Featured Birds

AMERICAN PIPIT

UNDER-THE-RADAR BIRD
WINTER

Often overlooked since it is a small, plain bird that stands upright and walks or runs but does not hop on the ground; can be found during the winter singly or in flocks usually on short-grass pasture, parks, or ball fields; has undulating flight pattern.

Identification: About 6.5 inches head to tail; brown above and buffy below with streaked or spotted breast; a pair of faint wing bars; delicate, thin bill.
Habitats: Old field, plateau prairie, river valley prairie, river, creek, tank, pond, and lake.
Cover: Mudflats of wetlands or ponds, short grasses, pastures, and disturbed areas such as plowed or recently burned fields.
Food: Aquatic and terrestrial invertebrates collected on the ground.

Stewardship: Control woody plant invasion, maintain short grass with sufficient bare ground, and encourage mudflats around ponds and tanks.
- Use light to moderate rotational grazing to maintain balance of grass, forbs, and bare ground.
- Control white-tailed deer and other browsing animals to increase forbs for insect diversity.
- Apply ant-specific insecticides to control red imported fire ants and increase insect diversity.

American Pipit

- Use summer or fall prescribed burns to control woody plants, reduce litter buildup, and reduce vegetation height. Late winter burns are probably too late to benefit this winter bird.

Other Potentially Useful Management:
- Avoid broad-spectrum insecticides to increase insect diversity.
- Use mechanical and chemical techniques to control woody plants, but they do not address litter buildup or vegetation height.

BLACK-THROATED SPARROW

PRIORITY BIRD

ALL YEAR

Attractive and popular, arid, grassland sparrow of the western Texas Hill Country; often overlooked when it comes to stewardship. Breeding bird surveys since 1966 show a troubling decline of the Black-throated Sparrow in Texas.

Identification: About 5.5 inches head to tail; black breast and chin contrast with white belly; bright white stripes radiating from above and below its bill; male's song from a high perch is easy to recognize.
Habitats: Plateau prairie, river valley prairie, shin oak savannah, and live oak savannah.
Cover: Thin ground cover of grasses and cactus with scattered trees and shrubs.
Breeding Territory: About 2–4 acres/pair; territory apparently contracts once incubation begins.
Nest: Open grass cup often built in a shrub about 14 inches above the ground.
Food: Insects gleaned from low branches but also on the ground where seeds and some plant material are collected.

Stewardship: Encourage diverse grasslands with sufficient bare ground while controlling woody plant invasion.
- Use light to moderate rotational grazing to maintain balance of grass, forbs, and bare ground.
- Control white-tailed deer and other browsing animals to encourage healthy shrubs for nesting and a diversity of forbs for food.
- Control tall woody plant invasion to reduce perches used by cowbirds during parasitism.

Black-throated Sparrow

- Use prescribed burns to control woody plants and reduce litter buildup.
- Avoid broad-spectrum pesticides to increase forbs and promote insect diversity.
- Control red imported fire ants with an ant-specific insecticide to increase insect diversity.
- Trap and dispatch cowbirds to reduce nest parasitism and increase nesting success.

Other Potentially Useful Management:
- Construct tepee-style brush piles where shrubs are lacking to increase shrub-level habitat.
- Control woody plants with appropriate mechanical or chemical methods to limit woody plant invasion; however, this does not remove litter buildup or increase bare ground.

CASSIN'S SPARROW

PRIORITY BIRD

ALL YEAR

Secretive sparrow found in western Texas Hill Country grasslands. Known to initiate nesting in response to substantial rainfall. Breeding bird surveys since 1966 show a disturbing population decline of Cassin's Sparrow populations in the Texas Hill Country.

Identification: About 5.5 inches head to tail; nondescript, pale gray-brown sparrow with faint yellow wash on wings; easiest to identify by male's loud, sweet song and territorial display; male sings as he flies up and then floats down on spread wings.

Habitats: Old field, plateau prairie, river valley prairie, pocket prairie, post oak savannah, shin oak savannah, and live oak savannah.

Cover: Dry grassland with scattered shrubs and low trees.

Breeding Territory: About 6.5 acres/pair.

Nest: Grass cup on ground or within a few inches of the ground.

Food: Mostly insects but some seeds usually picked up off the ground.

Cassin's Sparrow

Stewardship: Control woody plant invasion, and maintain adequate grass and forb (insect habitat) cover.
- Use light to moderate rotational grazing to encourage balance of forbs, grasses, and woody plants.
- Use prescribed burns with long rotations to control woody plants and reduce litter buildup and maintain adequate grass cover.
- Control white-tailed deer and other browsing animals to encourage woody plant recruitment, healthy shrubs, and a diversity of forbs.
- Limit pesticide use, and control red imported fire ants with an ant-specific insecticide to promote insect diversity and maintain food supply.
- Use mechanical and/or chemical techniques to control woody plant invasion, but this will not remove litter buildup or increase bare ground.

Other Potentially Useful Management:
- Construct tepee-style brush piles where lacking to increase shrub-level habitat.
- Control feral hogs to reduce possible nest predation.

EASTERN BLUEBIRD

FAVORITE BIRD

ALL YEAR

Beautiful blue bird frequently seen on fences and overhead wires; in the Hill Country, especially fond of open country with low ground cover adjacent to wooded riparian habitat; a cavity-nesting bird whose populations were once in decline but started to recover in 1960s and 1970s thanks to use of nest box design that excludes starlings.

Identification: About 7 inches head to tail; typical thrush shape with slender beak; males bright blue back and head, orange front, and white belly; females similar but with plain blue-gray back and head.

Habitats: Old field, plateau prairie, river valley prairie, pocket prairie, post oak savannah, shin oak savannah, live oak savannah, river, creek, and backyard.

Cover: Open grassland with adjacent woods or savannah.

Breeding Territory: Averages about 5 acres/pair.

Nest: Made of all the same kind of grass; in natural cavities and regularly uses nest boxes.

Food: Insects that live on a mix of forbs and grasses; also fruit, primarily during winter.

Eastern Bluebird

Stewardship: Maintain open, diverse grassland areas adjacent to woodlands or savannahs with adequate snags for nesting.
- Avoid broad-spectrum pesticides to increase forbs and promote insect diversity.
- Keep snags for cavities to increase nest sites. Pay special attention to potential of drought-killed trees.
- Use prescribed burns to encourage forbs and create snags suitable for nesting.
- Use light rotational grazing in fall or winter dormant season to promote forbs.
- Apply ant-specific insecticides to control red imported fire ants and increase insect diversity.
- Control white-tailed deer and other browsing animals to encourage a diversity of forbs.
- Control exotic species House Sparrows and European Starlings to minimize nest parasitism.
- Check boxes weekly, and remove exotic species to enhance reproduction.
- Add nest boxes placed approximately 100 yards apart and 100 feet from woodland edges to provide additional nest sites.
- Clean nest boxes after each nesting to allow additional nestings.

Other Potentially Useful Management:
- Mow periodically in fall or winter dormant season on no more than 30% of herbaceous area to promote annual forbs and insects.
- Maintain bird-friendly, natural or artificial water source to supply reliable and safe water.
- Keep cats indoors, and control stray and feral cats to reduce predation.

EASTERN MEADOWLARK

PRIORITY BIRD

ALL YEAR

A prairie emblem as it perches proudly on fences; common in fall and winter flocks but declining in most of its range; when flushed, flies away with quick wing beats followed by an "outer-spacecraft-like" glide. Breeding bird surveys since 1966 show a disturbing decline of Eastern Meadowlark populations in Texas.

Identification: About 9.5 inches head to tail; males larger than females; lemon-yellow front and black bib, thin beak, and short tail; except by song nearly impossible to distinguish from Western Meadowlark; song is plaintive, higher pitched, and flutelike but more varied than Western's; some say Western's song translates, "I come in peace, yes I do."

Habitats: Old field, plateau prairie, river valley prairie, pocket prairie, post oak savannah, shin oak savannah, and live oak savannah.

Cover: A variety of healthy grassland communities from rangeland to airports with balance of grasses, forbs, and bare ground.

Breeding Territory: About 7.0 acres/pair.

Nest: Grass lined; on the ground; often in a depression; sometimes with a top made of plant pieces placed over the nest.

Food: Mostly insects picked up from the ground.

Stewardship: Maintain diverse grassland with a balanced mix of grass for cover, forbs to attract insects, and bare ground where food is collected.
- Conduct prescribed burns to control invasion by woody plants, encourage forbs, and increase bare ground.
- Use appropriate mechanical and/or chemical techniques to control woody plant invasion.
- Employ light to moderate rotational grazing to encourage forbs and to maintain bare ground and grass cover at least 4–12 inches high.
- Apply ant-specific insecticides to control red imported fire ants and increase insect diversity.
- Avoid broad-spectrum pesticides to increase forbs and promote insect diversity.
- Control white-tailed deer and other browsing animals to improve forb diversity.
- Trap and remove cowbirds to reduce nest parasitism and increase nesting success.

Other Potentially Useful Management:
- Use infrequent (once every three years) mowing to increase forbs. Do *not* mow during the nesting season.
- Keep cats and dogs indoors, and control stray and feral cats to reduce predation.

Eastern Meadowlark

FIELD SPARROW

PRIORITY BIRD

ALL YEAR

Before the influence of European settlers and their domestic animals, probably benefited from the migratory grazing habits of huge bison herds and wildfire disturbances. Favorite habitat now is disturbed grassland or savannah and abandoned old fields. They live in these old fields from a year or so after abandonment for about 10 years when ground cover remains short and sparse enough for nesting. Breeding bird surveys since 1966 show a disturbing population decline of Field Sparrows in the Texas Hill Country.

Field Sparrow

Identification: About 5.75 inches head to tail; gray head with rusty cap and well-defined, white eye ring; conspicuous pink bill and legs; wing bars.

Habitats: Old field, plateau prairie, river valley prairie, pocket prairie, post oak savannah, shin oak savannah, and live oak savannah.

Cover: Short, sparse, and scrubby old fields; areas with a mix of grasses and low shrubs.

Breeding Territory: Average less than 2 acres/pair.

Nest: Grass cup on or near the ground, usually among spare clumps of short grasses.

Food: In fall and winter mostly grass seeds; in spring and summer a near equal mix of seeds and insects.

Stewardship: Maintain habitat in early- to mid-successional stages with mix of grasses and scattered shrubs.
- Use prescribed burns to minimize woody plant encroachment and maintain grassland structure in old fields.
- Use moderate grazing after the breeding season to help maintain early-successional stage.
- Do light disking every four to six years to keep old fields in an early-successional stage. Be sure to leave some shrubs.
- Apply ant-specific insecticides to control red imported fire ants and increase insect diversity.
- Avoid broad-spectrum pesticides to increase forbs and promote insect diversity.
- Build tepee-style brush piles about every 40 yards to help offset lack of shrub habitat.
- Control white-tailed deer and other browsing animals to improve forb diversity.
- Trap and remove cowbirds to reduce nest parasitism and increase nesting success.

Other Potentially Useful Management:
- Mow periodically in fall or winter dormant season on no more than 30% of herbaceous area to promote annual forbs and insects.

GRASSHOPPER SPARROW

PRIORITY BIRD

ALL YEAR

In spring and summer, recognized by its distinct "buzzy" song. Has an appetite for grasshoppers; both song and food may have contributed to its name. Declining throughout its range in the United States due to habitat loss, fragmentation, and degradation.

Identification: About 5 inches head to tail; flat head and stubby neck; chestnut-and-black back and wings; inconspicuous yellow lore (mark in front of eye).
Habitats: Old field, plateau prairie, river valley prairie, and shin oak savannah.
Cover: Grasses with bare ground and scattered shrubs.
Breeding Territory: About 2.0–4.5 acres/pair.
Nest: Grass cup on the ground with roof of vegetation and a side entrance.
Food: Mostly insects, especially grasshoppers, but grass seeds also important in winter diet; feeds on the ground.

Stewardship: Maintain larger tracts of grassland with a balance of bare ground for foraging, grasses for nesting, forbs for food, and some shrubs as cover.
- Limit or avoid insecticide use to maintain good supplies of insects.
- Increase insect diversity by controlling red imported fire ants with ant-specific insecticides.
- Control white-tailed deer and other browsing animals to encourage healthy scattered shrubs.

Grasshopper Sparrow

- Use light rotational grazing to encourage balance of forbs and grasses and woody plant diversity.
- Use prescribed burns to maintain desirable plant community of grasses and forbs and some bare ground to limit woody plant invasion.
- Use appropriate mechanical or chemical woody plant control techniques to control woody plant invasion.

Other Potentially Useful Management:
- Construct tepee-style brush piles where shrubs are lacking to increase shrub-level habitat.
- Control feral hogs, keep cats indoors, and control stray and feral cats to reduce predation.

LARK SPARROW

PRIORITY BIRD
ALL YEAR

In early summer, if you walk in a short grass field and flush several birds with bright white outer tail feathers, they may be a family or a group of lark sparrow families feeding on insects. Breeding bird surveys since 1966 show a disturbing decline of Lark Sparrow populations in the Texas Hill Country.

Identification: About 6.5 inches head to tail; males and females have striking rust, white, and black head markings plus white outer tail feathers.

Habitats: Old field, plateau prairie, river valley prairie, pocket prairie, post oak savannah, shin oak savannah, and live oak savannah.

Cover: Short to medium-tall grasses with good mix of grasses, forbs, and bare ground; adjacent to woody habitats; often in disturbed habitat such as old fields or after a fire.

Breeding Territory: Limited information, maybe 1 to about 15 acres/pair.

Nest: Lined with grass; usually in a shaded depression on the ground among short grasses.

Food: Primarily seeds and some insects picked up on the ground.

Stewardship: Maintain diverse grassland with mix of grass, forbs and some shrubs.
- Use prescribed burns to minimize woody plant encroachment and maintain grassland structure.
- Use moderate to heavy short-term grazing after the breeding season to help maintain early-successional stage.

Lark Sparrow

- Mow only in fall or winter dormant season to promote forbs. Do not mow during the nesting season.
- Control white-tailed deer and other browsing animals to encourage a diversity of forbs.
- Use limited cultivation such as occasional light disking to increase forbs and bare ground.
- Trap and remove cowbirds to reduce nest parasitism and increase nesting success.
- Apply ant-specific insecticides to control red imported fire ants and increase insect diversity.
- Avoid broad-spectrum pesticides to increase forbs and promote insect diversity.

Other Potentially Useful Management:
- Keep cats indoors, and control stray and feral cats to reduce predation.

LE CONTE'S SPARROW

PRIORITY BIRD

WINTER

Very shy winter visitor to wet meadows of the Hill Country; has unusually weak flight; when disturbed, flies out of the grass for a moment before dropping to run along the ground and hide.

Identification: About 4.75 inches head to tail; tiny sparrow; best identified by flight pattern and yellow-tan color when seen with the sun behind you.
Habitats: Old field, river valley prairie, pocket prairie, and post oak savannah.
Cover: Dense cover of mid- to tall grasses and forbs.
Breeding Territory: N/A
Nest: N/A
Food: Seeds of grasses and forbs.

Stewardship: Encourage forbs and maintain winter grass cover as high as possible, especially in pocket prairies.
- Use prescribed burns to minimize woody plant encroachment and maintain balance of grass and forbs.
- Employ light rotational grazing during growing season to improve balance of grass and forbs.
- Control white-tailed deer and other browsing animals to encourage a diversity of forbs.
- Avoid broad-spectrum herbicides to increase seed-producing forbs.

Other Potentially Useful Management:
- Mow infrequently (once every three years) to promote forbs, but probably has limited value.

Le Conte's Sparrow

LESSER GOLDFINCH

PRIORITY BIRD

ALL YEAR

Except when nesting, this striking finch is a wanderer, often moving in large flocks; recognized by its cheerful twittering. Breeding bird surveys since 1966 show a distinct decline of Lesser Goldfinch populations in the Texas Hill Country.

Identification: About 4.5 inches head to tail; males in this area have black head and back (occasionally green) with bright yellow front.
Habitats: Old field, river valley prairie, pocket prairie, post oak savannah, shin oak savannah, live oak savannah, river, creek, seep, springs, and backyard.
Cover: Weedy patches adjacent to wooded areas such as creeks or savannahs.
Breeding Territory: Little information, but males are reported to defend an area about 30 yards in diameter.
Nest: Small open cup, about 4 feet to almost 45 feet off the ground.
Food: Mainly seeds but also buds, flowers, and fruit.

Stewardship: Maintain a variety of seed-producing forbs for food and large trees for nesting.
- Use prescribed burns to encourage perennial forbs like Maximilian sunflower and maintain diverse savannahs.
- Create periodic disturbance by light disking for annual forb production such as annual sunflower.
- Control white-tailed deer and other browsing animals to improve or maintain plant diversity and structure.

Lesser Goldfinch

Other Potentially Useful Management:
- Maintain bird-friendly, natural or artificial water source to supply reliable and safe water.
- Use moderate to heavy short-term grazing in rotation system to encourage forb production.

MOURNING DOVE

PRIORITY BIRD

ALL YEAR

Commonly seen perching on telephone wires and fence lines or feeding on the ground; makes a familiar plaintive territorial call during March–September breeding season; wings whistle when they take off; flight is fast and straight; both parents share brooding and do not leave eggs exposed.

Identification: About 12 inches head to tail with male slightly larger than female; small head and bill; long, pointed tail; gray or tan with black spots on wings; short reddish legs and feet.

Habitats: Old field, plateau prairie, river valley prairie, pocket prairie, post oak savannah, shin oak savannah, live oak savannah, seep, springs, tank, pond, lake, and backyard.

Cover: Trees and tall shrubs for loafing and nesting; weedy patches for feeding.

Breeding Territory: Weakly defined; about 2.5 acres/pair.

Nest: Flimsy twig or grass nest 5–25 feet above the ground in trees, shrubs, or building ledges; two to three broods per season common; known to readily abandon nests at slightest threat.

Food: Seeds of grasses and forbs collected on the ground, including annual sunflower, croton, and ragweed, as well as crops such as milo and oats.

Stewardship: Maintain woody plants adjacent to grasslands or old fields for nesting, and promote annual forb production in grasslands or old fields.

- Plant annual food plot and follow with early spring disturbance.
- Disk lightly every other year to promote growth of annual forbs such as annual sunflower.

Mourning Dove

- Use prescribed burns every three to five years to maintain 10–20-acre openings in brush and wooded areas and to encourage perennial forbs like Maximilian sunflower.
- Control white-tailed deer and other browsing animals to encourage a diversity of forbs.
- Mow periodically in fall or winter dormant season on no more than 30% of herbaceous area to promote annual forbs.
- Avoid broad-spectrum herbicides to increase seed-producing forbs.

Other Potentially Useful Management:
- Keep cats and dogs indoors, and control stray and feral cats to reduce predation.
- Use light to moderate rotational grazing to encourage forbs and maintain bare ground.

RIO GRANDE TURKEY

PRIORITY BIRD

ALL YEAR

Three of the five subspecies of Wild Turkey (*Meleagris gallopavo*) in North America occur in Texas. The Eastern Wild Turkey (*M. g. silvestris*) is common throughout the eastern United States and in parts of East Texas; Merriam's Wild Turkey (*M. g. merriami*) is uncommon and found only in the mountains of West Texas; and the Rio Grande Turkey (*M. g. intermedia*) is found in much of the rest of Texas, including the Hill Country. Eastern Wild turkeys are the same species as the domestic turkey, but all subspecies of wild turkeys are far different in looks, behavior, and survivability. Traits of all wild turkey subspecies have been selected by and for living in its natural habitat. Wild Rio Grande Turkeys are sleek, alert, and very cautious but reluctant to take flight and prefer to run instead. Hybridization with other turkeys (including domestic) is very undesirable and should be avoided.

Rio Grande Turkey

Identification: Female about 37 inches and male about 46 inches head to tail; large dark bird with iridescent breast feathers, black-and-white wings, and long neck.

Habitats: Old field, plateau prairie, river valley prairie, post oak savannah, shin oak savannah, live oak savannah, river, creek, and canyon.

Cover: Native bunchgrasses at least 18 inches high needed for nesting; winter roosts often in large riparian trees.

Breeding Territory: Can travel 25 miles from winter roosts in search of good nesting cover; males display and breed in open fields with short cover.

Nest: Nests on the ground, constructed of nearby grasses, and well hidden in grasses 1–2 feet high within 0.5 mile of wooded area with understory 2–4 feet high. First few weeks of brood-rearing requires a 10-acre area with a mixture of trees, shrubs, and herbaceous ground cover per 100 acres.

Food: Grasses, insects and snails, mast (including pecans, acorns, and mesquite beans), green forbs, berries, grapes, wild onions, and seeds; annual diet = 36% grasses, 29% invertebrates, 19% mast, 16% forbs.
- Winter = mostly grasses and forbs
- Spring = mostly grasses, insects, and forbs
- Summer = mostly insects and grasses
- Fall = mostly mast, insects and grasses

Stewardship: The Rio Grande Turkey requires 1,000 acres or more for all phases of its life cycle; thus, small landowners cannot provide all their needs on one property. However, management of roost, nest, or brood-rearing habitat is valuable for these birds.
- Protect large winter roost trees from disturbance.
- Maintain low to moderate woody vegetation within 6 feet of roost trees.
- Provide native grass cover at least 18 inches high for nesting.
- Encourage ground cover of grasses and forbs for brood-rearing.
- Employ light stocking rates and rotation grazing systems that allow rested pastures during reproductive months of April to August.
- Use prescribed burns to encourage healthy grass and forb diversity.
- Mow to improve brood habitat, but avoid mowing in nesting areas.
- Control feral hogs to minimize possible nest predation.
- Provide permanent water in the form of a pond, guzzler, livestock tank, spring, or seep every square mile.

Other Potentially Useful Management:
- Move feeders and food plots frequently to reduce threats. Feeders and food plots may increase reproductive success, but they also concentrate birds and can increase predation, disease, and parasites. Thus, they should never replace habitat management.
- Control white-tailed deer and other browsing animals to encourage woody plant recruitment, healthy understory, and a diversity of forbs.

SAVANNAH SPARROW
UNDER-THE-RADAR BIRD
WINTER

A most abundant winter grassland sparrow; often seen in flocks but discounted as just another "little brown bird"; named for a specimen collected in Savannah, Georgia—not for its love of open, grassy habitat.

Identification: About 5.5 inches head to tail; small, brown sparrow with streaked back and upper breast; yellow eyebrows sometimes used as a key identification feature not always present.
Habitats: Old field, plateau prairie, river valley prairie, pocket prairie, post oak savannah, shin oak savannah, and live oak savannah.
Cover: Grassy areas with mix of forbs and bare ground.
Food: In winter eats forb and grass seeds and some insects collected on the ground.

Savannah Sparrow

Stewardship: Maintain large grassland blocks having a mix of grasses, forbs, and bare ground with cover throughout winter. Winter grass cover retention is a priority.
- Conduct prescribed burns to control invasion by woody plants, encourage forbs, and increase bare ground.
- Use appropriate mechanical and/or chemical techniques to control woody plant invasion.
- Employ light to moderate rotational grazing to encourage forbs and maintain bare ground.
- Use limited cultivation such as occasional light disking to increase forbs and bare ground.
- Control white-tailed deer and other browsing animals to improve forb diversity.
- Apply ant-specific insecticides to control red imported fire ants and increase insect diversity.
- Avoid broad-spectrum pesticides to increase forbs and promote insect diversity.

Other Potentially Useful Management:
- Mow infrequently (once every three years) to promote forbs, but may have limited value.

SCISSOR-TAILED FLYCATCHER

PRIORITY BIRD

SPRING, SUMMER, AND EARLY FALL

An insect-catching acrobat with a chatter call as it flies; found across the Hill Country from spring to late fall when it forms large flocks before flying south for the winter. Breeding bird surveys since 1966 show a modest decline of Scissor-tailed Flycatcher populations in the Texas Hill Country.

Identification: Female about 10 inches and adult male about 13 inches head to tail; long black-and-white forked tail; grayish back and head with pink sides and underwings.

Habitats: Old field, plateau prairie, river valley prairie, post oak savannah, shin oak savannah, live oak savannah, and river.

Cover: Open savannah or grassland with a mix of trees and/or adjacent woods or brush.

Breeding Territory: About 7–15 acres/pair.

Nest: About 7–30 feet above the ground in isolated trees surrounded by open, grassy habitat.

Food: Catches insects in the air or on the ground; hunts from a perch where it identifies potential prey.

Stewardship: Encourage insect and plant diversity, and control woody plant invasion in open areas.
- Apply ant-specific insecticides to control red imported fire ants and increase insect diversity.
- Avoid broad-spectrum pesticides to increase forbs and promote insect diversity.
- Use prescribed burns to minimize woody plant encroachment and maintain mix of grasses and forbs.

Scissor-tailed Flycatcher

- Control white-tailed deer and other browsing animals to encourage a diversity of forbs and replacement of woody plants.
- Use light to moderate rotational grazing to encourage balance of forbs, grasses, and bare ground.

Other Potentially Useful Management:
- Mow periodically in fall or winter dormant season to promote forbs and insects.
- Use limited cultivation such as occasional light disking to increase forbs and insects.

VESPER SPARROW

UNDER-THE-RADAR BIRD

WINTER

Abundant grassland bird usually seen in small winter flocks; when flushed, quickly identified by white outer tail feathers; in the Hill Country, beautiful song heard only occasionally in spring before migrating north to breeding territory.

Identification: About 6 inches head to tail; light brown with white outer tail feathers, streaked back, and distinct eye ring.
Habitats: Old field, plateau prairie, river valley prairie, pocket prairie, and shin oak savannah.
Cover: Grass, forbs, and some shrubs, with grasses averaging about 10 inches high.
Breeding Territory: N/A
Nest: N/A
Food: Mainly insects but also seeds of grasses and forbs collected on the ground.

Vesper Sparrow

Stewardship: Maintain grass cover with adequate forbs and bare ground for feeding.
- Conduct prescribed burns to control invasion by woody plants, encourage forbs, and increase bare ground.
- Use appropriate mechanical and chemical woody plant control techniques to stop woody plant invasion.
- Employ light to moderate rotational grazing to encourage forbs and maintain bare ground.
- Use limited cultivation such as occasional light disking to increase forbs and bare ground.
- Mow in fall or winter (dormant season) to promote forbs on no more 30% of grassland habitat.
- Control white-tailed deer and other browsing animals to improve forb diversity.
- Avoid broad-spectrum pesticides to increase forbs and promote insect diversity.
- Apply ant-specific insecticides to control red imported fire ants and increase insect diversity.

Other Potentially Useful Management:
- Add tepee-style brush piles where brushy habitat is lacking to increase shrub-level habitat.

3 Rivers and Creeks

Our relationship with riparian areas has come full circle from the early days of dependence, to destruction, to the renewed realization of the importance of riparian areas. Many now recognize and understand the connectivity of all things and that by altering one aspect of an ecosystem, everything and everyone is affected.—Ann Fallon, *Riparian Management Handbook*, 1998

INTRODUCTION

With elevations of 980 to 2,460 feet contributing to their character, Texas Hill Country streams arise in the hills and are fed by many springs. Because of changes in the underlying rock structure and soil, these streams alternate from top to bottom between straight and meandering, shallow and deep, high energy and low. They are renowned for natural variation, diverse vegetation, and high water quality.

Downstream, where smaller tributaries approach the larger stream, a Hill Country creek or river might have steep cut banks, often with a cliff on one side opposite a more open and flatter floodplain on the other. In addition, every stream has small-scale, natural variations in the fast-flowing riffle, sluggish pool, and quiet run sequences that alternate in stream reaches (short sections). These sometimes are so close together that a person can observe a riffle, pool, and run while standing in one place.

Ecologists call the area along a stream where water and land meet the riparian zone. This is the transition area between dry upland and the stream's water. Variations in the stream itself are reflected in the riparian vegetation that creates unique habitats for birds and other wildlife. It harbors most of the region's tallest trees, most notably cypress, sycamore, pecan, and cottonwood. These trees provide the essential high canopy that shades layers of smaller plants and the stream below. The combination of clear, clean water and many different native plant species creates a unique habitat that hosts the greatest variety of bird life in the Hill Country. One of the most eye-catching riparian birds common in the Hill Country is the Summer Tanager. It is, indeed, a summer bird and should be easy to see and hear along our streams.

DESCRIPTION: RIVERS AND CREEKS

When you think of rivers and creeks in the Texas Hill Country, do you see a mental image of cool, clear water flowing over flat rocks under tall cypress trees? To this picture, you can add a Green Kingfisher perched

---

FEATURED BIRDS OF RIVERS AND CREEKS

PRIORITY
Green Kingfisher
Louisiana Waterthrush
Red-shouldered Hawk
Summer Tanager
Yellow-billed Cuckoo

UNDER-THE-RADAR
American Wigeon
Bufflehead
Green Heron
Northern Parula
Yellow Warbler
Yellow-rumped Warbler

FAVORITE
Black Phoebe
Bullock's Oriole
Great Blue Heron
Wood Duck

OUTLAW
Brown-headed Cowbird

Note:
Priority = on at least one of the lists provided by TPWD or Oak and Prairies Joint Venture naming species in need of conservation assistance, or shown by USGS-sponsored Breeding Bird Surveys in the Edwards Plateau to be a species whose population is in decline
Under-the-radar = a regular bird in Texas Hill Country but often unnoticed
Favorite = popular with just about anyone who knows the bird
Outlaw = undesirable because of its behavior and impact on other bird species

Summer Tanager
(Photo by Lora L. Render)

Riparian habitat with understory and canopy

on a branch that hangs low over the water. This is accurate, but we can add more to the picture: much slower and deeper water flowing over a mud or sand bottom; water flow changing throughout the year; no water flow during an extreme drought. All of these images describe healthy Hill Country stream ecosystems. In addition, at higher elevations, the riparian zone is narrow with trees and shrubs hugging the creek. At the top of a watershed, streams often have carved out narrow box canyons.

SEASONAL OCCURRENCE OF BIRDS ON WELL-MANAGED RIVERS AND CREEKS

WINTER	SPRING AND FALL (MIGRANTS)	SPRING AND SUMMER (BREEDING BIRDS)	ALL YEAR (RESIDENTS)
American Goldfinch	American Redstart	Bell's Vireo	Belted Kingfisher
American Wigeon	Black-throated Green Warbler	Black-and-white Warbler	Bewick's Wren
Bufflehead	Blue-headed Vireo	Black-chinned Hummingbird	Black Phoebe
Cedar Waxwing	Clay-colored Sparrow	Blue Grosbeak	Black-bellied Whistling-Duck
Dark-eyed Junco	MacGillivray's Warbler (west)	Bullock's Oriole	Black-crested Titmouse
Gadwall	Mourning Warbler (east)	Golden-cheeked Warbler	Canyon Towhee
Green-winged Teal	Nashville Warbler	Green Heron	Canyon Wren
Hermit Thrush	Ruby-throated Hummingbird	Louisiana Waterthrush	Carolina Chickadee
Lesser Scaup	Swainson's Thrush	Northern Parula	Carolina Wren
Orange-crowned Warbler	Wilson's Warbler	Orchard Oriole	Chipping Sparrow
Ring-necked Duck	Yellow Warbler	Painted Bunting	Common Yellowthroat
Sedge Wren		Scissor-tailed Flycatcher	Eastern Phoebe
Sora		Summer Tanager	Golden-fronted Woodpecker
Yellow-rumped Warbler		Yellow-billed Cuckoo	Great Blue Heron
		Yellow-crowned Night Heron	Green Kingfisher
		Yellow-throated Vireo	Killdeer
			Ladder-backed Woodpecker
			Lesser Goldfinch
			Loggerhead Shrike
			Northern Bobwhite
			Red-shouldered Hawk
			Red-winged Blackbird
			Rio Grande Turkey
			Vermilion Flycatcher
			Wood Duck

west = western edge of Hill Country

White-eyed Vireos live in the habitat of these shady banks. Bewick's Wrens search for bugs on flood debris and plants growing near the ground. Summer Tanagers breed and rear their young in the upper tree canopy, whose shade supports the lush cool microhabitat below.

As a Hill Country creek continues downstream, its river valley expands into a wider floodplain with a larger riparian zone. Gradually, water slows as the stream begins to meander through alluvial soil deposits made by older streams. Here, because of a more consistent water level, food, and shelter, ducks such as wintering Gadwall, Bufflehead, American Widgeon, and Northern Pintail can be found as well as the occasional pair of Black-bellied Whistling-Ducks or Wood Ducks that nest here in the summer.

Green Kingfisher
(Photo by Ann Mallard)

Wooded riparian habitat with low perches over water

RIPARIAN ZONE OR RIPARIAN HABITAT

A stream's riparian zone is the area on both sides where the vegetation is dependent on the stream's moisture. Riparian vegetation varies depending on the geology and soils through which the stream flows. This area along a stream is also called riparian habitat and is a critical buffer to flooding. Texas Hill Country riparian habitat has some of the greatest bird diversity in all of the Hill Country. Riparian habitat is a magnet for winter birds such as Hermit Thrush and Spotted Towhee, for migrant

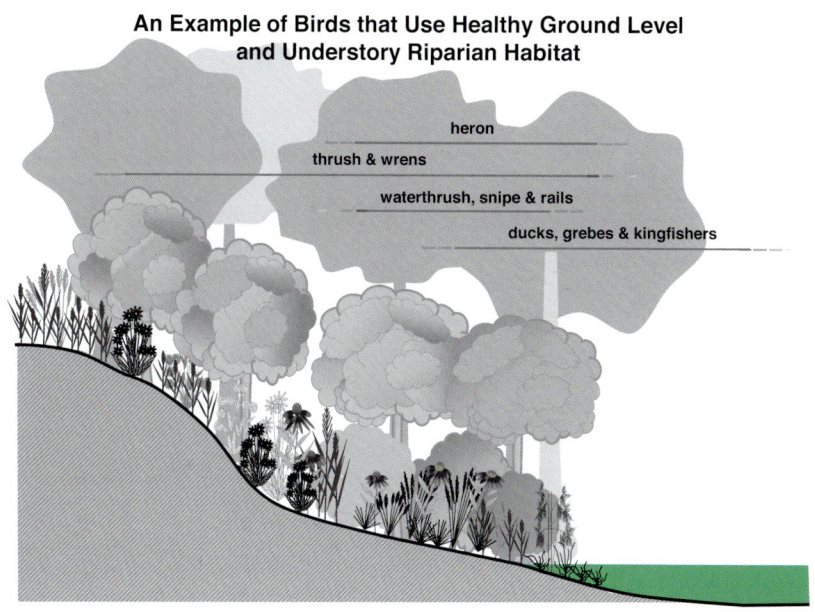

An Example of Birds that Use Healthy Ground Level and Understory Riparian Habitat

- heron
- thrush & wrens
- waterthrush, snipe & rails
- ducks, grebes & kingfishers

Figure 3.1

birds including the Mourning Warbler and Yellow Warbler, for resident birds such as the Black-crested Titmouse, and for breeding birds like the Bullock's Oriole.

What plants are dominant in Hill Country riparian habitat?

Bald cypress is one of the most well-known trees found in downstream Hill Country riparian zones. These tall trees grow along our larger and more perennial rivers and streams. Other large trees growing in riparian habitat include American sycamore, pecan, hackberries, cedar elm, and eastern cottonwood. But the riparian vegetation of a healthy creek or river is much more diverse and complex than a list of the large trees. Small trees and shrubs, including American beautyberry, roughleaf dogwood, elbow bush, and possumhaw, are important understory components that provide essential food and shelter for birds that live along Hill Country creeks and rivers.

BALCONES ESCARPMENT CANYONLANDS

As larger permanent streams, such as the Nueces, Guadalupe, and Colorado Rivers, approach and pass through the edge of the Balcones Escarpment, their channels are also carved by inflow from small and large tributaries. Some of these tributaries have carved true box canyons surrounded by high walls. The most diverse plant communities, often including rare plants, can be found in these tightly contained canyons or on the canyon walls inaccessible to deer, sheep, goats, and cattle. In turn,

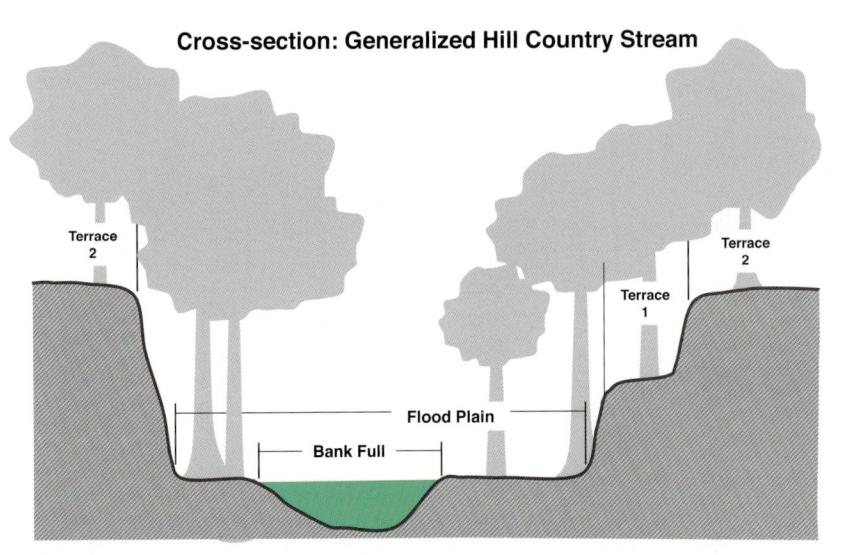

Cross-section: Generalized Hill Country Stream

[Diagram showing stream cross-section with labels: Terrace 2, Terrace 1, Flood Plain, Bank Full]

Figure 3.2

the diverse vegetation here supports many uncommon birds both resident and migrant, such as Bell's Vireo, Yellow-throated Vireo, Black-and-white Warbler, and Yellow Warbler.

The Golden-cheeked Warbler, the Hill Country's most charismatic endangered species, is especially fond of the tree and shrub mixture often growing in box canyons. Slope and aspect determine the existence of the plant communities that are critical for the Golden-cheeked Warbler as well as many other riparian birds. For more information, read about slope and aspect in chapter 1.

FLOODPLAINS

In the Hill Country, bankfull floods occur on the average every one to two years, but in any given year they may be more or less frequent. These are normal and important events that define the stream channel. When they happen, the stream should not experience excessive erosion or deposit. As floodwater leaves the stream's defined channel, it enters the hydrologic floodplain where the water is absorbed by the soil and recharges the alluvial aquifer, and it can reenter the stream. Thus, a flood can actually help sustain water flow through the year. As floodwater moves across the floodplain, its velocity slows and suspended soil particles fall to the ground. This floodplain deposition reduces the sediment moving downstream, improves downstream water quality, and lengthens the life of downstream reservoirs.

In addition, new sediments and vegetation in a floodplain increase its water-holding capacity, making it a better sponge for future "enrich-

Bullock's Oriole
(Photo by Lora L. Render)

Riparian vegetation with tree canopy and understory

ing" floods. Thus, floods, through the deposition of new soil and organic debris, are necessary for maintaining the health of this higher, flatter ground. More infrequently in larger floods, floodwater may also move onto the upper terraces above the hydrologic floodplain. Then, the higher vegetation also benefits from added moisture and sediment nutrients.

Yellow Warbler (Photo by Tom Rust)

Box canyon with dense vegetation and ephemeral stream

How do floods contribute to healthy streams and high-quality bird habitat?

- A thin mass of grasses, leaves, and small twigs deposited at the edge of the high-water mark, and often in other areas, is natural organic compost that decomposes to make nutrient-rich soil. Native and cultivated pecan trees thrive in some floodplains enriched by this organic material.

Stream flow variation between spring flood and late winter normal

- Floods move large debris around, creating downed deadfalls that become part of the stream-bank stabilization process that prevents erosion in future floods.
- The tangle of flood debris feeds and protects understory shrubs that are habitat for breeding songbirds such as Bewick's Wren, Louisiana Waterthrush, and the Common Yellowthroat.
- As high water recedes, the water-saturated soil is where American sycamore and bald cypress seeds germinate, replenishing and regenerating the essential upper canopy vegetation.

When do most Hill Country floods occur?

Because of highly variable seasonal rainfall and severe storms in the Hill Country, streams here can go from no flow to a raging torrent in a matter of hours. Floods can occur any time of year, but the biggest floods usually come in late spring and fall. Typically, lowest flow is in summer and winter.

How big a problem are Hill Country floods?

Every year lives are lost in the Hill Country as a result of floods. There are only a few places in the world with flood events that compare to the ones in the Hill Country, in which a placid creek or river can rise 20 to 30 feet in a matter of hours. An apparently tranquil low-water crossing can be impassable in a couple of minutes. Such crossings are extremely dangerous, even death traps, for the unwary driver. Turn around; don't drown! We cannot prevent huge floods or make every stream crossing safe. We always need to be aware and cautious.

What causes severe floods in the Hill Country?

Severe floods are caused either by hurricanes from the Gulf of Mexico or weather patterns across the Great Plains that collide with warm, moist air from the Gulf and produce sudden, intense thunderstorms. Another contributing factor is the steep gradient change of the streams that pass through the Hill Country.

CLIMATE CHANGE

Looking to the future, science presents a strong case for ongoing climate change that will produce more intense fluctuations of drought and flood. This suggests continuing flood risk to humans and change in riparian habitat that will intensify the need for informed land management to maintain healthy streams resilient to flood damage.

RIVERS AND CREEKS HABITAT MANAGEMENT GOALS

The goal of stream management is a healthy stream in dynamic equilibrium. Its streambed changes in modest amounts through erosion and deposition. Its watershed is covered with deep ground cover. Its riparian zone has vertical layers of deep ground cover, with low, middle, and upper levels made of native plants. These layers are home to many different species of migrant, breeding, and resident birds such as the Black Phoebe (as well as other native wildlife).

Western Hill Country stream with dense riparian vegetation

Black Phoebe (Photo by Lora L. Render)

PROBLEMS OF RIVERS AND CREEKS

Healthy Hill Country streams and their riparian zones provide important habitat for such a great diversity and number of birds that their problems deserve our best conservation efforts.

Problem 1: Native Plant Community Damage with Many Native Species Impaired or Missing

A healthy riparian zone has deep ground cover throughout its watershed and many different species of native plants that support diverse resident, breeding, and migrant bird species as well as other native wildlife. The natural native plant community along Hill Country rivers and creeks has been altered by deer, feral hogs, livestock, and human activities such as clearing, recreation, pond building, and roads.

TOO MANY DEER

The herd of deer you regularly see traveling through your property may be part of the problem. In many parts of the Hill Country, deer have removed the diverse collection of ground-covering plants that used to grow and hold the soil in drainages. With their loss, soil is washed away in every heavy rainstorm, as is the habitat that supports birds such as the Spotted Towhee and Louisiana Waterthrush. These birds eat the insects living on and around riparian, understory plants. They also need natural plant structure for protection.

Deer love to eat forbs and the leaves and new sprouts of many trees and shrubs that live near streams. These plants make up the brushy habitat where many birds find food, shelter, and nest space. In riparian habitat where there is no thick understory or a distinct browse line exists at the height of the deer's reach, there are too many deer.

What You Can Do: Manage Deer Density

To determine if you have a deer browse problem, use the deer browse evaluation chart in appendix 1. If you do have too many deer, reduce their numbers using an ongoing deer management plan. For information on reducing deer overpopulation, see chapter 8.

TOO MUCH CLEARING

Many riparian zones have been cleared to achieve a parklike look. Yours may have been cleared for picnicking or just so you can see the water without having to look through trees and brush. Unfortunately, the combination of tall trees over a lawn is terrible bird habitat and bad for the health of the stream.

Birds are displaced by too much clearing. Where water meets the shore is a plant community of sedges, rushes, broadleaf plants, and shrubs that tolerate and even thrive in consistently wet soil. These include aparejo muhly, spikerushes, Jamaican sawgrass (a sedge), water pennywort, and common buttonbush. Sometimes they are referred to as plants that "like to have their feet wet." Many of these plants provide essential food and cover to riparian birds such as Wood Duck, Gadwall, Blue- and Green-winged Teal, Great Egret, and Wilson's Snipe. Other birds often lost after too much clearing include the Hermit Thrush, Louisiana Waterthrush, Common Yellowthroat, and Green Kingfisher. All of these species need a shrubby layer and low branches.

Louisiana Waterthrush
(Photo by Thomas Collins)

Dense riparian understory and trees

Plants provide important food and cover for riparian birds in healthy riparian habitat. Trees that grow right on the edge of a stream include bald cypress, black willow, and American sycamore. Bald cypress trees are important nest sites for the Red-shouldered Hawk, Great Horned Owl, and Common Raven. Spreading branches of a mature American sycamore provide an excellent place for Great Blue Heron and Great Egret rookeries.

Moving away from the stream, the shoreline plant community gives way to a woodland mix of grasses, shrubs, and vines dominated by trees

Red-shouldered Hawk
(Photo by Lora L. Render)

Creek with high riparian tree canopy and ground cover

that grow on the drier adjacent floodplain. Common riparian trees are American sycamore, bald cypress, chinkapin oak, and pecan. Some of the grasses found in these floodplain terraces include eastern gamagrass, Canada wildrye, switchgrass, and broadleaf woodoats, which also provide important food and cover for riparian birds. Shrubs and vines include roughleaf dogwood, elbow bush, possumhaw, grape, poison ivy, and greenbrier. The fruit of these trees, shrubs, and vines are the backbone of the diet for many riparian birds.

Farther from the stream on higher terraces, the plant community often changes from woodland to savannah. These even drier soils support a more equal mix of trees and grasses. Trees and shrubs that may occur on the higher, drier terraces include hackberries, cedar elm, live oak, gum bumelia, and western soapberry. The Bell's Vireo is a priority species that nests in riparian understory habitat. It needs to be near reliable water, especially in arid areas on the western edge of the Texas Hill Country.

What You Can Do: Let It Go!

Letting a cleared area recover on its own is inexpensive and works well as long as you are not in too much of a hurry. If plants grow nearby, many local, site-specific species will return on their own, but it takes patience to wait for the plants to regenerate and bird species to return. After a few years, you might assess what native plants you have and which ones you want to add. Decide if there are important bird habitat species that are worth planting in limited numbers. They will act as seed sources for the future.

Under the right circumstances, planting can accelerate the restoration process, so birds like Common Yellowthroats, which need a shrubby understory layer, will return more quickly. How much floodwater goes down your stream will influence your chances of success with replanting. On rivers like the Guadalupe that have frequent high-velocity floods, new plants are often lost to floodwater before they can be fully established. Replanting has a greater chance of success on a seasonal creek where flooding is much less dramatic. Here new plants usually are able to survive the regular floods, and they can be planted to protect areas from erosion and accelerate the healing process.

RECREATIONAL HUMAN USE

Families often use the side of a stream for picnicking and fishing. Although we do not recommend these alterations for recreational use, we do know that creeks are sometimes dammed or dug out to make a deepwater pond for swimming.

What You Can Do: Manage Stream Access,
Pond Building, and Stream Crossings

1. Select new picnic areas.

If you see destabilization signs such as loss of ground cover or eroded banks, find a new picnic area and let the old one rest and recover. When you are selecting a new place for picnics and fishing, the most important decisions are the size and location. Fishing usually can be done from a natural opening and does not require much clearing. If clearing is necessary for your picnic area, make as small an opening as possible in a place that is unlikely to erode. A good location has little or no sign of active erosion and does not flood regularly (every year or two). Depending on the

WOODY PLANTS OF HEALTHY RIPARIAN HABITAT

TREES
American sycamore
Bald cypress
Black willow
Chinkapin oak
Elms
Hackberries
Pecan
Plateau live oak
Red mulberry

SHRUBS
American beautyberry
Brickell-bush
Common buttonbush
Elbow bush (spring herald)
Gum bumelia
Mexican buckeye
Possumhaw
Roughleaf dogwood
Western soapberry

VINES
Balsam gourd
    (Lindheimer's globeberry)
Carolina snailseed
Clematis
Grape
Greenbrier
Trumpet creeper
Virginia creeper

HERBACEOUS PLANTS OF HEALTHY RIPARIAN HABITAT

GRASSES
Broadleaf woodoats
Eastern gamagrass
Little bluestem
Switchgrass
Wildrye
Yellow indiangrass

FORBS
Black samson echinacea (purple cone-flower)
Blue mistflower
Brown-eyed Susan
Bundleflower
Bush sunflower
Cedar sage
Chile pequin
Common elderberry
Drummond's wild petunia
Engelmann daisy (cutleaf daisy)
Frostweed
Goldenrod
Goldenwave
Maximilian sunflower
Pigeonberry
Plateau goldeneye
Tropical sage
Turk's cap
Western ironweed
White boneset
White heath aster

topography of your property, the best picnic area may not be right next to the stream. The top choice for your family may be on an old floodplain terrace high above the stream with a nice view. You might be able to use this area without having to clear much understory and just add a trail down to the water for access and fishing. A well-selected picnic spot can be used safely for years, but watch for signs of overuse such as bare ground and erosion. They signal a time for rest and recovery. For more information, see chapter 4.

2. Protect stream vegetation during pond building.

Riparian plants add valuable habitat to the completed pond. Follow these guidelines as much as possible:

- Protect vegetation where water enters the pond to provide healthy wetland habitat.
- Confine heavy equipment to prevent unnecessary damage in the area.
- Keep the unaltered stream channel to ensure stability of the pond shape.
- Maintain the original stream waterway in the main portion of the pond to provide deep water for good fish habitat.

3. Limit stream crossings.

On many Hill Country ranches it is routine to drive through the streambed in a place where shallow water runs over smooth rock. Roads that cross streams can cause disruption of the streambed both upstream and downstream. Crossings should be limited in number and in locations without loose gravel or mud. Vehicles, especially four-wheelers, should *never* be driven in the streambed.

ROADS

Ranch roads in riparian zones cause disturbance, soil compaction, and erosion that destabilize riparian plants. These plants form the communities that are essential habitat for many birds.

What You Can Do: Keep Roads away from Riparian Zones
1. Avoid building a road that crosses a creek or runs parallel to a creek through its riparian zone.

Roads in riparian habitat have many disadvantages. They require high maintenance and are susceptible to erosion that takes years to restore. They also cause sedimentation that fills downstream pools and cause severe erosion known as down cutting that moves upstream and may be impossible to stop. While this discussion specifically addresses ranch roads, the same principles apply to trails built in riparian zones.

2. Keep roads that parallel a stream well away from the riparian zone.

3. Place a road built on a terrace well back from the edge of the active floodplain where it meets the terrace.

4. Avoid building a road near the outside bend of a stream where the hydraulics during a big flood are dramatically magnified.

5. Build the road at a low-water crossing with a natural rock base.
This is the best option if a ranch road must cross a stream. The next best choice is a crossing through a shallow riffle just below a pool.

6. Build the crossing so it approaches perpendicular to the stream flow.
Building the crossing in this way limits the area exposed to erosion. Sometimes approaching from a more parallel direction is unavoidable, but this should be minimized as much as possible.

7. Avoid using culverts.
A well-built low-water crossing is more stable than a culvert. A limited volume of water can pass through a culvert. Culverts concentrate stream flow, producing serious consequences in the Hill Country's frequent floods, including heavy silting and serious streambed erosion below the culverts. Ranch road culverts may also experience frequent washout.

LIVESTOCK WITH UNCONTROLLED ACCESS
The water, food, and shade of riparian habitat are very attractive to livestock. With uncontrolled access they hang out in the riparian area, overgrazing its plant community, physically damaging plants and stream banks, and reducing water quality.

What You Can Do: Limit Livestock Access
With careful management, livestock and birds can coexist on a stream. To prevent damage to riparian bird habitat, you must limit streamside grazing and watering.

- Allow livestock access to water in one spot, and move the access point before the area becomes bare and muddy.

Wood Duck (Photo by Lora L. Render)

River with diverse riparian vegetation and snag for nesting

- Pipe water away from the riparian area into a trough, and prevent livestock from getting water from the stream.
- Use flash grazing if livestock must graze along your riparian land. Flash grazing is moving livestock in and out quickly before they damage the plants that provide good riparian bird habitat.

Problem 2 : Streamside "Tidy" with Short, Thin Ground Cover

This may be considered an example or extension of problem 1. However, since the desire for a tidy appearance is so common and has a negative impact on nearly all riparian birds from the eye-catching American Redstart and Common Yellowthroat to the spectacular Wood Duck, we consider it a separate problem caused by a tendency to overmow, overgraze, or clean up too thoroughly.

OVERMOWING

You may want to mow near a stream to create a neat appearance. Unfortunately, neat lawns are *not* good bird habitat. Also, mowing in a riparian area is often responsible for excessive runoff and unnecessary streambank erosion.

What You Can Do: Refrain from Mowing in Riparian Zones

Depending on the size of the stream, maintain a 50–150-foot band with diverse understory plants along the edge of your stream. A natural mixture of native riparian plants will support wildlife, including some of our most valued birds, and their root systems will hold the soil and prevent excessive erosion in all but the most severe floods.

OVERGRAZING

On many ranches in the Hill Country, land is overgrazed and covered with a thin layer of short grass and a sprinkling of second-growth cedar. As occurs with overmowing, this unnatural combination produces excessive and rapid runoff after rainstorms. Heavy runoff leads to unnecessary erosion, bank erosion, and loss of soil. Damage by overgrazing in riparian habitat displaces many birds, including Hermit Thrush, Louisiana Waterthrush, Common Yellowthroat, Green Kingfisher, Bell's Vireo, Yellow Warbler, Green Heron, and Yellow-rumped Warbler.

What You Can Do: Rest the Land

Land may have been overgrazed for many years, and restoration requires patience. But a riparian zone has the benefit of groundwater and recovers faster than high, dry areas. Rest is the answer to restoration. When livestock are excluded, in most cases plants native to a riparian zone will return on their own. You should see new growth of riparian forbs and shrubs within a year or so.

NOTE: You may also have too many deer, and in this case reducing their numbers is also essential. Where there is an overpopulation of deer, they will eat up new sprouts as fast as they appear.

FLOOD DEBRIS

Floods carry and deposit jetsam (human-made debris like plastic and metal) and flotsam (plant material) left behind after a flood. Flotsam contains lots of seeds that will sprout and produce a nursery of seedlings. Many seedlings do not survive, but those that die are important, too,

Yellow-rumped Warbler
(Photo by Tom Rust)

River with riparian vegetation and flood debris

because they decompose and contribute to new soil that nourishes the seeds that do sprout. Regular deposition of flotsam is one of the reasons why riparian areas are so attractive to thrushes, thrashers, wrens, and warblers, including the Yellow-rumped Warbler, a winter resident. These birds live on the ground and in the lower levels of the riparian canopy.

What You Can Do: Clean Up the Jetsam but Keep the Flotsam

Cleaning up the ugly plastic and other human-made junk washed down by a flood will make the area look better. Leaving the natural flotsam might require a new appreciation of nature's mess. Think of this

Creek with diverse riparian trees and understory

Yellow-billed Cuckoo
(Photo by Lora L. Render)

decomposing plant material as rich natural food. It feeds many different insect species that in turn feed insect-eating songbirds. For example, fireflies and other insect species feed and lay their eggs on moist flotsam in contact with the soil. Their larvae live on the worms, snails, and other organisms also living in this rich decomposing material.

Problem 3: Excessive Erosion and Bare Ground

A stream may have good bird habitat and still experience significant change during a flood. A bank may collapse, and gravel bars may be deposited on the other side. Erosion is soil picked up and carried by

> **DEALING WITH EROSION AND BARE GROUND ALONG A STREAM**
>
> Regular monitoring is necessary to know what is normal and what is an extraordinary event for your stream. While sometimes the result of natural processes, excessive erosion and deposition can also be caused by severe overuse by wildlife and/or livestock, impervious cover, and major floods.
>
> - Visit your creek regularly. Take photos at identified locations to show water level during floods, normal flow, low flow, and severe drought.
> - Learn to recognize water level during bankfull events that occur on average every one to two years. These frequent floods, which lack intensity, are most important because they determine a stream's form and health.
> - Watch for erosion or deposition, and take note of the plant communities in this area.
> - Take photos and notes that record dominant plants. This will allow you to recognize major changes in vegetation types and conditions.
> - Document unusual changes in the amount of erosion or deposition.

fast-moving water. Deposition is soil or gravel dropped by water when it slows after running quickly during a flood. Both erosion and deposition are normal occurrences in the life of a stream. By their very nature, creeks and rivers change through time by these natural processes, which move or create new channels and build new floodplains. Therefore, erosion and deposition do not necessarily mean there is a problem. For example, a limited number of cut banks occur in a healthy riparian habitat and are used by Green Kingfishers for nest burrows.

SEVERE OVERUSE BY DEER, FERAL HOGS, AND/OR LIVESTOCK

Throughout the Hill Country, deer overpopulation and unlimited livestock access have degraded water quality and riparian habitat along streams and drainages for many years. The feral hogs are also increasing and spreading rapidly and do serious damage to riparian habitat wherever present.

What You Can Do: Manage Numbers of Deer, Hogs, and Livestock
1. Reduce deer overpopulation.

Good deer habitat, with plenty of forbs and low-growing woody plants, is good bird habitat. There is no contradiction between the two.

If you are a hunter or managing your land for deer hunting, be assured that land that supports many birds is good for deer. The key is to control deer density so deer are healthy but not so abundant that they damage the natural vegetation. To determine if you have a deer browse problem in woody habitats, use the deer browse evaluation chart in appendix 1.

A vigorous deer population in good Hill Country habitat is no more than about one deer per 10 acres. At this level, they will not damage the plant diversity that birds need, but damaged habitat must have only about one deer per 20 acres to accomplish recovery. If deer density in your area is very high, as it is in most of the Hill Country, you and your neighbors might require hunters to harvest two to three does before hunting bucks. If you do not hunt and are not comfortable with hunters on your place, you can at least support your neighbors' deer hunting. For more information on restoration of eroded bare ground, see sections on gardening and habitat recovery at end of chapter 8.

2. Control feral hogs.

Wherever you have excessive bare ground, the land does not support wildlife and there will be excessive and unnecessary erosion. Feral hogs root up their food and wallow in cool, wet places near streams. Thus, they disturb the plants that normally feed the birds and hold the soil, protecting it from being carried away by floodwater. Look for wallows in riparian habitat. If you see muddy bare spots, you have hogs. And because they are mostly nocturnal, you probably have many more than you might think. Your goal should be to get rid of them. This is a big challenge. For more information, see chapter 7.

3. Rest the land from livestock.

Extremely overgrazed riparian habitat has little ground cover, and floods will have caused serious erosion because of the bare ground. On flat alluvial plains, livestock disturbance may have been so great that floodwater cuts a second channel across the area so there are no roots to hold the soil in place. Though fairly common, this is an avoidable situation and one that can be corrected. Land with bare ground and excessive erosion needs to be rested so the natural native vegetation can recover. Restoration requires a number of steps. The good news is that when livestock are excluded, deer density controlled, and feral hogs reduced, bird habitat recovery will occur on its own. Expensive planting is not necessary. For more information, see problem 1 in chapter 2.

IMPERVIOUS COVER

Pervious cover is a community of ground-level plants like grass and forbs that allows the water to percolate into the soil. Pervious cover around our homes includes materials such as mulch, pavers set in sand, and decomposed granite. Impervious covers are materials that prevent water from soaking into the soil and force the water to run off, such as paved driveways, roads, and roofs. Before development, the land is permeable. As the land in a watershed is developed, some of it becomes impervious, and there is a corresponding increase and quickening in runoff that must be handled by streams draining the watershed. The increase in water running quickly into the stream contributes to bigger floods that may remove trees, shrubs, forbs, and grasses that are home to birds and other wildlife.

What You Can Do: Limit Impervious Cover

While it is impossible to eliminate all impervious cover in a development, there are many practices that help moderate runoff. These include rain catchment systems, use of pervious material for sidewalks and driveways, and retention or rain ponds that hold and delay storm runoff.

MAJOR FLOODS

A healthy riparian zone will protect a stream from erosion during modest bankfull floods that occur regularly (on average, every one to two years). However, even well-managed streams can experience flood damage from extreme high-water events. Lately these severe floods seem to be happening more often, which may be due to changes in our climate and alterations in the watershed.

What You Can Do: Repair Erosion

When repairing erosion in the riparian zone, there are two areas to address:

- Land immediately adjacent to an eroded bank may have lost much or all of the topsoil. In any case, it is best to let this area heal on its own. Leave most, preferably all, of the debris piles deposited by floodwater. These piles provide protection and organic matter for replacement plants (trees, shrubs, forbs, and grasses). In addition, they are great nest and feeding areas for wrens and other understory birds.
- Fields near the newly eroded area should have adequate buffer (50–150 feet) of natural riparian vegetation. Good buffers include tall trees with vines, midlevel shrubs, and low-level forbs and grasses. If the area does not have a healthy buffer, planting or

River with high riparian canopy

Northern Parula
(Photo by Lora L. Render)

allowing one to grow naturally is an important protective measure to minimize further erosion and ultimately restore a healthy riparian habitat. The migrant Black-throated Green Warbler forages for insects and poison ivy berries in riparian canopy trees. The Northern Parula is a beautiful songbird that nests and feeds in tall riparian trees.

Problem 4: Invasion by Exotic Plants and Animals and/or Cedar

Many exotic animals and plants are common in and along Hill Country streams. Some of these are smallmouth bass, Asian clam, house mouse, axis deer, vinca, common four o'clock, ligustrum, Chinese tallow, chinaberry, and salt cedar. They are displacing native species that have been here for centuries. Because many birds depend on specific native species for survival, invasive exotics produce inferior bird habitat.

A diverse mixture of woodland or savannah species such as elms (cedar, American, slippery), pecan, hackberries, plateau live oak, western soapberry, grapes, elbow bush, switchgrass, yellow indiangrass, little bluestem, Maximilian sunflower, Illinois bundleflower, and Engelmann daisy, is normal for riparian terraces. When cedar invades and becomes dominant to the point of displacing these plants, it limits bird food production in the form of seeds, fruits, and insects that occur naturally. It also reduces bird nest sites, especially in the understory and at ground level. The problem of invasive exotics has been caused by both humans and livestock, intentionally and unintentionally.

HUMAN INTRODUCTION

You or the people who owned your property in the past have probably introduced exotic plants or animals for landscaping, recreation, or just by accident. Since they may have been there for a long time and you have gotten used to them, you might want to know if their removal is necessary to restore healthy bird habitat. The short answer is yes.

EXOTIC PLANTS

Some birds readily eat the seeds and fruit of exotic plants, so you might think that they are good for the birds. Unfortunately, invasive exotic plants like Chinese tallow, chinaberry, elephant ear, ligustrum, and giant reed outcompete and displace native plants. Thus, they are responsible for decreasing native plant diversity. This means that these exotic plants are responsible for many bird species losing the native plant nutrients and structure that they need. Also, lower plant diversity provides a vulnerable, less stable food supply, and when the native species are gone, most of the insects that feed on them are also lost.

What You Can Do: Remove Exotic Plants

If you want more birds, get rid of the invasive exotic plants. In the fall, Chinese tallow leaves turn gorgeous bright red and orange colors that rival those of the most beautiful maples. Do not be fooled by this temporary allure. Chinese tallows are spreading rapidly along riparian corridors because they are prolific seed producers and their roots release an allelopathy into the soil that inhibits the growth of other naturally occurring plants. Flag Chinese tallow trees in the fall when they are easy to spot. Get rid of them in the spring when they are easiest to kill. Cut and paint their stumps with an herbicide. Paint rather than spray the stump; this method is effective and protects the sensitive riparian zone.

EXOTIC FISH

Birds such as kingfishers, herons, and egrets all depend on healthy and diverse native fish populations. No matter what anyone tells you, exotic game fish compete with and reduce native fish populations. For example, Guadalupe bass, the Texas state fish, has been severely impacted by the introduction of nonnative smallmouth bass. Most significantly, smallmouth bass have hybridized with Guadalupe bass and made their continued existence in many streams uncertain.

What You Can Do: Do Not Stock Nonnative Fish
Never add exotic fish species to any Hill Country stream or lake. There is little you reasonably can do to eliminate those already present, and the best option is simply to not make the situation any worse.

EXOTIC DUCKS AND GEESE

Once introduced, exotic waterfowl are present year-round and reduce shoreline vegetation that provides food for species like Wilson's Snipe, Green Heron, and dabbling ducks, including Gadwall, American Wigeon, Northern Pintail, Green-winged Teal, and Blue-winged Teal. Most native ducks are migratory, and even large flocks allow extended periods of rest and recovery for the plant community that makes up their habitat. Exotic ducks and geese may also spread infectious and fatal diseases to native waterfowl.

What You Can Do: Remove Exotic Waterfowl
Fortunately, these introduced species are much easier to deal with than feral hogs and exotic fish. Trap and remove domestic waterfowl as soon as possible, and no matter how much you might like them, do not add any more. Once the exotic waterfowl are gone, simply allow the area to recover naturally. It is not necessary to spend money on planting.

LIVESTOCK INTRODUCTION

Exotic plant seeds are introduced when livestock are fed supplemental hay or grain and also by birds that that frequent these livestock feeding areas and then spread the hay or grain seeds to other areas. Often the animals' two favorite places are where they eat and drink. Thus, when they pick up seeds either on their bodies or by eating them, livestock spread exotic seeds to the riparian zone when they go to drink.

What You Can Do: Limit Livestock Access
If possible, place stock tanks far away from riparian habitat. At least, limit livestock to a small portion of a stream. Monitor the area regularly, and remove exotics as soon as they appear.

American Wigeon
(Photo by Jeff Forman)

Shallow stream pool with emergent vegetation

Problem 5: Diminished Water Quality and Quantity

Poor water quality impacts riparian birds that depend on clean, clear water. For example, as water quality declines, fish and aquatic invertebrate populations are altered so that various insect-eating birds such as the Black Phoebe and the Common Yellowthroat may not find adequate food. Muddy water also impacts fish and aquatic invertebrates, and silt smothers sensitive riparian plants, decreasing the quality of aquatic habitat. Muddy water even makes it difficult for kingfishers, egrets, and herons, including the Green Heron, to see and catch fish.

Green Heron (Photo by Lora L. Render)

Stream with shallow edge, vegetation, and clear water

Obviously, a dry stream means loss of fish and aquatic invertebrate food for ducks, herons, egrets, and kingfishers. Also, when water levels fluctuate beyond the norm, there are drastic changes in the surrounding riparian habitat. For example, when riparian areas dry up, cedar and/or exotic plants move in, replacing the more diverse native plant community. Without these plants, riparian birds lose their natural food and shelter. Huge, damaging floods that erode soil and wash out plants can have a similar effect by removing a portion of the plant community riparian birds live on.

POOR WATERSHED MANAGEMENT

How the land is managed upstream from you has a huge impact on the health of your stream. Deep ground cover in a watershed enriches water quality by producing a stream resistant to erosion and silting. Deep ground cover also slows runoff and absorbs water to penetrate the soil, move slowly through the soil, and finally into the stream. Thus, a healthy watershed helps stabilize stream flow.

Poor watershed management produces an unhealthy stream with diminished water quality and quantity. A stream fed by a watershed with little or no ground cover is subject to severe erosion and silting as well as the extremes of extra-high floods and very low flow during dry periods. Waterfowl such as the black and white Bufflehead that winter in the Hill Country require consistent water and benefit from good watershed management. This stunning duck and others needing good water conditions and stable flow will make any landowner proud.

What You Can Do: Maintain Deep Ground Cover and Reduce Runoff

Your stream must be fed by a healthy watershed, where the ground cover acts like a sponge and a filter. The ground cover absorbs rainwater, and the rainwater is filtered as it moves slowly through the soil and finally into the stream. Thus, landowners need to work together to create and protect healthy, watershed-wide ground cover. Follow these guidelines for effective watershed management:

- Be a good neighbor, and work with your neighbors for the common good of the stream.
- Model watershed stewardship on your property by maintaining healthy plant communities in all habitat types.
- Practice ongoing brush management, control deer overabundance, and use rotational grazing with appropriate stocking levels for maximum plant production and diversity.
- If you live in a housing development, reduce runoff with pervious cover, retention or rain ponds, and rain catchment systems.

LOSS OF HEALTHY RIPARIAN HABITAT

Healthy riparian habitat has many levels of diverse plant species and dense ground cover that promote good water quality and quantity, provide a good home for birds, and protect a stream from unnecessary erosion and soil loss. How neighbors along a stream manage their riparian zone impacts the quality of this habitat downstream. You may be allowing your riparian habitat to grow naturally, but others upstream may be doing just the opposite.

Bufflehead (Photo by Lora L. Render)

Secluded stream pool with healthy aquatic habitat

Do not forget that feral hogs damage riparian bird habitat. Hog wallows reduce water quality and cause loss of fish species that are dependent on clean water. Consequently, fish-eating birds such as kingfishers, egrets, and herons, including the Great Blue Heron, lose a food source. Hogs also root up and kill riparian plants. Their loss can ruin ground-nesting bird habitat.

What You Can Do: Restore Watershed Condition

Be a model for good land stewardship. Make sure you have healthy riparian vegetation on your own property. This helps moderate erosion during floods and can inspire others to do the same. Do what you can to increase watershed awareness in your community. Talk to your neighbors about these solutions:

Great Blue Heron
(Photo by Scott Swearingen)

Stream pool with clear water and riparian vegetation

- Maintain deep ground cover.
- Thin dense cedar.
- Harvest deer annually.
- Limit livestock access to streams.
- Trap/dispatch feral hogs and shoot on sight.
- Do not clear in the riparian zone.
- Be a good watershed neighbor. Every well-managed property in your watershed benefits every downstream property.

Above all, be patient. Changing habits formed over a lifetime is not easy, but everyone benefits from improved land stewardship.

SUMMARY OF HABITAT PROBLEMS AT RIVERS AND CREEKS

HABITAT PROBLEMS	STEWARDSHIP SOLUTIONS	AFFECTED BIRDS
Problem 1: Native plant community damaged with many native species impaired or missing	Reduce deer overpopulation Stop clearing riparian plants and replant in some areas Allow rest and recovery in damaged areas Limit livestock access Keep roads away from stream	Black Phoebe Common Raven Common Yellowthroat Gadwall Great Blue Heron Great Egret Green-winged Teal Louisiana Waterthrush Red-shouldered Hawk Wood Duck
Problem 2: Streamside "tidy" with short, thin ground cover	Allow 50–150-foot undisturbed buffer zone along stream Exclude livestock or allow only limited access to stream Leave flotsam in place Reduce deer overpopulation	Bell's Vireo Common Yellowthroat Green Kingfisher Green Heron Louisiana Waterthrush Yellow Warbler Yellow-rumped Warbler Wood Duck Wrens
Problem 3: Excessive erosion and bare ground	Reduce deer overpopulation Encourage deer hunting Maintain long-term cooperation with neighbors to manage deer Do long-term feral hog removal Rest the land from livestock Allow 50–150-foot undisturbed buffer zone along stream Install rainwater catchment Add rainwater retention ponds	Black-bellied Whistling-Duck Louisiana Waterthrush Northern Parula Sora Yellow-billed Cuckoo
Problem 4: Invasion by exotic plants and animals and/or cedar	Survey and remove invasive exotic plants regularly Do not stock nonnative fish Trap and remove exotic waterfowl Water livestock far from riparian habitat or limit their access Conduct long-term cedar management	American Wigeon Eastern Phoebe Gadwall Green Heron Green Kingfisher Green-winged Teal Warblers Wrens
Problem 5: Diminished water quantity and quality	Use pervious ground cover Add rain retention ponds Install rainwater catchment Maintain deep ground cover Thin dense cedar Harvest deer annually Limit livestock access to stream Remove feral hogs Maintain healthy riparian zone	Black Phoebe Belted Kingfisher Bufflehead Common Yellowthroat Great Blue Heron Great Egret Green Heron Green-winged Teal Lesser Scaup

Bird Summaries: Featured Birds

AMERICAN WIGEON
UNDER-THE-RADAR BIRD
FALL, WINTER, AND EARLY SPRING

Attractive dabbling duck especially adapted to foraging on plants; known to steal and eat plants brought to the surface by coots and diving ducks.

Identification: About 20 inches head to tail; male has white forehead and iridescent green mask extending backward from around his eye; female has grayish, somewhat speckled head and no mask; when wigeon is taking to flight, a black patch is visible on upper surface of the trailing edge of the wing with a white patch toward the forewing.

Habitats: River, creek, tank, pond, and lake.

Cover: Emergent or submerged aquatic vegetation growing in water 6–12 inches deep.

Breeding Territory: N/A

Nest: N/A

Food: Almost exclusively aquatic and wetland vegetation, including roots, stems, and seeds of duckweed, widgeon grass, spike rush, and bulrush; possibly wetland grasses and clover.

Stewardship: Maintain shallow-water habitat with healthy aquatic vegetation.
- Drop pond water depth to expose mudflats and encourage growth of wetland vegetation during late spring and early summer.
- Maintain water depth less than 12 inches to provide habitat for dabbling ducks during fall and winter, when birds are present.
- Maintain adequate aquatic vegetation to protect essential food for wigeons present in winter when controlling aquatic vegetation either by chemical or biological (grass carp) methods.
- Limit shoreline access by livestock to maintain deep, clean water and healthy aquatic vegetation for feeding.

Other Potentially Useful Management:
- Manage for healthy watersheds to maintain water quality and quantity.
- Use pond design features, such as shallow benches, to promote good winter waterfowl habitat when building or maintaining tanks or lakes.
- Install water-control structures to facilitate management of water depth during dam construction or repair.
- Control feral hogs to maintain shoreline vegetation and water quality in tanks or lakes and to promote plant diversity in the watershed.

American Wigeon

BLACK PHOEBE

FAVORITE BIRD

ALL YEAR

Found on the western side of the Hill Country and closely associated with water; nonmigratory but wanders in winter; active with constantly wagging tail when perched and frequent flights to catch aerial insects.

Identification: About 6.25 inches head to tail; charcoal above, including head, and white on lower breast and belly.

Habitats: River, creek, canyon, seep, springs, tank, pond, and lake.

Cover: Small to medium-sized trees and shrubs adjacent to water.

Breeding Territory: About 130–525 feet between nests.

Nest: Mud nest 3–10 feet high; often on rock ledges; also readily uses human structures such as buildings and culverts.

Food: Insects caught on the wing from a perch usually about 6.5 feet above the ground.

Stewardship: Encourage and maintain insect and plant diversity in riparian habitat.
- Control white-tailed deer and other browsing animals to maintain healthy riparian vegetation.
- Exclude livestock from riparian areas to maintain shoreline vegetation and water quality.
- Control brush encroachment by salt cedar (tamarisk) by appropriate methods to promote insect and plant diversity.
- Minimize or avoid the use of insecticides to maintain food supply.
- Control feral hogs to maintain riparian vegetation and water quality across the watershed.

Other Potentially Useful Management:
- Manage for healthy watersheds, including deep and sustainable ground cover to maintain riparian health and water quality.
- Use deer exclosures to encourage woody plant recruitment.

Black Phoebe

BUFFLEHEAD
UNDER-THE-RADAR BIRD
FALL AND WINTER

Strikingly beautiful black and white diving duck with a big head; easily flushed and springs up quickly without running across the water; common in small winter flocks and very alert to danger.

Identification: About 14 inches head to tail; male larger than female; male black and white with a large white patch from the eye to back of head; female dark brown above, paler below with a small white cheek patch.
Habitats: River, creek, tank, pond, and lake.
Cover: Open-water lakes and tanks with some emergent and aquatic vegetation and undisturbed streams with deep pools.
Breeding Territory: N/A
Nest: N/A
Food: Aquatic insects, snails, crayfish, and aquatic plants.

Stewardship: Maintain deep, clean water with healthy aquatic vegetation with diverse aquatic invertebrates.
- Limit livestock access to the shoreline to maintain deep, clean water and healthy aquatic vegetation for feeding.
- Do not use insecticides that might leach into the water to protect the Bufflehead's food sources.
- Be extremely cautious when using aquatic herbicides to avoid unnecessary damage to aquatic insect habitat.

Bufflehead

- Maintain protective vegetation along shorelines to foster a friendly environment for this flighty species.
- Avoid deep stream pools in fall and winter to prevent unnecessary disturbance to Bufflehead flocks.

Other Potentially Useful Management:
- Use pond design features, such as shallow benches, when building or maintaining tanks or lake—to promote good winter waterfowl habitats.
- Install water-control structures during dam construction or repair to facilitate water-depth management.
- Control feral hogs to maintain shoreline vegetation and water quality in tanks and across the watershed.

BULLOCK'S ORIOLE

FAVORITE BIRD

SPRING, SUMMER, AND FALL

Gorgeous western oriole; usually in tall riparian trees or mature mesquite; male and female sing similar songs; early in nesting female may sing more than the male. Significant, long-term population declines have been recorded in western North America; population appears stable in Texas.

Identification: About 7.75 inches head to tail; males bright orange contrasted with black shoulders, back, chin, and cap and a black stripe through the eye; black wings with white wing patch; females yellow with grayish back and belly and dark wings.

Habitats: Old field, pocket prairie, post oak savannah, shin oak savannah, live oak savannah, river, creek, seep, springs, and backyard.

Cover: Canopy of riparian trees or other mature trees with herbaceous ground cover.

Breeding Territory: May defend a combined nesting and feeding territory; multiple nests have been observed in one tree; food availability may be important in territorial behavior.

Nest: Long, hanging, woven bag 6–25 feet above the ground.

Food: Mainly insects but also fruit and some nectar, usually foraged in bushes or the tree canopy.

Stewardship: Maintain healthy riparian and woodland habitat with well-developed overstory and diverse herbaceous ground cover.
- Control white-tailed deer and other browsing animals to encourage woody plant recruitment, healthy understory, and a diversity of forbs.
- Exclude livestock from riparian areas to maintain diverse understory.
- Use light to moderate rotational grazing to provide a diverse plant community in other areas.

Bullock's Oriole

- Avoid broad-spectrum pesticides, and control red imported fire ants with an ant-specific pesticide to promote insect diversity and maintain food supply.
- Control spread of regrowth cedar with appropriate mechanical methods to maintain diverse riparian and savannah habitats.

Other Potentially Useful Management:
- Control feral hogs to maintain healthy riparian vegetation.
- Use deer exclosures to enhance overstory tree recruitment.
- Control cowbirds to limit nest parasitism.
- Maintain natural or artificial bird-friendly water source to supply reliable and safe water.
- Use prescribed burns in savannahs to increase forbs and maintain diverse woody plant communities.

GREAT BLUE HERON
FAVORITE BIRD
ALL YEAR

Common and popular wading bird found around most Hill Country streams, ponds, and lakes; largest heron in the Texas Hill Country; has a wingspan of 6 feet.

Identification: About 47 inches head to tail; large grayish bird with long legs and neck; white head with black above the eye extending back and ending in a long crest.
Habitats: River, creek, canyon, seep, springs, tank, pond, lake, and backyard.
Cover: Shoreline or wetland vegetation.
Breeding Territory: Colonial in rookeries, but occasionally a pair will nest singly.
Nest: Large jumble of sticks, usually in rookery tree that may be up to 100 feet tall.
Food: Varied but mainly fish; includes crustaceans, mammals, and insects; food often found and captured by slowly stalking or waiting in shallow water.

Stewardship: Encourage high water quality and diverse stream and wetland vegetation.
- Maintain large riparian trees such as American sycamore and cedar elm to foster possible rookery sites.
- Limit roads and prevent disturbance within approximately 1,000 feet of an active rookery to promote successful reproduction and prevent abandonment.
- Control white-tailed deer and other browsing animals to maintain woody plant recruitment and healthy understory.
- Exclude livestock from riparian areas to maintain habitat quality.

Great Blue Heron

Other Potentially Useful Management:
- Control feral hogs to maintain healthy shoreline vegetation and high water quality.
- Exclude livestock from ponds or lakes to protect feeding habitat.
- Manage for healthy watershed, including deep and sustainable ground cover to maintain riparian health and water quality.

GREEN HERON

UNDER-THE-RADAR BIRD

ALL YEAR

Small heron that silently stalks its prey in shallow water during mornings and evenings; often perches in trees and usually roosts on or near the ground; most abundant as migrant in spring and fall; a few pairs nest in places having ideal wooded riparian habitat; has a loud alarm call when disturbed; locally known as "fly-up-the-creek" because it follows the path of the stream when flushed.

Identification: About 18 inches head to tail; dark, gray-green back and wings; throat and neck white with rust color on sides of neck and head.

Habitats: River, creek, springs, tank, pond, and lake.

Cover: Dense shrubs, wooded shoreline, and tall herbaceous vegetation near streams, wetlands, or ponds.

Breeding Territory: Calls to announce territory around its nest but may nest in colonies and defend only the nest itself.

Nest: Stick platform, usually concealed in riparian tree or shrub from ground level to 30 feet.

Food: Minnows, frogs, crayfish, aquatic invertebrates, and insects often caught when wading.

Stewardship: Maintain dense vegetation near water on your property.
- Control white-tailed deer and other browsing animals to encourage riparian woody plant recruitment and healthy understory.
- Control feral hogs to maintain riparian vegetation, shoreline vegetation, and water quality across the watershed.
- Limit livestock access to the shoreline to maintain healthy shore and aquatic vegetation and water quality.

Green Heron

- Do not eliminate shallows through stream channelization to foster shallow feeding habitat.
- Do not use insecticides that might leach into the water to maintain good water quality and abundant aquatic life.
- Maintain adequate aquatic vegetation to protect aquatic food sources when controlling aquatic vegetation either by chemical or biological (grass carp) methods.
- Control regrowth cedar using appropriate mechanical methods to maintain riparian plant diversity.

Other Potentially Useful Management:
- Install water-control structures during dam construction or repair to facilitate water-depth management.
- Build ponds with shallow benches to promote feeding habitat.

GREEN KINGFISHER

PRIORITY BIRD

ALL YEAR

Uncommon and solitary kingfisher of eastern and southern Hill Country; jets fast and straight just above stream surface; hunts food while sitting on a convenient low perch; rarely in ponds or lakes. Habitat damage in the past caused this species to decline in Texas but apparently has increased in recent decades.

Identification: About 8.75 inches head to tail; dark green back, head, and wings; white collar; male has rufous-colored breast.

Habitats: River, creek, tank, pond, and lake.

Cover: Overstory and understory foliage of trees and shrubs on rivers and streams with clean, clear water; occasionally on confluence of stream with lake pond.

Breeding Territory: Approximately one pair per mile of stream; low water levels could increase territory size.

Nest: Burrow dug in stream bank about 2 inches in diameter and up to 3 feet deep with entrance covered by vegetation; one brood/year.

Food: Dives from a low perch for small fish, tadpoles, and aquatic insects spotted in clear water.

Stewardship: Maintain riparian understory along streams and ponds and promote high water quality.

- Manage for healthy watersheds, including deep and sustainable ground cover to maintain riparian health and water quality.
- Avoid unnecessary clearing and trimming to maintain low (5 feet or less) perches.
- Control white-tailed deer and other browsers to maintain riparian understory and woody plant recruitment.
- Control regrowth cedar using appropriate mechanical or hand clearing to maintain diverse riparian woodlands.
- Control feral hogs to prevent excess runoff and maintain high-quality water for food.
- Exclude livestock to maintain healthy riparian habitat and high water quality.
- Use ant-specific bait to control red imported fire ants and limit nest predation.
- Avoid broad-spectrum pesticides to maintain diverse and healthy aquatic life.

Other Potentially Useful Management:

- Use deer exclosures to encourage woody plant recruitment.

Green Kingfisher

LOUISIANA WATERTHRUSH

PRIORITY BIRD

SPRING AND SUMMER

Uncommon thrushlike warbler that nests in riparian habitat; known for its tail wagging; its presence usually indicates high water quality.

Identification: About 6 inches head to tail; light brown above with white below and breast streaked with brown; white eyebrow and incomplete eye ring.
Habitats: River, creek, and springs.
Cover: Well-developed tree canopy and healthy streamside understory, including downed logs and clear, clean water.
Breeding Territory: Linear following stream course, ranging from 650 feet to more than 3,000 feet (over 0.5 mile), but averages about 1,300 feet.
Nest: Nests along creeks usually with a current; open cup made of leaves and mosses usually placed in a depression on the creek bank or at the base of a downed log.
Food: Forages primarily for invertebrates along the bank or in the water.

Louisiana Waterthrush

Stewardship: Maintain and encourage healthy, undisturbed riparian areas with stable banks and high water quality.
- Exclude livestock from riparian areas to maintain understory and bank integrity.
- Control feral hogs to maintain and improve water quality and ground-level herbaceous cover and reduce nest destruction.
- Control white-tailed deer and other browsing animals to encourage healthy woody understory.
- Trap and remove cowbirds to reduce nest parasitism and increase nesting success.
- Avoid broad-spectrum pesticides to maintain diverse and healthy aquatic life.
- Use ant-specific bait to control red imported fire ants and limit nest predation.

Other Potentially Useful Management:
- Control regrowth cedar using appropriate mechanical or hand clearing to maintain diverse riparian woodlands.
- Manage for healthy watersheds, including deep and sustainable ground cover to maintain riparian health and water quality.
- Use deer exclosures to encourage woody plant recruitment.

NORTHERN PARULA

UNDER-THE-RADAR BIRD

SPRING AND SUMMER

Uncommon to rare, shy warbler of riparian habitats; easily identified by its call; known to build its hanging nest in Spanish moss, but whether or not it uses ball moss (a very common Hill Country epiphyte) is unknown.

Identification: About 4.5 inches head to tail, blue-gray with yellow throat and breast; small yellow patch on back and two small white wing patches.
Habitats: River, creek, and canyon.
Cover: Diverse mid- to upper riparian canopy.
Breeding Territory: Averages about 0.75 acre per pair.
Nest: Usually 10–20 feet high; hanging nest often in a bundle of Spanish moss or cluster of leaves in wooded riparian habitat.
Food: Insects and spiders collected from the tips of branches.

Northern Parula

Stewardship: Encourage diverse tree canopy, and maintain epiphytes such as Spanish moss and ball moss.
- Control white-tailed deer and other browsing animals to encourage healthy understory and woody plant recruitment.
- Exclude livestock from riparian areas to maintain understory.
- Control regrowth cedar using appropriate mechanical or hand clearing to maintain diverse riparian woodlands.
- Avoid broad-spectrum insecticides, and control red imported fire ants with an ant-specific insecticide to promote insect diversity and maintain food supply.

Other Potentially Useful Management:
- Use deer exclosures to encourage woody plant recruitment.
- Manage for healthy watersheds, including deep and sustainable ground cover to maintain riparian health and water quality.
- Trap and remove cowbirds to reduce nest parasitism and increase nesting success.

RED-SHOULDERED HAWK

PRIORITY BIRD

ALL YEAR

Breeds late January through May; very vocal; loud, screaming calls heard during late winter, spring, and summer; young fledge when about six weeks old but depend on parents for another 4.5 months; pair of Great Horned Owls may take over their nest; home range approximately 400 acres.

Identification: About 20.5 inches head to tail; rust spot on shoulders; long, narrow wings; long tails with narrow black and white bands and white tips; yellow legs.

Habitats: River valley prairie, mixed wooded slope, post oak savannah, shin oak savannah, live oak savannah, river, creek, canyon, tank, pond, and lake.

Cover: Tall trees and dense shrubs near water.

Breeding Territory: Spacing of nests averages about 2,200–4,200 feet; a pair commonly uses same territory for many years.

Nest: Usually 20–60 feet above the ground; big stick nest in major fork of a tall tree, lined with bark, leaves, moss, down, and feathers.

Food: Looks and listens for prey (small mammals, snakes, lizards, frogs, fish, birds, insects, and crayfish) from a perch or on the wing over fields near wooded stream or wetland.

Stewardship: Maintain tall, healthy overstory and understory in riparian areas as well as adjacent open habitat for hunting territory.

- Control white-tailed deer and other browsing animals to encourage woody plant recruitment.

Red-shouldered Hawk

- Control regrowth cedar using appropriate mechanical or hand clearing to maintain plant and animal diversity.
- Minimize or avoid the use of pesticides to maintain stable food supply.
- Use brush piles to provide important cover for small mammals where brushy habitat is lacking.
- Use prescribed burns in uplands to promote plant and animal diversity in nearby feeding grounds.

Other Potentially Useful Management:

- Use deer exclosures to encourage woody plant recruitment.
- Exclude livestock from riparian areas to maintain understory.
- Manage for healthy watersheds, including deep and sustainable ground cover to maintain riparian health and water quality.

SUMMER TANAGER

PRIORITY BIRD

SPRING AND SUMMER

Gorgeous bird that specializes in eating bees and wasps; breeding males defend their territories by singing incessantly throughout the day.

Identification: About 8 inches head to tail; male all red like a male cardinal but no crest; female yellow-green.
Habitats: Mixed wooded slope, post oak savannah, river, creek, canyon, and backyard.
Cover: Nests in understory to upper canopy, open woodlands with relatively closed canopy.
Breeding Territory: About 22–27 acres/pair.
Nest: Open cup, in tree 10–35 feet above the ground.
Food: Insects mostly, especially bees and wasps; kills its prey by knocking it against a branch.

Summer Tanager

Stewardship: Encourage diversity in both understory and overstory plants and maintain a fairly closed canopy.
- Control white-tailed deer and other browsing animals to encourage woody plant recruitment and healthy understory.
- Exclude livestock from riparian areas to maintain diverse understory.
- Use light to moderate rotational grazing to provide a diverse plant community in other areas.
- Trap and dispatch cowbirds to reduce nest parasitism and increase nesting success.
- Control regrowth cedar using appropriate mechanical or hand clearing to maintain diverse riparian woodlands.
- Avoid broad-spectrum insecticides, and control red imported fire ants with an ant-specific insecticide to promote insect diversity and maintain food supply.

Other Potentially Useful Management:
- Use deer exclosures to encourage woody plant recruitment.
- Manage for healthy watersheds, including deep and sustainable ground cover to maintain riparian health and water quality.
- Control feral hogs to maintain riparian vegetation and related insects.

WOOD DUCK

FAVORITE BIRD

ALL YEAR

One of the most beautiful ducks in the Texas Hill Country; may be found along large rivers of the southern and eastern Hill Country, where it nests in tree cavities and nest boxes. Feeds by dabbling or short shallow dives; has strong, fast flight and elaborate courtship behavior.

Identification: About 20 inches head to tail; males have sharply contrasting colors of rusty breast, black head framed with white stripes on chin and eyebrow, and iridescent green crest; females generally drab with crest and white eye patch.

Habitats: River, creek, springs, tank, pond, and lake.

Cover: Diverse shoreline vegetation of streams, ponds, lakes, and wetlands having shallow water and emergent vegetation.

Breeding Territory: Where nest sites are abundant, will nest in close proximity; male is nonterritorial but will chase other males away from its mate.

Nest: Large cavities; readily uses nest boxes.

Food: Various seeds, fruits, acorns, and vegetation of wetland and riparian plants along with a significant amount of invertebrates.

Wood Duck

Stewardship: Maintain or encourage diverse shoreline and shallow-water habitats, including healthy riparian habitat with mature trees containing nest-site cavities to promote reproduction.
- Retain snags and trees with cavities within about 1.2 miles of suitable aquatic habitat to provide nest sites.
- Add nest boxes with predator guards where natural nest cavities are limited to promote reproduction.
- Control lake water levels and encourage emergent wetland plants (while maintaining relatively constant water levels when possible) to foster food production.
- Protect beavers as a beneficial companion species to create diverse shallow-water habitat and food for wood ducks.
- Exclude grazing from lake and riparian shorelines to maintain or encourage dense aquatic and terrestrial vegetation for food.
- Control regrowth cedar using appropriate mechanical or hand clearing to maintain diverse shorelines and riparian woodlands.
- Avoid broad-spectrum pesticides to maintain diverse and healthy aquatic life.

Other Potentially Useful Management:
- Manage for healthy watersheds, including deep and sustainable ground cover to maintain riparian health and water quality.
- Use deer exclosures to encourage woody plant recruitment.
- Control white-tailed deer and other browsing animals to encourage healthy woody understory.
- Control feral hogs to maintain and improve water quality and ground-level herbaceous cover.

YELLOW WARBLER
UNDER-THE-RADAR BIRD
SPRING AND LATE SUMMER

Active, bright yellow; found almost exclusively in dense riparian thickets; regular Texas Hill Country migrant that does not breed or winter here.

Identification: About 5 inches head to tail; yellow all over; males often with rusty streaks on breast.
Habitats: Mixed wooded slope, river, creek, canyon, seep, and springs.
Cover: Shrubby or regrowth woody thickets, often willow.
Breeding Territory: N/A
Nest: N/A
Food: Almost exclusively insects gleaned from tips of branches.

Stewardship: Encourage low-growing woody plants, especially along and adjacent to wet areas for good cover and insect diversity.
- Control white-tailed deer and other browsing animals to encourage woody plant recruitment and healthy understory for cover.
- Control regrowth cedar using appropriate mechanical or hand clearing to maintain diverse riparian woodlands.
- Exclude livestock from riparian areas to maintain diverse understory.
- Use light to moderate rotational grazing to provide a diverse plant community in other areas.
- Avoid the use of broad-spectrum insecticides to protect insects for food.
- Apply ant-specific insecticides to control red imported fire ants and increase insect diversity.

Yellow Warbler

- Control feral hogs to maintain and improve water quality and ground-level herbaceous cover.

Other Potentially Useful Management:
- Manage for healthy watersheds, including deep and sustainable ground cover to maintain riparian health and water quality.
- Use deer exclosures to encourage woody plant recruitment.
- Use prescribed burns in savannahs to increase forbs and maintain diverse woody plant communities.
- Construct tepee-style brush piles to increase shrub-level habitat.
- Keep cats indoors, and control stray and feral cats to reduce predation.

YELLOW-BILLED CUCKOO

PRIORITY BIRD

SPRING AND SUMMER

Large, shy bird of Hill Country riparian woods; heard much more often than seen; called "raincrow" by old-timers because its loud, clucking call is thought to forecast rain. West of Rockies its numbers have dropped dramatically, creating conservation concerns; Texas Hill Country breeding bird surveys show a slight downward trend but still fairly common across the region. Voracious appetite for caterpillars makes it an important biological control for webworms.

Identification: About 11 inches head to tail; gray-brown above and white below; long tail with prominent black and white pattern on lower surface that appears as white dots when perched; yellow lower bill and black upper bill.
Habitats: Post oak savannah, shin oak savannah, live oak savannah, river, creek, canyon, and springs.
Cover: Trees with diverse thickets, especially near streams.
Breeding Territory: Unknown. May avoid competition with other nesting pairs by staggered time of egg laying.
Nest: Platform of short twigs set on a horizontal branch 4–10 feet above the ground.
Food: Forages for insects (over 90%), especially caterpillars and cicadas; also eats lizards and berries.

Yellow-billed Cuckoo

Stewardship: Maintain undisturbed and diverse habitat of trees and brush along streams as essential habitat for food, cover, and nesting.
- Control regrowth cedar using appropriate mechanical or hand clearing to maintain diverse riparian woodlands.
- Control white-tailed deer and other browsing animals to encourage woody plant recruitment and healthy understory.
- Exclude livestock from riparian areas, and in other areas use rotational grazing at light to moderate levels to maintain a diverse plant community.
- Avoid the use of broad-spectrum insecticides to protect insects for food.
- Use prescribed burns in savannahs to increase forbs and maintain diverse woody plant communities.
- Manage for healthy watersheds, including deep and sustainable ground cover to maintain riparian health and water quality.

Other Potentially Useful Management:
- Use deer exclosures to encourage woody plant recruitment.
- Control feral hogs to maintain ground-level herbaceous cover.
- Apply ant-specific insecticides to control red imported fire ants and increase insect diversity.

YELLOW-RUMPED WARBLER

UNDER-THE-RADAR BIRD

WINTER

Common warbler is a welcome sight feeding in the dull winter palette of edge habitats; often actively hunts for food on tips of branches and piles of wood.

Identification: Averages 5.5 inches head to tail; yellow on rump, sides, and often with yellow cap.

Habitats: Old fields, pocket prairie, mixed wooded slope, post oak savannah, shin oak savannah, live oak savannah, river, creek, canyon, seep, springs, tank, pond, and backyard.

Cover: Woody and shrubby edges.

Breeding Territory: N/A

Nest: N/A

Food: Invertebrates and fruit gleaned from branches.

Stewardship: Maintain or encourage shrubby or regrowth habitat edges along fencerows, woodland borders, and riparian areas with debris and a variety of fruiting trees, shrubs, and vines.
- Avoid the use of broad-spectrum insecticides to protect insects for food.
- Control white-tailed deer and other browsing animals to encourage woody plant recruitment, understory, and healthy fruiting trees, shrubs, and vines.
- Use prescribed burns in uplands to maintain and encourage shrub and diverse woody plant communities.
- Control regrowth cedar using appropriate mechanical or chemical methods to maintain diverse woodlands.
- Exclude livestock from riparian areas to maintain diverse understory.

Yellow-rumped Warbler

- Use light to moderate rotational grazing to provide a diverse plant community in other areas.

Other Potentially Useful Management:
- Use deer exclosures to encourage woody plant recruitment.
- Apply ant-specific insecticides to control red imported fire ants and increase insect diversity.
- Construct tepee-style brush piles to increase shrub-level habitat.

4 Canyons, Springs, and Seeps

When European explorers entered Texas, Indians often guided them over well-worn trails from one group of springs to another.

Springs are often found in ravines, canyons, and valley trenches. These places are also convenient dumping grounds, and as a result are frequently filled with old car bodies, refrigerators, tires, and other trash. More care should be taken to preserve Texas' beautiful springs. —Gunnar M. Brune, 1981

INTRODUCTION

Canyons, springs, and seeps are sprinkled throughout the Hill Country. Steep topography with rugged canyons provides the geologic features that allow groundwater to spill out at seeps and springs with unique, relatively uncommon plant communities important to birds and other wildlife. In otherwise dry country, moisture is especially important to the uncommon Bell's Vireo because moisture supports the type of dense vegetation this bird needs. The Rio Grande Turkey also depends on the vegetation at springs and seeps, where it feeds on green plants in an otherwise dry environment.

These three habitats are closely linked to each other, with canyons often containing springs and seeps that add plant diversity, resulting in extremely productive bird habitat. It may be difficult to separate the value of moist bird habitat at a spring or seep from that of a canyon and its natural vegetation. Therefore, it is important to keep in mind the complementary relationship between these three habitats. Slope and aspect of the land also are important in determining the vegetative composition of these critical microhabitats throughout the Hill Country.

DESCRIPTION: CANYONS

Canyon location is determined by the consistency and porosity of underlying rock, size of the watershed draining to a specific point, and the amount of elevation change. Where the land falls off quickly and the underlying rock is soft and porous, water cuts steep canyons. These canyons are often "V" shaped with a wide mouth and narrow upper interior. Steep, narrow canyons are shaded and relatively cool. They collect and hold moisture and support many deciduous trees and shrubs that do not grow in the hot, dry climate above the canyon. When these shaded canyons are steep enough to exclude deer, they have high plant and wildlife diversity, including wintering birds such as the Golden-crowned Kinglet.

FEATURED BIRDS OF CANYONS, SPRINGS, AND SEEPS

PRIORITY
Bell's Vireo
Canyon Wren
Common Yellowthroat
Orange-crowned Warbler
Sedge Wren
Yellow-throated Vireo

UNDER-THE-RADAR
Black-and-white Warbler
Golden-crowned Kinglet
Hermit Thrush
Song Sparrow
White-eyed Vireo
Wilson's Snipe

FAVORITE
Cedar Waxwing
Eastern Phoebe
Indigo Bunting

OUTLAW
Brown-headed Cowbird

Note:
Priority = on at least one of the lists provided by TPWD or Oak and Prairies Joint Venture naming species in need of conservation assistance, or shown by USGS-sponsored Breeding Bird Surveys in the Edwards Plateau to be a species whose population is in decline
Under-the-radar = a regular bird in Texas Hill Country but often unnoticed
Favorite = popular with just about anyone who knows the bird
Outlaw = undesirable because of its behavior and impact on other bird species

Bell's Vireo (Photo by Lora L. Render)

Dense understory in dry habitat near water (Photo by Patsy Inglet)

DESCRIPTION: SPRINGS

Springs are formed when rainwater soaks into the ground, reaches an impermeable layer, and is forced out onto the surface. Springs have more stored water than seeps and can run long after soaking rains have ended. Since spring flow depends on rainfall and the Hill Country has frequent droughts, spring flow fluctuates and depends on weather and natural recharge, land management practices, and amount of groundwater removed by wells that tap into the same aquifer.

SEASONAL OCCURRENCE OF SOME BIRDS ON WELL-MANAGED CANYONS, SPRINGS, AND/OR SEEPS

WINTER	SPRING AND FALL (MIGRANTS)	SPRING AND SUMMER (BREEDING BIRDS)	ALL YEAR (RESIDENTS)
American Goldfinch	American Redstart	Bell's Vireo	Bewick's Wren
Cedar Waxwing	Black-throated Green Warbler	Black-and-white Warbler*	Black-crested Titmouse
Dark-eyed Junco	Blue-headed Vireo	Black-chinned Hummingbird	Cactus Wren*
Golden-crowned Kinglet	Clay-colored Sparrow	Golden-cheeked Warbler*	Canyon Wren
Hermit Thrush	MacGillivray's Warbler (west)	Indigo Bunting	Carolina Chickadee
Orange-crowned Warbler	Mourning Warbler (east)	Louisiana Waterthrush**	Carolina Wren
Sedge Wren**	Nashville Warbler	Orchard Oriole	Chipping Sparrow
Song Sparrow	Ruby-throated Hummingbird	Painted Bunting	Common Yellowthroat
Sora	Swainson's Thrush	Summer Tanager*	Eastern Phoebe
Spotted Towhee	Tennessee Warbler	White-eyed Vireo	Golden-fronted Woodpecker*
White-crowned sparrow	Wilson's Warbler	Yellow-billed Cuckoo	Lesser Goldfinch**
Wilson's Snipe	Yellow Warbler	Yellow-crowned Night Heron**	Red-winged Blackbird
Yellow-rumped Warbler		Yellow-throated Vireo*	Rio Grande Turkey*
			Rufous-crowned Sparrow
			Vermilion Flycatcher

*Uses canyons but not springs or seeps **Uses seeps and springs but not canyons
west = western edge of Hill Country

Texas Parks and Wildlife aquatic biologist and Hill Country springs research scientist Chad Norris reports, "Springs and associated aquatic, wetland, and riparian ecosystems are among the most productive, biologically diverse, ecologically unique, and threatened habitats. Most, if not all, of the springs I visit have been impacted or disturbed in some way in the past. The protection and restoration of spring ecosystems should be a high priority for land management and conservation agencies."

As icons of the Texas Hill Country, our springs are an integral part of the character and culture of the region. They vary greatly in their basic characteristics.

- Some discharge a tremendous volume of water; others have a small but steady flow.
- Some issue from a large defined opening; others emerge from many inconspicuous openings.
- Some bubble up vertically; others trickle out horizontally from the walls of a canyon or cliff.
- Some feed rivers; others flow naturally onto relatively level, fertile ground and create wetland habitats called fens. Experts agree that fens are threatened habitats in need of active conservation.

Wildlife and many resident, migrant, and breeding birds depend on reliable, safe spring water in areas where no other water is available;

Golden-crowned Kinglet
(Photo by Lora L. Render)

Canyon with water

some come to drink and others to feed at Hill Country springs. These birds include one of our favorites—the Canyon Wren, which does not need to drink but specializes in nesting and feeding in the rock crevices where the moisture of the spring keeps insects alive.

DESCRIPTION: SEEPS

Hill Country seeps are found on hillside slopes, hillside benches, and canyon walls and near streams. A seep results when water emerges from the ground slowly and less consistently than water from a spring. In the

Dry rocky canyon with spring nearby

Canyon Wren (Photo by Ruben F. Ayala)

Hill Country, seeps have less water than springs and normally do not flow year-round. They depend on saturated soils and rock and recent rain. Seeps are formed as rainwater soaks into the ground, reaches a layer that is relatively impermeable (often clay or shale), and is forced to move sideways until it leaks from the side of a hill or canyon wall. Thus, sections of an impermeable layer that allow water to leak or "seep" out are often visible at the same elevation. And additional seeps may be found by following this layer along a hillside.

Seeps are important microhabitats used by birds and other wildlife. A few birds, including the Common Yellowthroat, use seeps and springs as their principal habitat in the Hill Country. Many other birds use seeps as an important source of water or as way stations during long migrations. The wetland plant community found at a seep also maintains and improves water quality by filtering the water.

FERN-COVERED SEEPS

These seeps tend to be on steep, shaded slopes or rocky outcrops. They usually have limited but consistent flow that supports water-loving plants specially adapted to this microhabitat. Botanists call this microhabitat the Southern Maidenhair-Herbaceous Vegetation Association. This fern-covered seep plant community is uncommon and occurs on cliff faces and lower slopes of forested box canyons of the Hill Country. Fern seeps are usually in narrow horizontal bands where water seeps from exposed limestone or where water from perennial or nearly perennial creeks provides consistent moisture. Southern maidenhair fern (*Adiantum capillus-veneris*) is usually common on cliff-face seeps. Lindheimer's maiden fern, thin-leaf brookweed, and Lindheimer rosettegrass are more common on moist marly rubble at the foot of the same cliffs.

HILLSIDE OR BENCH SEEPS

These seeps are sunny, often treeless, and support many grasses and other herbaceous plants adapted to growing in wet soil. This marshy microhabitat is important to birds such as wrens and phoebes. As a water source for wildlife, seeps are usually ephemeral or seasonal. However, they provide extra moisture that supports the associated plant communities that are able to survive frequent dry periods. Thus, they add plant diversity to an otherwise dry landscape. Lindheimer muhly, seep muhly, and aparejo muhly often are cornerstones of these areas. Bushy bluestem, spikerushes, and switchgrass are also often found at seeps.

Woody plants are usually not found at seeps because these areas may dry out for extended periods. In a dry cycle, cedar may encroach on a seep, but it usually does not survive in the water-saturated soil of wet years. The wet habitat of a seep varies from a few square feet to several acres. Smaller seeps are much more common than larger.

How do birds use the microhabitat of springs and seeps?

In the relatively dry landscape of the Hill Country, moisture in these microhabitats benefits birds in several ways:

- Water at springs and seeps provides essential drinking places for many birds, including doves, sparrows, kinglets, and orioles.

- Insects living on the diverse array of water-loving plants are food for insectivorous birds such as warblers and flycatchers.
- Flowering plants like Turk's cap and salvia attract hummingbirds and butterflies.
- Fruit of some vines such as Virginia creeper and Carolina snailseed are eaten by Cedar Waxwing, Hermit Thrush, and Northern Mockingbird.
- In wet seasons, the microhabitat of a seep becomes an oasis for wintering birds such as Sedge Wren, Savannah Sparrow, and Le Conte's Sparrow. During cold months, these birds use seep habitat as a place to escape north winds and to eat the insects and the seeds produced by the plants living there.

PROBLEMS OF CANYONS, SPRINGS, AND SEEPS

As in all Hill Country habitats, rooting and wallowing by feral hogs, excessive grazing and trampling by livestock, improper cedar management, invasive nonnative plants, and browsing by too many deer or exotic animals all damage the native plant communities and reduce the quality of canyon, spring, and seep habitat for nesting, wintering, and migrating birds. For example, the brilliant blue Indigo Bunting is one of the area's most beautiful birds. It finds water, food, and shelter in healthy canyons, although it may go undetected because its spring and fall migratory visits are brief and breeding pairs are fairly rare in the Hill Country.

Problem 1: Habitat Damage and Loss Caused by Feral Hogs

Feral hogs are a huge, growing problem in the Hill Country. Their numbers have risen dramatically in recent years for several reasons:

- Hogs have few predators.
- More residents are feeding deer, and hogs also thrive on this food.
- Although hog hunting is gaining popularity, it is not sufficient to address the environmental problems hogs create.
- Feral hog meat continues to be undervalued.
- Hogs are intelligent, well adapted, elusive, and difficult to control.

Do not expect to get rid of them once and for all. You must trap and dispatch hogs on a continual basis. If they are on your property, they probably have been there longer and are more plentiful than you realize.

CANYONS

Feral hogs are attracted to cool, moist areas of Hill Country canyons where they hide, feed, and wallow. This is especially true in the warmest months when hogs use cool, wet areas to prevent overheating. Their

Indigo Bunting (Photo by Lora L. Render)

Spring-fed canyon and shrubs with berries

feeding and wallowing habits are destructive to plants in these areas. Although not a major cause of cedar invasion, the feeding and wallowing habits of feral hogs contribute to this problem. They disturb the ground and make more space for cedar seeds to take root.

SPRINGS

Feral hog rooting and wallowing behaviors inevitably destroy native plants growing near springs and replace the native plant community with bare, muddy ground. The space may remain devoid of plants, or the natural community may be replaced by exotic invasive species. In either case, birds such as Wilson's Snipe that normally live in the area have lost

Spring damaged by feral hogs (Photo by Kory Perlichek)

the food and shelter they need to survive. For example, after feral hogs move in, the gnatcatchers, warblers, and wrens that live on the insects and berries typically found at springs must go elsewhere for food.

Hog wallowing can also have a serious impact on spring water quality. Chad Norris found feral hog damage at about half of the Hill Country springs he has visited and significant impairment at about 25% of these sites. The worst damage is where spring water runs across fairly level soil and supports wetland vegetation. Feral hogs damaging spring water quality is a difficult problem to solve. It is nearly impossible to exclude hogs from a spring site with fencing. If the site is fairly small, fencing might be done with hog panels, but this is expensive. Limited resources may best be spent on intensive feral hog removal.

SEEPS

Feral hogs do enormous damage to natural seep vegetation, especially during hot, dry months when they wallow in wet places to control their body temperature. Birds, including the Song Sparrow, feed on the vegetation found at healthy seeps, and when hogs wallow in a seep, the water, soil, plants and birds are all impacted. It is difficult to exclude hogs from a wet seep, and the best option is reducing hog numbers with regular, long-term control measures.

Whenever you see a seep with disturbed ground and plants removed, suspect feral hogs to be the cause. Ask yourself these questions: Does it

Song Sparrow (Photo by Lora L. Render)

Small seep with healthy vegetation

look like someone has been out with a spade turning over the soil? Does your seep now have the appearance of a depressed muddy pigpen? If so, you probably are looking at hog damage or a hog wallow. The severity of your problem depends on the extent of the damage. If the disturbance is old, limited to one spot, and not too large, you may not have too many animals and trapping a few hogs could take care of it.

What You Can Do: Control Feral Hogs

- Use a long-term trap and dispatch program. Currently, corral traps are considered a major improvement over the old box-style traps. Corral traps catch many more hogs and do not trap deer. Contact your local AgriLife Extension office for a good publication that discusses the pros and cons of each technique.
- Look for damage regularly to catch feral hog activity early so that you can act before the problem becomes intractable and expensive. If you cannot decide, take pictures from different angles and send them to your local TPWD wildlife biologist or your AgriLife Extension agent for expert advice.
- Hogs may be hunted and killed at any time of year in the Hill Country; all you need is a hunting license.

For more information on feral hog control, see chapter 7.

Problem 2: Habitat Overgrazed and Trampled by Livestock

Canyons and other areas with springs and seeps contain fragile plant communities that provide food, water, and shelter to breeding and migrant birds. If at all possible, exclude livestock to protect the native plants, birds, and other wildlife that inhabit these special places.

CANYONS

All but the steepest canyons of the Hill Country have been overgrazed and overbrowsed by cattle and goats for many years. Canyonland that has been overstocked with goats will be missing the shrubs, forbs, and tree branches within their reach. Canyons heavily overbrowsed by goats also may have limited understory dominated by cedar, agarita, and Texas persimmon. It may lack the low-level nest cover needed by birds such as Bell's Vireo, Painted Bunting, and towhees. This habitat damage can also have a significant impact on birds such as Cedar Waxwings that feed on the fruit of possumhaw, Carolina snailseed, American beautyberry, and other berry-producing plants that should be growing there.

Shallow canyons and the openings of steeper canyons overgrazed by too many cattle will have very short ground-covering plants lacking the seeds, flowering forbs, and cover needed by birds that live near the ground. This also contributes to bare ground and runoff. When goats, cattle, sheep, or horses have been allowed uncontrolled access to a canyon for a long time, overuse may have done extreme damage to the plant community and you will see large areas of bare ground and signs of ongoing erosion and soil loss. This kind of long-term damage requires immediate restoration measures, and recovery will take years.

What You Can Do: Rest the Land and Limit Livestock Numbers and Access

- In areas of severe damage to the ground cover and wide areas of bare ground, it is very important to cover the bare ground and deter further soil loss. In fall or early winter, one option is to plant a light mix of native wildrye and some forbs like firewheel and purple prairie clover.
- Cattle are preferred livestock, if you need income from ranching and also want to manage your land for bird diversity. Convert to a rotational grazing system with plenty of rest time to restore sustainable grassland and healthy bird habitat.
- For cattle grazing purposes, you can treat the relatively flat land with decent soil in some canyons as grassland or savannah. After plants recover in these shallow canyons, restocking at a sustainable level will also improve habitat for birds. Rio Grande Turkey may nest among bunches of little bluestem, and Painted Buntings are attracted to grass seeds. Wintering wrens and sparrows can also find shelter from winter winds.
- Rotational grazing with long and varied periods of rest will also allow the natural regeneration of hardwood trees and understory shrubs and vines.

For more information on restoring overgrazed land, see chapter 1; for information on rotational grazing, see chapter 2.

SPRINGS

Poor land management practices can reduce the quantity and quality of water in springs. When livestock are allowed access to a spring, they drink, trample, and lounge in a comfortable place. Thus, their presence may damage the spring opening. In severe cases, the spring can be totally silted in. Livestock have a particularly deleterious effect on the native plants found at a spring. In the Hill Country's hot, dry climate, livestock spend more time at a spring than elsewhere in their pasture. Thus, they overeat the plants that are convenient and trample and often wipe out the plants that once grew there.

When allowed access to a spring, livestock beat down banks with their hooves. The ensuing erosion adds silt to the water and reduces its natural clarity. Silt combined with livestock excrement changes the aquatic invertebrate community, and those species that cannot tolerate this degradation are lost. Common Yellowthroats, Louisiana Waterthrushes, Green Herons, Wilson's Snipes, Northern Cardinals, Eastern Bluebirds, and Say's, Black, and Eastern Phoebes all eat the flying adults of aquatic

Overflowing livestock tank

mayfly, dragonfly, and damselfly nymphs. Waterbirds such as egrets and herons eat crawfish common in springs. Crawfish are more tolerant of poor water quality, but their numbers also may be reduced by livestock damage at springs.

When ranch managers allow improper stocking levels on upland pastures, damage to the land that recharges a spring can be severe. Too many cattle for too many years cause soil compaction and loss of dense ground cover. This means heightened runoff and less water being absorbed by the soil. Mismanagement of this sort is common in the Hill Country and often contributes to the loss of spring water quality and quantity.

> **Remember: Spring water is naturally clear and free of silt and contains aquatic invertebrates that are an important food source for many birds.**

What You Can Do: Manage Livestock for Bird Habitat

Landowners can be major players in the conservation of small springs that are recharged locally.

- To maintain the best bird habitat provided by springs, use good fencing to exclude livestock and protect the fragile plant communities needed by the birds for food and cover.
- If you must water animals from a spring, use PVC pipe and a solar pump to move the water to a tank set far enough away that the cattle can drink without disturbing the spring or its wetland habitat. Make sure that the water tank is not above the springs where the cattle's waste filters through the ground and contaminates the subsurface spring water.
- If your spring has silted in so that flow is greatly reduced or stopped, little wetland habitat will be present. In this case, it may be possible to remove much of the silt material, replenish spring flow, and restore wetland habitat. A restored wetland will have healthy, native, water-loving plants for food and shelter as well as shallow water where birds can drink.
- Use rotational grazing and limit stocking levels throughout the watershed so livestock do not damage the natural structure of ground-cover plants. Healthy ground cover acts to capture rainfall, allowing it to penetrate the soil and prevent runoff. When rain enters the ground, it can percolate through the subterranean geologic structures to recharge the spring below.
- Do not allow continuous overflow on any livestock trough. Although trickling water attracts birds and other wildlife, overflow is preventable water loss. To water wildlife, run the water into a small, shallow ground-level container where little evaporation can occur and the birds can wade and drink safely. Exclude livestock and feral hogs from the water container, and place it near undergrowth cover where the birds can escape predators.

SEEPS

Cattle spot-graze their favorite plants. This results in overgrazed seeps and is especially harmful to lush vegetation at the beginning of a dry season. Spot-grazing also causes trampling that may eventually eliminate the more sensitive native plants. Without its natural plant community, the microhabitat of a seep can no longer support birds such as the Common Yellowthroat and the Eastern Phoebe.

Continuous grazing, high stocking levels, and use of a seep as drinking water can destroy its natural plant community. This overuse often

Sedge Wren (Photo by Tom Mast)

Meadow seep with dense vegetation

leaves an expanse of bare ground that is no longer habitat for winter birds such as the Sedge Wren and Le Conte's Sparrow. These are special birds that depend on this unique Hill Country habitat in their winter range. The Sedge Wren has disappeared from much of its breeding range in the northeastern United States due to habitat loss.

Livestock can also impair the quality and quantity of water from seeps:

- Their feet and waste cloud the water and add bacteria.
- They damage the microhabitat associated with the seep, so wetland plants no longer filter and clean the water as it comes to the surface.

- If there are too many animals on the land, they will reduce the ground cover above a seep. Thin ground cover means more runoff, less water absorbed by the soil, and not as much water coming out of the seep.
- Less water at a seep and loss of its natural plant community are keys to loss of the birds that normally drink, eat, hide, and nest in this microhabitat.

What You Can Do: Remove Livestock or Reduce Stocking Levels

- Fence the seep to exclude livestock.
- Use alternative livestock water sources set away from seeps.
- At least, limit livestock access to a seep with a rotational grazing system.
- Always maintain stocking levels appropriate for the condition of your land.

If you must use a seep to water livestock, pipe the water downhill to a trough. Fence the trough's wet overflow area to exclude livestock and provide moist habitat where birds and other wildlife can drink and hide from predators. Overflow at the downhill trough will replace some of the original seep habitat *only* if the livestock do not ruin it. This does not fully restore the natural seep habitat but is much better than leaving the overflow area unprotected.

Problem 3: Improper Cedar Management

For more information, see chapter 9.

CEDAR INVASION IN CANYONS

Cedar invasion is a problem especially common on the hot, dry west-, southwest- and south-facing canyons. A canyon choked with cedar in place of its natural diverse plant community lacks the space and food for many birds, including insect eaters such as the Black-chinned Hummingbird and Eastern Phoebe. On the bottom of many canyons, browsers (goats, exotics, or white-tailed deer) have eaten up their preferred plants, and cedar has gotten a big advantage. Thus, cedar is now much more abundant, and many natural canyon communities are no longer diverse and balanced.

What You Can Do: Clear Cedar and Rest Land

Conduct selective cedar removal by avoiding steep slopes and maintaining plant diversity. Then, rest the land so the grasses and native perennials can recover. How long this takes depends on the severity of

Canyon spring with plant diversity

Eastern Phoebe
(Photo by Lora L. Render)

overgrazing and the length of time it has occurred along with the timing and amount of rainfall. Plants cannot recover without water during their growing season. To assess the status of ground cover and presence of birds and other wildlife, visit the area regularly. Take pictures using fixed photo points at regular intervals.

CEDAR INVASION IN SPRINGS

Because cedar is intolerant of moist soil, invasive cedar is usually not a problem at Hill Country springs. But cedar growing everywhere is often targeted as the cause of springs and creeks going dry. It is difficult to separate fact from fiction here. Many examples presented to support this argument are incidental with no scientific study that uses controls for seasonal variation and rainfall. For example, a story of springs flowing after cedar clearing might be due to plentiful rain following a dry spell rather than the absence of cedar.

Current data seem to support the argument that cedar clearing over a very large area is necessary to produce a significant increase in groundwater and consequent spring flow. In other words, clearing cedar off a single property may improve grassland habitat but will not make a measurable improvement in spring flow. One possible exception might be a small spring fed by a watershed within property boundaries, where cedar can be removed across its entire watershed.

What You Can Do: Participate in Large-Scale Watershed Cedar Management

If you own all the land above your spring and the cost is not prohibitive, cedar clearing could be worth a try. However, because the subsurface pathways feeding a spring are often complicated and to a large extent unknown, there is no guarantee that cedar clearing will significantly improve spring flow.

CEDAR INVASION IN SEEPS

Cedar is an especially xeric plant that cannot grow in the wet areas of a seep. It often circles the margins of a seep and tends to encroach on the wetter areas during especially dry years. Cedar also spreads quickly across disturbed areas with plenty of bare ground. If this happens over the watershed of a seep, it is possible that a dense thicket could reduce the amount of water flowing into and out of the seep.

What You Can Do: Control Cedar Both at the Seep and throughout Its Watershed

Get rid of cedar as soon as it encroaches on a seep. Ongoing cedar clearing over the watershed of a seep may promote natural water content.

INDISCRIMINATE CEDAR CLEARING

Many landowners have an "all or nothing attitude" toward cedar. They try to clear all the cedar, often using methods that also remove other trees and shrubs valuable to healthy bird habitat. In areas of high deer density, cedar clearing opens up space that becomes especially vulnerable to deer browse and further plant loss. Some seeds and seedlings of plants preferred by deer are probably still present. When the seeds sprout, deer move right in and snap them up. In the Hill Country, sudden and serious erosion is usually due to excessive cedar clearing, which exposes soil that is washed away in the next big rainstorm. If heavy rain falls before grasses and forbs have a chance to grow, erosion is inevitable. Loss of precious soil produces a less productive habitat when it does recover.

Wherever there are erosion and bare ground, habitat for birds and other wildlife is impaired and in need of restoration. For example, the tiny Black-and-white Warbler nests on the ground. It cannot successfully rear its young on eroded bare ground lacking native plants to hide its nest. How you clear cedar, manage livestock, and control deer and exotics is important for a healthy canyon plant community and the presence of this lovely little bird.

INDISCRIMINATE CEDAR CLEARING IN CANYONS

Cool, north-facing canyons where Golden-cheeked Warblers come to nest and rear their young are especially important. This endangered species depends on a mixture of old cedar trees for nest-building material and hardwoods for food. In narrow, steep-sided canyons, cedar clearing may not be desirable. If the cedar here is old and mixed with hardwoods such as Spanish oak and cedar elm, you may have valuable Golden-cheeked Warbler habitat that should be managed differently to encourage this endangered species. You could remove some small, pole-sized cedar to open the canopy and allow sun to reach the ground in a few places to encourage the growth of many different woody plants. Especially in remote canyons that are wet and cool, it might be best to exclude livestock altogether and protect the uncommon plant assemblages important to the Golden-cheeked Warbler, other breeding birds such as Carolina Chickadee or Black-crested Titmouse, and migrants such as Hermit Thrush and Indigo Bunting.

Many other birds also lose habitat when an area is cleared of cedar. Indiscriminate cedar clearing removes the natural mix of trees and understory plants (including rare and endemic bigtooth maple, canyon mock orange, sycamore-leaf snowbell, and big red sage). These plants produce food such as seeds and berries as well as a place for birds to forage for insects. When they are removed, many birds are affected: Black-chinned Hummingbirds feed on nectar and the tiny insects attracted to big red sage flowers; Carolina Chickadees eat the seeds and buds of bigtooth maples; Golden-cheeked Warblers, Summer Tanagers, and Hermit

Black-and-white Warbler
(Photo by Lora L. Render)

Canyon with mature trees and understory

Thrushes glean insects from maple, snowbell, and canyon mock orange plants.

What You Can Do: Use Selective Cedar Clearing

- Use cedar clearing methods that protect deciduous trees and old cedar trees for maximum protection of Golden-cheeked Warbler habitat.

- In cool, rugged canyons, remove some cedar to open the canopy and allow sunlight to reach the ground and provide energy for a diverse woody plant community.
- On the moderately sloped ground of shallow canyons and at the openings of others, remove regrowth cedar on a regular basis. This will allow a ground cover of grasses and forbs to flourish and prevent erosion.
- Watch for and protect uncommon or endemic Hill Country plants such as canyon mock orange, sycamore-leaf snowbell, big red sage, and bigtooth maple.

INDISCRIMINATE CEDAR CLEARING IN SPRINGS AND SEEPS

Indiscriminate cedar clearing on flat or gently rolling land, but especially around seeps associated with canyons, can reduce the natural plant diversity of these moist microhabitats and impact the bird populations using them. Cedars at the edge of a seep are places where birds forage for berries and insects, and cedars also give birds natural shelter from bad weather and predators. Indiscriminate cedar clearing over a large area impacts many species using the habitat: Birds such as Cedar Waxwings and Western Scrub-Jays lose cedar berries that are an essential part of their winter diet. Other birds such as kinglets and phoebes lose shelter and insects needed in the winter.

What You Can Do: Use Selective Cedar Clearing

When landowners are managing habitat for birds and wildlife, cedar clearing requires a thoughtful, measured approach.

- As you remove cedar near a seep on flat or rolling land, leave some cedar trees around the seep as cover and a place for birds to forage for food.
- Near a canyon seep, it is best to remove only small regrowth cedar and leave the larger trees needed by nesting canyon birds such as the Golden-cheeked Warbler.
- Take your time. Watch carefully to see how the plant community around the seep is responding and what birds are using the area.
- Many of the native plants that grow at seeps and springs are not readily available from seed companies, but fortunately most will return on their own with good land management practices.
- Be patient and enjoy the recovery process.

Remember: Some birds and other wildlife use cedar for food and shelter. Cedar clearing for wildlife management must leave adequate natural habitat for their benefit.

Problem 4: Invasion of Nonnative Plants

Invasive exotic plants introduced by humans have had a dramatic negative impact on native plant diversity. They may be woody trees or shrubs or annual or perennial grasses and forbs. An invasion of exotic plants reduces native plant populations. Thus, wherever you see many exotic plants, you can assume that this habitat is not producing its natural combination of bird food, shelter, and nest space. Invasive nonnative plants will not go away by themselves, and many well-adapted exotic species have become abundant and are very difficult to eradicate.

OVERSTOCKING IN CANYONS THAT HAVE INVASIVE NONNATIVE PLANTS

Canyons stocked with too many domestic animals have damaged habitat, often with many invasive exotic plants unpalatable to the livestock. The longer overstocking has occurred, the less diversity and the more undesirable plants, such as Chinese tallow, chinaberry, and ligustrum, will be present.

What You Can Do: Remove Invasive Exotic Species

If only a few exotic plants are present in a limited area, they may be removed by hand, but if you have a large area and many exotics, make a long-term control plan. You will need to remove them annually for some time, but the longer you work at them, the easier it will become. Treatment methods fall into five categories:

1. Mechanical: Use tools to cut, pull, lop, or shear unwanted plants.
2. Cultural: Use overseeding, mowing, or prescribed burning to foster vigorous native plant growth.
3. Biological: Release insects, mammals, or pathogens to stress unwanted plants.
4. Chemical: Apply synthetic herbicides to kill or severely stress unwanted plants.
5. Other: Use low-risk methods such as applications of orange oil, vinegar, soap solutions, and plastic sheets. The effective use of plastic sheets is limited to annuals and some biennials.

The abundance and type of invasive plant determine the preferred removal methods. Remove large, woody invasive shrubs and trees like ligustrum, chinaberry, and Chinese tallow annually during their growing season. Cut down woody plants and paint their stumps with herbicide. If there are not too many forbs, grasses, and seedlings present, hand-pull the seedlings. When treating a large area, spray with herbicide. Be sure to follow directions on the label.

Remember: Many native plants found at Hill Country springs are relatively uncommon and limited to these special sites because of their dependence on consistent water. All the problems found at springs work together to impair and eliminate these species.

Spring-fed wetland with dense vegetation

Common Yellowthroat
(Photo by Lora L. Render)

INVASION OF TERRESTRIAL PLANTS AT SPRINGS

The invasion of exotic terrestrial plants at springs is a serious and pervasive management problem closely related to other problems (human recreation, livestock damage, excessive deer browse, feral hog wallows) that build on one another to cause the loss of native plant communities. Severely disturbed habitat is vulnerable to invasion by exotic plant species. When humans, livestock, feral hogs, and deer or wildlife (includ-

Remember: Exercise extreme caution when using herbicides because they can kill aquatic invertebrates and fish when applied near springs and seeps.

ing birds) introduce exotic plant seeds to an area that does not have its natural plant community, there is no competition or natural control to their spread, so they often take over the site. Thus, a spring can be completely changed into habitat unattractive to the birds that lived in the original community. One of these is the Common Yellowthroat, a stunning bird that may be common where a spring has natural wetland habitat and is sometimes found in other Hill Country locations where dense vegetation hangs over water. According to Chad Norris, invasive terrestrial plants commonly found at Hill Country springs include ligustrum, chinaberry, Chinese tallow, Japanese honeysuckle, fig, watercress, mint, and elephant ear.

What You Can Do: Exclude Large Animals and Practice Ongoing Invasive Terrestrial Plant Removal

Control feral hogs, reduce deer overpopulation, and exclude livestock. As you control the damaging large animals, remove exotic plants in a manner that can be maintained over many years. Break the work into manageable pieces completed annually before the targeted plant species go to seed. Removal may be mechanical or chemical. Size of the infestation and age of the plants determine which technique is preferred. For small infestations or young plants, use mechanical techniques such as hand pulling or a mechanical lever such as Pullerbear, Extractigator, Up-Rooter, or Root Jack. Extensive stands or large plants often require herbicide treatment or a combination of mechanical and herbicide control (cut and treat stump with herbicide). When using chemicals, treat most invasive species during the growing season when the plants are most susceptible. Read and follow all herbicide labels.

For more information on the best way to remove invasive exotic species on your land, consult your local Native Plant Society of Texas chapter, Texas A&M AgriLife Extension agent, or the Texas Invasives Database produced by a collaboration of government agencies and non-profit organizations (www.texasinvasives.org/invasives_database/).

INVASION OF AQUATIC PLANTS AT SPRINGS

According to Chad Norris, in the Hill Country exotic aquatic plants such as hydrilla, water lettuce, watercress, and elephant ear frequently invade spring ponds. Their introduction may have been deliberate or accidental. Invasive aquatic species often proliferate to such an extent that they displace the existing native plant community, destroy the natural food chain, reduce fish populations, and interfere with fishing and swimming.

How do invasive aquatic plants harm kingfishers, ducks, herons, and egrets?

By reducing visibility and supporting fewer fish, invasive aquatic plants may reduce the number of birds that normally find their food in a spring pond. Many Hill Country wading birds and kingfishers use their eyes to hunt and find their prey. For example, if the surface of the water is covered with plants, kingfishers that hunt from a perch above the water cannot see their prey in the water. If the plants are so thick that they reduce the amount of light penetrating the water, they also harm fish and other aquatic organisms that live in spring ponds. If the pond supports fewer fish, it has less food for diving ducks, herons, and egrets. Large springs are especially vulnerable to this problem. Some sites such as Aquarena Springs in San Marcos may be virtually impossible to restore. Here invasive aquatic plant species have been present for generations, and restoration to its natural state would be extremely difficult and expensive.

What You Can Do: Use Ongoing Invasive Exotic Plant Removal

When controlling invasive aquatic plants, keep some vegetation for fish habitat. Fish live among aquatic plants where they feed and hide. A pond with "weeds" will have many more fish than one with no weeds. Both mechanical and chemical methods are available for invasive aquatic plant control:

- Use aquatic weed rakes for manual removal.
- Apply aquatic herbicides such as Rodeo for chemical control. Be careful with chemical treatment. If too successful, you may also lose your fish. The decomposition of the dead plant material reduces dissolved oxygen. When there is an excessive amount of plant decomposition, dissolved oxygen can become so low that a fish kill occurs.
- Protect noninvasive, native aquatic plants such as coontail, which creates good aquatic habitat for waterfowl and other wildlife.

Problem 5: Habitat Browsed by Too Many Deer and/or Exotic Animals

White-tailed deer overpopulation is common in most areas of the Hill Country. Exotic axis deer are of special concern in the eastern half of the Hill Country. Aoudad and moufflon sheep can cause serious overbrowsing in canyons of Bandera, Real, and other counties in the western part of the Hill Country. Deer control is important when improving bird habitat because the plants damaged by deer are needed by birds for food, shelter, and nest building. For example, if you want Cedar Waxwings to use your property during their spring and fall migration, you will have to have the trees, shrubs, and vines that produce their favorite berries. Red mulberry, possumhaw, and Carolina snailseed are especially valuable for these and other migratory birds that eat fruit.

WOODY PLANT RECRUITMENT

White-tailed deer and other browsers eat the tender, young sprouts and seedlings of many deciduous trees and shrubs. As a result, canyons all over the Hill Country are losing species that can no longer be replaced by young plants when the mature plants die. Foresters call this a lack of recruitment.

Upper canyon walls that are steep enough to exclude browsers have more plant diversity. Lower down, where the large animals have easy access, browsers eat the seedlings of their favorite woody plants so none can grow into saplings or mature trees. A mat of seedlings all the same very small size under a large deciduous tree is evidence of lack of recruitment. If you check the area and do not find many saplings (5–15 feet high), this means that there are no young trees growing up to replace the old one when it dies. Some experts refer to this as the "slow train wreck" of woodlands in the Hill Country: slow because the death of mature trees coupled with the loss of replacement trees takes many years to play out. And it will have devastating consequences for the natural habitats that birds and other wildlife depend on. (Extended, severe droughts may accelerate this "train wreck" by killing mature trees.)

In some places lack of recruitment under deciduous trees is also accompanied by a general loss of understory. In these places, the area is open and free of almost all small woody plants and has become useless to birds such as Hermit Thrush and Bell's Vireo, which depend on dense understory.

OVERBROWSING IN CANYONS

Canyons with too many deer and/or other browsers often exhibit lack of recruitment for oak trees and other deciduous hardwoods. Loss of hardwood species limits the number of mature trees that normally provide a stable supply of food for birds and other wildlife. For example, the Yellow-throated Vireo needs mature trees common in healthy canyons of the southern and eastern Edwards Plateau. This bird lives high in tall, mature trees, where it nests and forages for insects to feed itself and its young. Also, different oak species respond differently to climate conditions and produce substantial acorn crops in different years. Thus, species such as Golden-fronted Woodpeckers and Western Scrub-Jays

Canyon and shrubs with berries

Cedar Waxwing
(Photo by Lora L. Render)

that feed on acorns need a variety of mature oak tree species to provide an acorn crop that is reliable from year to year.

Heavy browsing by too many deer has produced more bare ground than is natural in many Hill Country canyons. Canyons overbrowsed by deer and overgrown with cedar experience gradual soil loss. Too many deer and too much cedar result in inferior bird habitat in side canyons and small drainages where deer have eaten up their favorite plants, the natural understory and ground cover are missing, and bare ground is open to slow and insidious erosion for many years. Many species, including the Orange-crowned Warbler, depend on midlevel understory that is routinely damaged by too many deer. This warbler is a fairly com-

Yellow-throated Vireo
(Photo by Tom Mast)

Canyon with water and mature trees

Remember: Loss of deciduous saplings to browsers and loss of mature deciduous trees to drought is leading to a "a slow train wreck" for many important Hill Country trees.

mon Hill Country winter bird, whose population is in a long-term downward trend.

OVERBROWSING AT SPRINGS AND SEEPS

Heavy browsing by deer is also a problem for Hill Country springs and seeps accessible to deer. The diversity of trees, shrubs, vines, and forbs found at healthy springs is one of the characteristics that makes them

Canyon with low and midlevel understory

Orange-crowned Warbler
(Photo by Jeff Forman)

productive and inviting places for birds. Since too many deer damage most of these plants and contaminate the springs with their waste, preserving this bird habitat requires reducing deer overpopulation.

At seeps, tall grasses, forbs, and some woody plants also produce important Hill Country habitat used by birds such as wrens, warblers and sparrows. Although seeps have less water and plant diversity than most healthy springs, they produce important moist microhabitat for these birds in an otherwise dry landscape. The amount of water and plant growth at a seep varies from year to year depending on rainfall and temperature, and this makes them particularly vulnerable to damage by deer browse. The seeps found on gently rolling hills are especially vulnerable to damage by deer overpopulation because they are so easily accessible.

What You Can Do: Reduce Deer Density and Remove Exotic Browsers
It may have taken many years of dense cedar and deer overpopulation to reduce native plant communities, drastically expose the soil, and create erosion. But if deer are radically reduced so they no longer eat every sprout that appears, the natural plant community may be able to restore itself. The following steps will be helpful:

- Use the deer browse evaluation chart in appendix 1 to determine if you have a deer browse problem.
- Commit to long-term, annual deer removal, which is essential to meaningful deer population control.
- Restore plant communities.
- Leave the restoration site alone for a few years after reducing the deer population.
- Learn what plants should be growing in this habitat in your part of the Hill Country.
- Watch what happens, and take pictures at fixed points to document the changes.
- Reintroduce desirable native plants if necessary after three to five years by planting a few as seed source.
- Surround new plants with deer exclosures to make sure that the plants are not eaten. Exclosures will protect important plants found around seeps such as grapes, slippery elm, Texas mulberry, hawthorn, and Carolina buckthorn. Contact your local NRCS office about its cost-sharing program that funds large-scale deer exclosures.
- Reduce exotic animal (axis, aoudad, sika, and fallow deer) numbers as much as possible.
- Identify existing plants preferred by deer. Place exclosures around these plants to protect them from browsers.

For more information on deer management, see chapter 8.

Problem 6: Modifications for Human Use
In the Hill Country, springs produce highly desirable water in an otherwise dry landscape. Consequently, they are used by humans as well as livestock, feral hogs, deer, birds, and other wildlife. The changes made

Spring-fed wetland with dense ground cover

Wilson's Snipe (Photo by Lora L. Render)

by this use inevitably decrease the value of a spring as wetland habitat, where birds normally drink and feed on insects, seeds, and berries. Wilson's Snipe, Hermit Thrush, and Sedge Wren are a few of the birds that depend on this habitat in their Hill Country wintering grounds.

SPRINGS MODIFIED FOR HUMAN USE

In a dry climate fresh, cold water is so scarce and desirable that wherever accessible, Hill Country springs have been altered by human use in the following ways:

- Drained with perforated pipe to adjacent or downhill water tanks for livestock
- Filled to change boggy habitats into pasture
- Sculpted into water holes for livestock
- Landscaped or dammed and made into amenity lakes
- Channelized to concentrate the spring's flow or direction
- Covered in a spring box

However, improving a spring for human use does *not* improve the native plant and wildlife habitat. Where water from many springs spreads out in a wide, shallow wetland, the natural dense herbaceous vegetation is often perfect cover for the Common Yellowthroat, Sora, and Wilson's Snipe. Altering such a site to retain and deepen the water removes an important habitat component for these birds.

What You Can Do: Limit Human Use and Restore as Much as Possible to a Natural State

- If you own a spring that remains untouched, with its natural structure and plant community intact, you are very fortunate. Leave it alone. If you have a spring that runs across an area unchanged by turf lawn, a dugout channel, or a pond, you have a jewel worth protecting. Check it regularly for invasive species, and carefully remove any that you find.
- Do not modify the habitat of a natural spring to build a tank or pond. If your spring has already been modified to make a deep pool, with edges so steep that the birds cannot reach the water safely, you can fill the pool with rocks or reshape its edge to make shallow-water access that is safe for birds and other small creatures.
- If you want to restore the natural flow of a spring that has been ditched or channelized, consult a natural resource agency such as NRCS or TPWD for the best methods to accomplish your goal.
- If your spring is not too radically altered by human use, protect the native plant community, as well as birds and wildlife, and control human use.
- You may limit human use to occasional visits by a few individuals. Avoid the area when birds are nesting or when many birds are using the area during migration.
- For bird-watching in this area, you can observe from a spot far enough away so the birds are not disturbed by your presence.
- If you want to visit the site frequently without disturbing birds and other wildlife, you can set up a small but comfortable blind for closer viewing.

SEEPS MODIFIED FOR HUMAN USE

A natural seep is not just the water but a microhabitat that provides essential food, water, and shelter for migrant, wintering, and breeding birds. Seeps support many Hill Country birds, including one of the more common but rarely seen songbirds, the White-eyed Vireo, and a much larger, more well-known bird, the Rio Grande Turkey. Seeps may have been modified for human use in the following ways:

- Landscaped for aesthetics or recreation
- Planted with exotic grasses
- Filled to change their boggy habitat
- Drained to provide water elsewhere
- Dug out to make a pool of water

Rock seep with native understory plants

White-eyed Vireo
(Photo by Lora L. Render)

What You Can Do: Remove Modifications, Restore, and Limit Human Use

Return natural flow to modified seeps by plugging diversions, removing drains, restoring the native plant community, controlling invasive exotic plants (see problem 4), and restoring the seep so it can be used safely by small birds for drinking and bathing. Seeps associated with rock ledges or shelves may be especially difficult to restore. The problem is how to collect the water on the lower end where it can be held in a small, shallow receptacle appropriate for birds. Sometimes using a narrow, deep container filled with rocks allows safe access for the birds. Also, consider using native flat stone that is slightly concave to retain just enough water for drinking.

Problem 7: Loss of Natural Flow and/or Poor Water Quality

There are many potential causes for a decline in water quantity and quality of Hill Country springs and seeps. They include poor land management in general and the already discussed problems of feral hog damage and cedar encroachment as well as groundwater table decline and drought vulnerability.

LOSS OF NATURAL FLOW IN SPRINGS

Because scientific data are lacking on Texas' springs, it is difficult to make an evidence-based statement on the cause of spring flow decline in many places in Texas. Gunnar Brune, who did fieldwork in the mid- to late 1970s, reported that 65 of the 281 major and historical springs in the state were dry. However, some of these years had little rainfall, and some of the springs he reported as dry are flowing today. Current research on Hill Country springs being done by TPWD, along with groundwater data being collected by Texas' groundwater conservation districts, will provide much more information for future planning and management.

LOSS OF NATURAL FLOW IN SEEPS

If not enough is known about changing water quantity and quality at Hill Country springs, even less is known about what has and is happening to seeps. Seeps are small and fed locally, and as far as we know, no research has been done on their condition. We do know that seeps usually are not protected when land is developed or ranched, the two major causes of alteration to Hill Country seeps. Many birds depend on moist habitats, including seeps that support the plants and insects they need to survive. One such bird is the Hermit Thrush, a winter Texan that escapes the frozen north for a warmer place where it can find insects to eat until the next breeding season.

GROUNDWATER TABLE DECLINE

In certain areas, increased groundwater pumping has reduced groundwater levels and availability. Lower groundwater levels result in diminished spring flow. As the human population grows, the number of wells pumping Hill Country groundwater continues to increase, and we are seeing spring flow decline in counties where population growth is highest.

The principal aquifer under the Hill Country is the Trinity-Edwards Aquifer. There are places where groundwater use has increased to the point that the amount of water flowing from certain springs has been reduced.

- Jacobs Well, a large spring near Wimberley, is fed by the Middle-Trinity Aquifer through a widespread system of underground faults northwest of the spring. Woodcreek, the largest development in this area, pumps water from community wells in a major underground path that supplies water to Jacobs Well spring. When Woodcreek wells are on, the flow from Jacobs Well declines; when they are off, the spring increases. These daily oscillations can be seen on the USGS hydrograph, available online at http://waterdata.usgs.gov/usa/nwis/uv?08170990.
- Comanche Springs is near Fort Stockton and once was one of the largest springs in Texas. In the 1970s, Brune reported that Comanche Springs had gone dry. In fact, the springs had gone dry in 1961 due to extensive agricultural groundwater pumping. In the winter of 1996, after years of decreased pumping and some wet years, the springs started flowing again. They now flow when agricultural irrigation pumping ceases each winter from late December to March.

What You Can Do: Get Involved and Limit Groundwater Pumping

Milan Michalec, member of the board of directors of Cow Creek Groundwater Conservation District, reports that groundwater in Texas is governed by the "rule of capture," which grants landowners the right to pump and use water beneath their property. In practice, the Texas legislature limits the right of capture by enabling citizens to manage local groundwater through groundwater conservation districts. In 2011, the Texas legislature passed Senate Bill 332, affirming ownership of groundwater and the authority and responsibility of groundwater conservation districts to manage groundwater. Comal, Llano, Travis, and Williamson Counties and part of San Saba County are not in a groundwater conservation district and remain subject only to the rule of capture.

If you live in a groundwater conservation district and want to have a voice in conserving groundwater as a sustainable resource, attend the meetings and be an informed conservation advocate. Groundwater management plans and associated rules may vary from district to district. Groundwater conservation district directors are typically elected to serve four-year terms. Change in a district's directors may be followed by a change in how many and where wells may be drilled. As David K. Langford, retired CEO of the Texas Wildlife Association and a Hill Country

Tank with Floating Cover for Bird Access

Top View

Floating Cover

Side View

Water Level

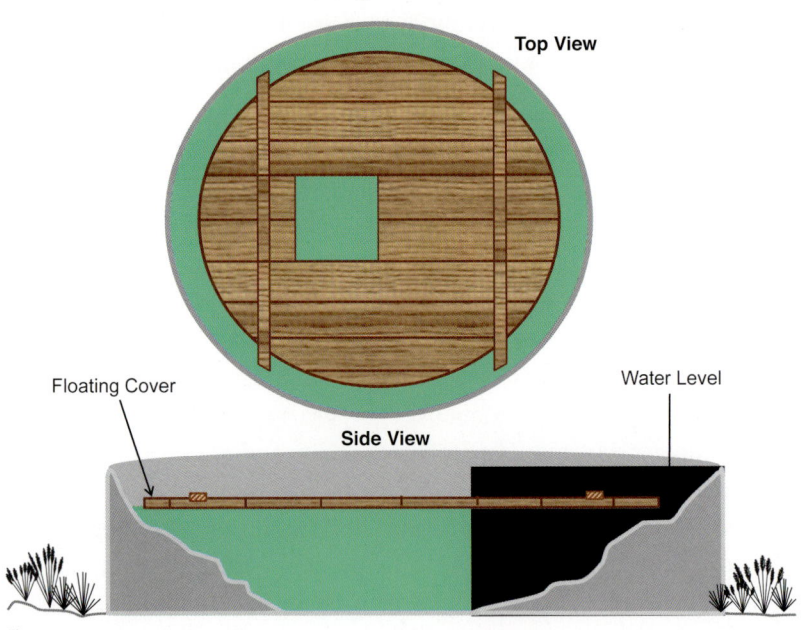

Figure 4.1

resident says, "Everyone's life and economic investment depend on the sensible management of our groundwater resources. All citizens should be involved in the process. Meet your elected groundwater conservation district officials and hold them accountable. They are charged with protecting your groundwater." Let your groundwater conservation district directors know that you support conservation of sustainable groundwater levels, so springs and streams continue to flow.

The biggest consumption of groundwater is landscape watering, and the springs most affected are those where water is being tapped consistently and heavily. Perhaps large springs (such as Jacobs Well) that draw water from a wide area are in more trouble than small springs fed by a small recharge area on one or two ranches. But smaller springs are important, too, and there are several ways to limit groundwater pumping on your property that will protect your springs and seeps.

Never use groundwater to maintain a pond. Putting groundwater in a surface pond puts the water up where it can evaporate. Today, this is an unethical and unsustainable land management practice. Minimize evaporation from livestock and wildlife watering devices. On structures especially for birds and other wildlife, make the watering spot as small as possible. On a livestock trough, use a floating cover to limit excessive evaporation from the surface. The large animals and birds can drink from around the edges.

Hermit Thrush
(Photo by Lora L. Render)

Seep with healthy ground cover

Do not allow livestock troughs to overflow. Although persistent trickling water attracts birds and other wildlife, overflow is preventable water loss. Use xeric plants in your landscaping and limit watering to a minimum. Plant in appropriate habitat and at the appropriate time. Water only to get them started and then let them survive or fail on their own. Save water; do not water to keep your landscape plants green in a dry season.

DROUGHT

Spring flow is dependent on rainfall, and some springs are especially vulnerable to drought. During times of plentiful rainfall, most Hill Country springs flow, but in dry seasons and drought years, decreased water flow

Remember: Wet and dry cycles are regular Hill Country climatic variations that have gone on for millennia. Seeps, their plant and insect communities, as well as the birds that use seeps have adapted to these harsh conditions.

at many springs is natural. During an extended period, many springs may stop flowing altogether until the drought breaks and adequate rain falls again. These periods depend on large-scale weather patterns and are somewhat predictable.

Seeps depend on impermeable strata, saturated soil, and plenty of rain. Water from a seep is rain that has been soaked up by the ground, pulled down by gravity until it reaches an impermeable clay or shale layer, and then pushed sideways until it leaks out onto the surface. Because of their relatively low water flow even during normal years, it is common for Hill Country seeps to dry up during a drought. When and how often a seep dries up is related to its location and area weather patterns, which are fairly predictable.

What You Can Do: Plan and Conserve

Obviously, we cannot prevent the inevitable droughts that occur regularly in the Hill Country. The best we can do is to be prepared for them. Do not be fooled by a big rain, wet season, or rainy year.

- Always conserve water on your land—all year—every year. Do *not* pump well water to make or maintain a pond for livestock or recreation.
- Maintain deep ground cover that will absorb every drop of rain, even in heavy rains, so all the water can penetrate the soil, percolate down, and recharge the groundwater that springs and seeps depend on.
- Place floating covers on the surface of livestock troughs to avoid unnecessary water loss.
- Harvest rain for watering wildlife and livestock. Consider all the rainwater harvesting possibilities on your property. Rainwater harvesting can range from a small wildlife guzzler to large catchments such as a barn roof or the land itself. For more information on guzzlers, see chapter 5.

SUMMARY OF HABITAT PROBLEMS

Problem 1: Habitat damage and loss caused by feral hogs

HABITAT TYPE	STEWARDSHIP SOLUTIONS	SOME BIRDS AFFECTED
Canyons	Use long-term trap and dispatch plan	Gnatcatchers
Springs	Check regularly for hog damage	Wrens
Seeps	Hunt and kill hogs anytime	Warblers

Problem 2: Habitat overgrazed and trampled by livestock

HABITAT TYPE	STEWARDSHIP SOLUTIONS	SOME BIRDS AFFECTED
Canyons	Maintain deep ground cover	Bell's Vireo
	Use rotational grazing	Painted Bunting
	Perhaps stop livestock access and rest land	Rufus-crowned Towhee
	Maintain sustainable livestock levels	Cedar Waxwing
		Rio Grande Turkey
		Sparrows
		Wrens
Springs	Exclude livestock with fence	Common Yellowthroat
	Use rotational grazing with sustainable stocking levels in watershed	Eastern Bluebird
	Remove silt	Green Heron
	Perhaps pipe water to distant tank	Louisiana Waterthrush
	Prevent livestock trough overflow	Northern Cardinal
		Phoebes
		Wilson's Snipe
Seeps	Exclude livestock from seep with fence	Le Conte's Sparrow
	Use rotational grazing with sustainable stocking levels in watershed	Sedge Wren
	Provide livestock water elsewhere	Song Sparrow
	Perhaps pipe water downhill to livestock trough	
	Fence trough overflow to exclude livestock	

Problem 3: Improper cedar management

HABITAT TYPE	STEWARDSHIP SOLUTIONS	SOME BIRDS AFFECTED
Canyons	Conduct selective cedar removal that protects deciduous trees and old cedar	Black-chinned Hummingbird
	In cool canyons, okay to open canopy somewhat	Black-crested Titmouse
	Rest the land	Carolina Chickadee
	In shallow canyons, remove regrowth cedar regularly	Eastern Phoebe
	Protect uncommon and endemic plants	Golden-cheeked Warbler
	Visit and assess area regularly	Hermit Thrush
		Indigo Bunting
		Summer Tanager
Springs and seeps	Cedar control throughout watershed might help	Black-and-white Warbler
	Be cautious with cedar removal near seep or spring	Cedar Waxwing
	Remove regrowth cedar encroaching on seep but leave some in surrounding area	Phoebes
		Ruby-crowned Kinglet
	Assess response of plant community near spring or seep and be patient	Western Scrub Jay

Problem 4: Invasion of nonnative plants

HABITAT TYPES	STEWARDSHIP SOLUTIONS	SOME BIRDS AFFECTED
Canyons and springs	For invasive terrestrial exotic plants: Control feral hogs Reduce deer overpopulation Exclude livestock Use appropriate methods for plant removal	Common Yellowthroat
Springs	For invasive aquatic plants: Use aquatic weed rake Limit aquatic herbicide use to protect noninvasive, native aquatic plants Keep some native aquatic vegetation for healthy habitat	Common Yellowthroat Ducks Egrets Herons Kingfishers

Problem 5: Habitat browsed by too many deer and/or exotic animals

HABITAT TYPES	STEWARDSHIP SOLUTIONS	SOME BIRDS AFFECTED
Canyons, springs, and seeps	Remove or reduce exotic animals as much as possible Use deer browse evaluation chart to determine extent of problem Practice long-term, annual deer removal to control population Restore plant communities Reduce or remove exotics Use exclosures to protect plants preferred by deer	Cedar Waxwing Golden-fronted Woodpecker Orange-crowned Warbler Western Scrub-Jay Yellow-throated Vireo

Problem 6: Modifications for human use

HABITAT TYPES	STEWARDSHIP SOLUTIONS	SOME BIRDS AFFECTED
Springs	Check undamaged springs regularly for invasive species and remove as soon as they appear Do not modify springs to build pond or tank Restore altered spring flow to natural state Limit human use to protect native plant community at springs	Common Yellowthroat Hermit Thrush Sedge Wren Sora Wilson's Snipe
Seeps	Plug diversions Remove drains Restore native plant community Control invasive, exotic plants to restore altered or damaged microhabitat	Rio Grande Turkey White-throated Vireo

Problem 7: Loss of natural flow and/or poor water quality

HABITAT TYPE	STEWARDSHIP SOLUTIONS	SOME BIRDS AFFECTED
Springs and seeps	Be involved with your groundwater conservation district so it supports conservation of sustainable groundwater to maintain springs and seeps Never use groundwater to fill pond Minimize watering trough evaporation with floating cover Prevent water troughs from overflowing Landscape with xeric plants Maintain deep ground cover Harvest rainwater for wildlife and livestock and other watering needs	Hermit Thrush

Bird Summaries: Featured Birds

BELL'S VIREO
PRIORITY BIRD
SPRING AND SUMMER

Uncommon small, plain bird often overlooked as it climbs and flits through low foliage in search of insects. It depends on dense vegetation near water; numbers declining due to clearing of wetlands and riparian zones.

Identification: About 4.75 inches head to tail; gray to faint green above and white to pale yellow below; one clear wing bar with another faint one above; faint, broken, white eye ring. May be easiest to identify by its unmistakable and unmusical scratchy two-phrase song: *Cheedle cheedle chee? Cheedle cheedle chew.*

Habitats: Old field, post oak savannah, shin oak savannah, live oak savannah, river, creek, canyon, seep, and springs.

Cover: Dense, often regrowth vegetation near water.

Breeding Territory: About 1.5 acres/pair.

Nest: Deep cup hung from a branch about 3 feet off the ground in dense shrubs and small trees or occasionally in herbaceous vegetation.

Food: Almost exclusively (99%) insects and spiders gleaned as it moves though vegetation 10–20 feet off the ground. Climbs and flits through foliage in search of insects, sometimes catching flushed insects in midair, gleaning them from foliage while hovering, or grabbing them while climbing through the brush.

Stewardship: Maintain dense, diverse understory, especially adjacent to water, for nest sites and insect food.
- Control white-tailed deer and other browsing animals to maintain dense and diverse shrubland.
- Exclude livestock from riparian areas.
- Use rotational grazing at light to moderate levels to maintain a diverse woody plant community.
- Use prescribed burns in upland areas to promote dense regrowth woody habitat.

Bell's Vireo

- Control regrowth cedar using appropriate mechanical or hand clearing to maintain diverse woodlands.
- Trap and remove cowbirds to reduce nest parasitism and increase nesting success.
- Control stray and feral cats to reduce predation.
- Avoid broad-spectrum insecticides, and control red imported fire ants with an ant-specific insecticide to promote insect diversity.

Other Potentially Useful Management:
- Use deer exclosures to encourage woody plant recruitment.
- Manage for healthy watersheds, including deep and sustainable ground cover to maintain riparian health and water quality.
- Build tepee-style brush piles to help offset lack of shrub-level habitat.

BLACK-AND-WHITE WARBLER
UNDER-THE-RADAR BIRD

SPRING, SUMMER, AND FALL

Somewhat common breeding bird, often seen feeding in canyon and riparian trees; picks insects from tree bark, even hanging upside down like a nuthatch.

Identification: About 4.75 inches head to tail; unmistakable black and white stripes; creeps busily along tree trunks and large branches.
Habitats: Mixed wooded slope, river, creek, and canyon.
Cover: Healthy understory; dense and diverse canopy of mature trees.
Breeding Territory: About 3.5–9.0 acres/pair.
Nest: Cup nest lined with bark and set directly on the ground in moist canyons and riparian woods.
Food: Eats insects foraged from tree bark; often on dead branches, large branches, and trunks.

Stewardship: Encourage healthy understory and diverse mature canyon and riparian woodlands.
- Control white-tailed deer and other browsing animals to improve or maintain plant diversity and woody plant recruitment.
- Limit control of cedar in woodlands to regrowth/pole-sized cedar using mechanical methods while retaining mature trees to maintain balance of deciduous and evergreen (cedar) woody plants and relatively closed canopy.
- Do not clear cedar in wet canyons and riparian habitat during mid-April to mid-June breeding season to protect nesting birds.
- Trap and remove cowbirds to reduce nest parasitism and increase nesting success.

Black-and-white Warbler

- Exclude livestock from mature woodlands and riparian areas to maintain woody plant diversity.
- Limit mature tree trimming, and retain dead branches to protect insect-foraging habitat.
- Avoid broad-spectrum insecticides to promote insect diversity.

Other Potentially Useful Management:
- Use deer exclosures to encourage woody plant recruitment.
- Control red imported fire ants with an ant-specific insecticide to increase insect diversity and reduce possible nest predation.
- Maintain healthy watersheds with deep and sustainable ground cover to protect riparian health.
- Control feral hogs to reduce nest predation.
- Control stray and feral cats to reduce predation.

CANYON WREN

PRIORITY BIRD

ALL YEAR

Has an unforgettable song, a series of beautifully liquid, descending notes. It depends on its extra large feet and long claws to creep across rock walls in search of food; not known to drink water; gets moisture from its food. Breeding bird surveys since 1967 show a disturbing decline of Canyon Wren populations in the Texas Hill Country. In otherwise suitable habitat, known to nest around human habitation, often in locations where cats can be deadly.

Identification: About 5.25 inches head to tail; rusty-brown with white throat and breast; brown belly with white spots; long, thin, down-curved beak.

Habitats: Mixed wooded slope, river, creek, canyon, seep, and springs.

Cover: Rock crevices, canyon walls, and rocky cliffs.

Breeding Territory: Averages about 2.25 acres/pair.

Nest: Shallow cup made of moss, plant fuzz, and feathers; built in rock crevice or sometimes in human habitations near suitable habitat.

Food: Insects and spiders gleaned from rocks and crevices.

Stewardship: Minimize disturbance of cliffs and steep rocky slopes and encourage invertebrate diversity.
- Avoid cedar control and other management practices where birds are present from April through August to minimize disturbance of nest birds.
- Avoid broad-spectrum pesticides to increase forbs and promote insect diversity.

Canyon Wren

Other Potentially Useful Management:
- Maintain water in this bird's natural habitat near cliffs and in rocky canyons to support the invertebrates it eats that require moisture.
- Control white-tailed deer and other browsing animals to maintain plant diversity used by insects.
- Control feral and stray cat populations, and keep cats indoors to reduce predation.

CEDAR WAXWING

FAVORITE BIRD

WINTER AND SPRING

Common, social, and handsome winter bird; Texas Hill Country has highest concentration of winter flocks in the United States; travels in large flocks that feed on plentiful fruit and move on when it is consumed; Cedar Waxwing flocks easily identified by their high-pitched whistles.

Identification: About 6 inches head to tail; sleek tan body; black mask and throat; crest; yellow tail band; short black bill and legs; male and female similar; immature with light streaks and usually no mask.

Habitats: Old field, river valley prairie, pocket prairie, mixed wooded slope, post oak savannah, shin oak savannah, live oak savannah, river, creek, canyon, seep, springs, tank, pond, lake, and backyard.

Cover: Canopy and understory of trees, shrubs, and vines with abundant fruit.

Breeding Territory: N/A

Nest: N/A

Food: An extraordinary fruit specialist; can survive on fruit alone for several months but will eat some insects gleaned or caught on the wing like a flycatcher.

Stewardship: Maintain healthy habitat that supplies a plentiful food source.
- Control spread of regrowth cedar using appropriate mechanical or chemical methods to maintain plant diversity.
- Control white-tailed deer and other browsing animals to encourage plant and insect diversity.

Cedar Waxwing

- Use prescribed burns in uplands to maintain shrubs and good plant diversity.
- Restore and protect fruit-bearing trees, shrubs, and vines to enhance their food supply.

Other Potentially Useful Management:
- Maintain bird-friendly, natural or artificial water source to supply reliable and safe water.
- Use rotational grazing to encourage plant diversity.
- Use deer exclosures to encourage fruiting plant recruitment.
- Avoid broad-spectrum insecticides to promote insect diversity.

COMMON YELLOWTHROAT

PRIORITY BIRD

ALL YEAR

A masked warbler of wet places usually heard before seen. Loud, clear call of *wichity, wichity, wichity*. Nests are frequently parasitized by Brown-headed Cowbirds. Yellowthroat parents hide the location of their nest by walking across the ground to and from the nest when entering and leaving.

Identification: About 4.75 inches head to tail; male olive or brown and bright yellow throat with bright black mask; female same but no mask or very faint one.

Habitats: River valley prairie, pocket prairie, live oak savannah, river, creek, canyon, seep, springs, tank, pond, lake, and backyard.

Cover: Dense wetland vegetation and brushy thickets, often near water.

Breeding Territory: About 0.5–3.0 acres/pair, occasionally larger.

Nest: Bulky cup sometimes with partial roof made of a variety of plants such as grass, sedge, and forbs lined with soft material; built on or near the ground, usually in clump of grass or cattails, in reeds, or in low shrub up to 3 feet from the ground.

Food: Insects and spiders gleaned from dense wetland vegetation near the ground; also will fly out to catch flying insects and occasionally eat seeds.

Stewardship: Conserve their limited habitat by protecting and restoring wetland vegetation at springs, ponds, and seeps.
- Maintain riparian habitat along rivers and creeks to promote cover and nest sites.
- Avoid broad-spectrum pesticides to increase forbs and promote insect diversity.
- Control red imported fire ants with an ant-specific insecticide to increase insect diversity.

Common Yellowthroat

- Exclude livestock or use light grazing for short durations to maintain plant diversity and herbaceous cover.
- Control feral hogs to maintain marshy and shoreline vegetation.
- Do not mow to maintain deep wetland vegetation.

Other Potentially Useful Management:
- Control white-tailed deer and other browsing animals to encourage a diversity of forbs.
- Maintain healthy watersheds with deep and sustainable ground cover to protect riparian health.
- Control invading woody plants by appropriate mechanical methods to maintain wetland vegetation.
- Use pond design features, such as shallow benches and water-control structures, that benefit shorebirds and waterfowl to promote dense wetland vegetation.

EASTERN PHOEBE
FAVORITE BIRD

ALL YEAR

Most common and familiar flycatcher in the Hill Country; named for its distinct and often repeated two-note *phee-bee* territorial call; benefits from human development because of nest-site preference and tolerance of humans nearby. Usually seen singly, perched upright and "hawking" insects from a small branch.

Identification: About 6.5 inches head to tail; dark gray or brown back and head; no conspicuous wing bars; constantly wags its tail.

Habitats: Pocket prairie, mixed wooded slope, post oak savannah, shin oak savannah, live oak savannah, river, creek, canyon, seep, springs, tank, pond, lake, and backyard.

Cover: Wooded areas with understory and shrubs, usually near water.

Breeding Territory: About 3 acres/pair.

Nest: Mud and grass lined with moss and hair; usually placed on ledge of cliff, building, or bridge; reuses same nest year after year with some cleaning and repair.

Food: Flying insects, typically caught in short flight from a perch on branch of shrub or tree; supplements food with fruit of plants such as sumac and poison ivy, especially during cold winter months.

Stewardship: Maintain healthy diverse woodland edges with adjacent herbaceous vegetation.
- Thin or remove cedar thickets, and promote a mixture of woody shrubs in canyons and near streams, springs, and seeps to maintain healthy woodlands.
- Use light to moderate rotational grazing to encourage balance of forbs, grasses, and bare ground in addition to woody plant diversity.
- Use prescribed burns in savannahs to increase forbs and maintain woody plant communities.
- Trap and remove cowbirds to reduce nest parasitism and increase nesting success.

Eastern Phoebe

- Control white-tailed deer and other browsing animals to encourage woody plant recruitment and a diversity of forbs.
- Avoid broad-spectrum pesticide use to increase forbs and insect diversity.
- Do not remove nests in order to protect a preferred nest location and promote reproduction since birds return to old nests.

Other Potentially Useful Management:
- Control red imported fire ants with an ant-specific insecticide to increase insect diversity.

GOLDEN-CROWNED KINGLET

UNDER-THE-RADAR BIRD

WINTER

Tiny, very active bird that seems never to stop flitting and flicking its wings as it searches for insects in shrubs and low trees near moisture.

Identification: About 4 inches head to tail; olive-green above and pale below; bold black-and-white stripes on face; two white wing bars; crown patch orange on males and gold on females.

Habitats: Mixed wooded slope, post oak savannah, shin oak savannah, live oak savannah, river, creek, canyon, seep, springs, tank, pond, lake, and backyard.

Cover: Lower branches of trees and understory of small trees and shrubs; various moist woodlands.

Breeding Territory: N/A

Nest: N/A

Food: Gleans small insects such as aphids, bark beetles, and scale insects and insect eggs from trees and shrubs.

Stewardship: Maintain diverse woodlands with healthy understory.
- Control regrowth cedar using appropriate mechanical or chemical methods to maintain diverse woodland with healthy understory.

Golden-crowned Kinglet

- Control white-tailed deer and other browsing animals to encourage deciduous woody plant recruitment and diversity.
- Avoid broad-spectrum insecticides to promote insect diversity.
- Maintain bird-friendly, natural or artificial water source to supply reliable and safe water.

Other Potentially Useful Management:
- Use deer exclosures to encourage woody plant recruitment.

HERMIT THRUSH
UNDER-THE-RADAR BIRD
WINTER

Somewhat shy and reclusive yet charming relative of the robin; its glorious song not often heard in the winter; usually seen by itself hunting for food among litter where understory is open but sheltered.

Identification: About 6.75 inches head to tail; rich brown head and back; rusty rump and tail with smudgy spots on breast; slender beak.
Habitats: Mixed wooded slope, post oak savannah, shin oak savannah, live oak savannah, river, creek, canyon, seep, and springs.
Cover: Low, dense shrubs and ground-cover vegetation in moist woodlands.
Breeding Territory: N/A
Nest: N/A
Food: Insects, including beetles, ants, caterpillars, grasshoppers, and crickets, as well as snails and earthworms foraged from ground litter.

Stewardship: Maintain woodlands with dense understory and diverse ground cover including leaf litter.
- Protect or restore water quality and quantity at springs, seeps, and streams.
- Avoid use of herbicides and pesticides in low, moist areas to protect natural vegetation and insects in essential winter habitat.

Hermit Thrush

- Control white-tailed deer and other browsing animals to encourage woody plant recruitment and healthy understory.
- Maintain bird-friendly, natural or artificial water source to supply reliable and safe water.
- Control regrowth cedar using appropriate mechanical or chemical methods to maintain diverse woodland.

Other Potentially Useful Management:
- Use deer exclosures to encourage woody plant recruitment.

INDIGO BUNTING

FAVORITE BIRD

EARLY SPRING, SUMMER, AND FALL

Brilliant blue bird sometimes seen in early spring under bird feeders during a "fallout" when hungry migrants, knocked down by strong winds, feed voraciously before they take off for the rest of their trip.

Identification: About 5 inches head to tail; male brilliant blue all over; female plain brown; beak conical like a sparrow's.

Habitats: Old field, pocket prairie, post oak savannah, shin oak savannah, live oak savannah, river, creek, canyon, seep, and springs.

Cover: Shrubs and small trees often in moist locations.

Breeding Territory: About 3.5 acres/pair.

Nest: Sturdy cup of grass, leaves, and bark strips lined with downy material; 1.0–3.5 feet above ground in shrub, small tree, or tangle.

Food: Forages on the ground or in low foliage for insects and spiders; also eat some seeds, wild berries, and grain.

Indigo Bunting

Stewardship: Maintain understory, dense shrubs, and vines, especially around streams, seeps, and springs.

- Avoid broad-spectrum pesticides to increase forbs and promote insect diversity.
- Encourage berry-producing shrubs and vines to protect food sources.
- Control spread of regrowth cedar using appropriate mechanical or chemical methods to maintain savannah and shrubland.
- Use prescribed burns in uplands to control cedar and maintain a relatively open shrubby habitat with diverse grassland community.
- Control white-tailed deer and other browsing animals to encourage woody plant recruitment and healthy shrubs.
- Trap and dispatch cowbirds to reduce nest parasitism and increase nesting success.

Other Potentially Useful Management:

- Use deer exclosures to encourage woody plant recruitment and protect fruit-producing plants.
- Control feral hogs to maintain healthy riparian vegetation.
- Apply ant-specific insecticides to control red imported fire ants and increase insect diversity.
- Use light to moderate rotational grazing to encourage balance of forbs and grasses and woody plant diversity.
- Keep cats indoors, and control stray and feral cats to reduce possible predation.
- Remove some regrowth cedar in canyons, which may be beneficial, but only in limited amounts, to encourage restoration of diverse understory.

ORANGE-CROWNED WARBLER
PRIORITY BIRD

FALL AND WINTER

Tiny, plain, and common winter songbird often seen moving quickly from perch to perch in low foliage. One of only a handful of warblers that winter in the Texas Hill Country. Occurs in most of North America; species has been slowly decreasing throughout its range for 45 years.

Identification: About 5 inches head to tail; pale gray-green above and dull yellow below with faint streaks. Orange spot on head rarely visible; usually appears to have no distinctive markings.

Habitats: Old field, mixed wooded slope, post oak savannah, shin oak savannah, live oak savannah, river, creek, canyon, seep, springs, and backyard.

Cover: Dense understory and midlevel foliage of trees, shrubs, and vines.

Breeding Territory: N/A

Nest: N/A

Food: Actively gleans insects and spiders from twigs, moss, and leaves.

Stewardship: Conserve structure and diversity of understory trees, shrubs, and vines.
- Restore native plant communities to prevent soil and habitat loss.
- Control white-tailed deer and other browsing animals to encourage woody plant recruitment, healthy understory, and a diversity of forbs.
- Control regrowth cedar using mechanical or chemical methods to promote diversity of trees, shrubs, and vines for cover.
- Avoid broad-spectrum herbicides and insecticides, and control red imported fire ants with an ant-specific insecticide to increase forbs and promote insect diversity.
- Use prescribed burns in savannahs to increase forbs and maintain diverse woody plant communities.

Other Potentially Useful Management:
- Remove feral hogs to protect natural vegetation.
- Use light to moderate rotational grazing to encourage balance of forbs and grasses and woody plant diversity.
- Mow periodically in fall or winter (dormant season) on no more than 30% of herbaceous area to promote annual forbs.
- Construct tepee-style brush piles to increase shrub-level habitat.

Orange-crowned Warbler

SEDGE WREN

PRIORITY BIRD

WINTER

Secretive wren of healthy wet meadows associated with seeps or springs. When flushed, has a weak, almost crippled flight; may be easy to see where it landed but is very difficult to flush a second time. Listed here as a priority species because it has experienced population declines through its breeding range to the north due to drained wetlands and consequent loss of habitat.

Identification: About 4.25 inches head to tail; light brown to rusty with dark streaks on back; weak eye line; short bill; identification aided by its weak up-and-down flight pattern.

Habitats: River, creek, seep, springs, tank, pond, and lake.

Cover: Tall herbaceous vegetation of marshes or wet meadows; also marshy areas adjacent to streams or ponds.

Breeding Territory: N/A

Nest: N/A

Food: Not well studied but probably feeds exclusively on invertebrates collected on the ground.

Stewardship: Maintain tall herbaceous vegetation adjacent to water in association with seeps, springs, creeks, ponds, tanks, and lakes.

- Exclude livestock from grazing in these areas during winter months to maintain necessary cover and food.
- Limit grazing to early in the growing season, and allow plenty of regrowth before winter to promote wintering habitat structure.
- Control invasive brush and exotic woody plants with appropriate and cautious mechanical or chemical methods to conserve herbaceous cover.
- Control feral hogs to protect natural vegetation for cover and food.
- Avoid broad-spectrum pesticides, and control red imported fire ants with an ant-specific insecticide to increase forbs and promote insect diversity.
- Practice watershed management to maintain water flow and healthy wet meadows or marsh habitat.

Other Potentially Useful Management:

- Use pond design features, such as shallow benches, that benefit waterfowl when building or maintaining tanks or lakes to promote good winter habitat.
- Install water-control structures during dam construction or repair to facilitate water-level management.

Sedge Wren

SONG SPARROW
UNDER-THE-RADAR BIRD
WINTER

Winters in the Texas Hill Country and is somewhat uncommon. A streaky, "little brown bird" found in a variety of open habitats. Scratches with both feet as it searches for seeds and insects on the ground.

Identification: About 5.75 inches head to tail; has a long, rounded tail and dark brown or chestnut-streaked back; pale breast with dark streaks often with central spot; whitish throat.

Habitats: Old field, river valley prairie, pocket prairie, post oak savannah, shin oak savannah, live oak savannah, river, creek, canyon, seep, and springs.

Cover: Shrubby fields, dense understory along woodland or stream edges, and marshy vegetation.

Breeding Territory: N/A

Nest: N/A

Food: In winter, mainly seeds of forbs and grasses collected on the ground; also some insects.

Stewardship: Encourage healthy mix of herbaceous cover and shrubs, especially adjacent to wet areas.
- Control white-tailed deer and other browsing animals to improve forb diversity and maintain a healthy shrub community.
- Exclude livestock from riparian areas, seeps, and springs, and in other areas use rotational grazing at light to moderate levels to maintain a diverse plant community.
- Control regrowth cedar using appropriate mechanical or chemical methods to maintain shrubs and woody plant diversity.
- Control feral hogs to maintain healthy wetland vegetation.
- Use prescribed burns to control cedar and maintain a relatively open shrubby habitat with diverse grassland community.
- Avoid broad-spectrum pesticides to increase forbs and promote insect diversity.

Other Potentially Useful Management:
- Use tepee-style brush piles to increase shrub-level habitat where brushy habitat is lacking.
- Keep cats indoors, and control stray and feral cats to reduce predation.
- Maintain healthy watersheds with deep and sustainable ground cover to protect riparian health.
- Control red imported fire ants with an ant-specific insecticide to maintain healthy supply of insects.
- Use deer exclosures to encourage woody plant recruitment.

Song Sparrow

WHITE-EYED VIREO
UNDER-THE-RADAR BIRD
SPRING AND SUMMER

Secretive bird of wooded areas; hard to see but easy to locate by its incessant song heard all day long during breeding season; adaptations of its bill and feet allow it to specialize in eating relatively large insects for its size.

Identification: About 5 inches head to tail; muted-yellow with bright yellow spectacles, white eyes, and two white wing bars; immature birds have brown eyes.
Habitats: Mixed wooded slope, post oak savannah, shin oak savannah, live oak savannah, river, creek, canyon, seep, springs, and backyard.
Cover: Shrubs and thickets; moist canyons, riparian areas, and woodland edges; brush near springs and seeps.
Breeding Territory: About 3 acres/pair, but higher density is likely in good-quality habitat.
Nest: Cup in low shrub, up to 6 feet above the ground and made of leaves, grass, and pieces of bark.
Food: Insects, spiders, snails, and tiny lizards.

Stewardship: Maintain understory trees and shrub habitat, especially near springs or seeps and along creeks and rivers.
- Control regrowth cedar using appropriate mechanical or chemical methods to maintain shrubs and woody plant diversity.
- Avoid broad-spectrum insecticides to promote insect diversity
- Exclude livestock from riparian areas, and in other areas use rotational grazing at light to moderate levels to maintain a diverse plant community.

White-eyed Vireo

- Control white-tailed deer and other browsing animals to encourage woody plant recruitment and healthy understory and shrub community.
- Trap and dispatch cowbirds to reduce nest parasitism and increase nesting success.

Other Potentially Useful Management:
- Use prescribed burns in savannahs to maintain diverse woody plant communities.
- Use deer exclosures to encourage woody plant recruitment.
- Apply ant-specific insecticides to control red imported fire ants and increase insect diversity.
- Maintain healthy watersheds with deep and sustainable ground cover to protect riparian health.

WILSON'S SNIPE
UNDER-THE-RADAR BIRD
WINTER

Shy, well-camouflaged wetland bird; usually not seen until flushed; takes off with quick zigzag flight; wings make a startling loud sound; feeds by probing soft mud with its long bill. To get a good look, do not disturb it if you spot it on the ground. Bird with a beautiful pattern and interesting to watch.

Identification: About 11 inches head to tail; stocky body; long bill; short legs; brown color with clear stripes on back and head.
Habitats: River, creek, canyon, seep, springs, tank, pond, and lake.
Cover: Marshy areas associated with water, consisting of dense grass and sedges.
Breeding Territory: N/A
Nest: N/A
Food: Soil insects, earthworms, and other invertebrates that live underground plus some seeds and leaves.

Stewardship: Restore and maintain deep wetland vegetation associated with water, such as springs, seeps, ponds, tanks, lakes, and canyons for food and cover.
- Control feral hogs to maintain wetland and shoreline vegetation and water quality.
- Limit livestock access to the shoreline and wetland areas to maintain healthy vegetation for food and cover.
- Do not mow in marshy areas to maintain deep winter vegetation.
- Avoid broad-spectrum pesticides to protect wetland vegetation and invertebrate diversity.

Wilson's Snipe

- Manage for healthy watersheds, including deep and sustainable ground cover to maintain water quality and quantity.
- Manage water levels during fall and winter to maintain damp soils and healthy vegetation.
- Manage for saturated soils during growing season to encourage the growth of wetland vegetation.

Other Potentially Useful Management:
- Control woody plants by appropriate mechanical methods to maintain marshy vegetation.
- Use pond design features, such as shallow benches, that benefit shorebirds and waterfowl when building or maintaining tanks or lakes to promote good winter habitat.
- Install water-control structures during dam construction or repair to facilitate water-level management.

YELLOW-THROATED VIREO

PRIORITY BIRD

SPRING AND SUMMER

Most colorful of all vireos in Texas but relatively uncommon and little known; a boreal edge species that forages slowly and methodically overhead; nests in Lost Maples State Park.

Identification: About 5.5 inches head to tail; olive head, cheek, and back; yellow throat, chest, and around dark eyes; two white wing bars.

Habitats: Post oak savannah, live oak savannah, river, creek, canyon, and backyard.

Cover: Upper story and midlevel vegetation of tall deciduous trees, wooded riparian area, and wet canyons.

Breeding Territory: Probably about 7 acres/pair in suitable riparian habitat; as an edge species is highly susceptible to cowbird parasitism.

Nest: Small open cup held together by caterpillar and spider silk; suspended from fork of narrow branch usually near trunk and 3–60 feet above the ground; prefers tall, deciduous, riparian trees.

Food: Mostly insects and spiders gleaned from bark, branches, and leaves.

Stewardship: Maintain and manage large tracts of diverse, multilayered, deciduous hardwoods with a relatively closed canopy.
- Control regrowth cedar using mechanical or hand clearing to maintain diverse riparian woodlands.
- Control white-tailed deer and other browsing animals to encourage woody plant recruitment and healthy understory.
- Exclude livestock from riparian areas, and in other areas use rotational grazing at light to moderate levels to maintain a diverse plant community.
- Manage for healthy watersheds, including deep and sustainable ground cover, to maintain riparian health and water quality.

Yellow-throated Vireo (Photo by Tom Mast)

- Avoid the use of broad-spectrum insecticides to protect insects for food.
- Use prescribed burns in savannahs to increase forbs and maintain diverse woody plant communities.
- Trap and remove cowbirds to reduce nest parasitism and increase nesting success.

Other Potentially Useful Management:
- Use deer exclosures to encourage woody plant recruitment.
- Apply ant-specific insecticides to control red imported fire ants and increase insect diversity.

> **FEATURED BIRDS OF CONSTRUCTED TANKS, PONDS, AND LAKES**
>
> PRIORITY
> Killdeer
> Lesser Scaup
> Red-winged Blackbird
> Sora
> Spotted Sandpiper
> Vermilion Flycatcher
> Yellow-crowned Night-Heron
>
> UNDER-THE-RADAR
> Black-bellied Whistling-Duck
> Gadwall
> Red-breasted Merganser
>
> FAVORITE
> Barn Swallow
> Belted Kingfisher
> Blue-winged Teal
> Dark-eyed Junco
> Green-winged Teal
> Ring-necked Duck
>
> OUTLAW
> Brown-headed Cowbird
> Egyptian Goose
> Muscovy Duck
>
> Note:
> Priority = on at least one of the lists provided by TPWD or Oak and Prairies Joint Venture naming species in need of conservation assistance, or shown by USGS-sponsored Breeding Bird Surveys in the Edwards Plateau to be a species whose population is in decline
> Under-the-radar = a regular bird in Texas Hill Country but often unnoticed
> Favorite = popular with just about anyone who knows the bird
> Outlaw = undesirable because of its behavior and impact on other bird species

5 Constructed Tanks, Ponds, and Lakes

Continuity is at the heart of conservatism: ecology serves that heart.
—Garrett Hardin

INTRODUCTION

In the Hill Country, there are spring-fed ponds and pondlike pools that form in streambeds. All other tanks, ponds, and lakes are artificially constructed. In this chapter, we limit our discussion to human-made water-retention structures. Since water is often a limiting factor for wildlife, and surface water seems to be declining, both natural and human-made water-collecting structures have the potential to create valuable habitat for birds, including the stunning Vermilion Flycatcher. These natural and artificial water bodies are especially important for wintering and migrant waterfowl in drought years when freshwater marshes along the Texas coast are reduced. During these years, waterfowl that usually winter along the coast must find wetland habitat elsewhere, and Texas' inland ponds, tanks, and lakes become critical winter habitat.

What are the differences between constructed ponds, tanks, and lakes?

In the Texas Hill Country, artificial ponds are often called "tanks." Here we will use the words "pond" and "tank" interchangeably to refer to a body of water about one surface acre or less. Tanks are built primarily to water livestock and are generally bowl shaped with a regular, sloping shoreline that allows easy access for livestock. Livestock tanks attract many species of wildlife, including resident, wintering, migrant, and breeding birds.

A big tank (more than one surface acre) is often called a lake. Lakes are frequently made by building a dam on a stream or drainage, and its natural features define the shoreline. Marsh habitat along flat, shallow edges is a common bonus when building a lake. Constructed lakes are usually designed and built for recreation or flood control but commonly serve livestock and wildlife as well. In winter, the deep water of a good-sized lake provides open-water habitat for diving ducks such as Lesser Scaup, Bufflehead, Redhead, and Red-breasted Merganser. Areas of deep water also maintain larger fish populations through dry years. Overflow elements such as a large emergency spillway are common on human-made lakes, and many also have a drain pipe/principal spillway.

Pond with woody vegetation and perches

Vermilion Flycatcher
(Photo by Tom Rust)

They protect the dam during large floods and serve as a way to maintain a more consistent water supply for livestock and wildlife.

DESCRIPTION: CONSTRUCTED TANKS, PONDS, AND LAKES

Constructed tanks, ponds, and lakes can, but often do not, provide suitable habitat for birds, and many landowners who own a tank wonder why they do not attract migratory ducks such as the Green-winged Teal and Ring-necked Duck. The difference between a tank that birds use and one that they cannot use often depends on the type of vegetation and water depth at or near the shoreline. Shoreline shape and depth are im-

Lesser Scaup (Photo by Lora L. Render)

Large deep lake with healthy aquatic habitat

portant habitat features in both natural and constructed ponds. Pond water depth and shoreline vegetation fluctuate naturally from season to season. Human-made structures that mimic good natural features are the best for attracting birds.

SHORELINE AND WETLAND VEGETATION
Shorebirds, waterfowl, and other birds need plenty of plants for food and shelter around a constructed water source. These birds do best if

SEASONAL OCCURRENCE OF SOME BIRDS ON WELL-MANAGED TANKS, PONDS, AND LAKES

WINTER	SPRING AND FALL (MIGRANTS)	SPRING AND SUMMER (BREEDING BIRDS)	ALL YEAR (RESIDENTS)
Bufflehead	American Wigeon	Barn Swallow	Belted Kingfisher
Dark-eyed Junco	Least Sandpiper	Blue-winged Teal	Black-bellied Whistling-Duck
Gadwall	Pectoral Sandpiper	Common Nighthawk	Chipping Sparrow
Green-winged Teal	Sandhill Crane	Great Egret	Common Yellowthroat
Lesser Scaup	Wilson's Phalarope	Orchard Oriole	Eastern Phoebe
Northern Pintail		Painted Bunting	Great Blue Heron
Pied-billed Grebe		Purple Martin	Green Heron
Redhead		Red-shouldered Hawk	Green Kingfisher
Red-breasted Merganser		Yellow-crowned Night-Heron	Killdeer
Ring-necked Duck			Red-winged Blackbird
Sedge Wren			Spotted Sandpiper
Sora			Vermilion Flycatcher
Wilson's Snipe			
Yellow-rumped Warbler			

the plant community growing around a pond or lake is diverse and includes grasses, forbs, shrubs, rushes, and sedges. Dabbling ducks such as wigeon, teal, and Gadwall feed on wetland vegetation and seeds both in shallow water and on land nearby.

RELIABLE WATER AND WETLAND VEGETATION

Open water is valuable habitat where the super-aerial Chimney Swift, Barn Swallow, and Purple Martin feed, drink, and bathe in flight. Perennial water means that bulrushes and giant rush can grow at the edges of lakes and larger tanks. Both provide excellent nest habitat for Red-winged Blackbirds in early spring and summer. The females build sturdy nests with rushes or sedges and line them with fine grass. Fairly reliable water supports shoreline woody plants such as roughleaf dogwood, possumhaw, and common buttonbush. These shrubs create perches at water's edge and shade for fish populations. Perches, fish, and clear water are essential to attracting the Belted Kingfisher. Sparrows, vireos, wrens, warblers, and finches eat insects, fruit, and seeds of shoreline shrubs and forbs such as goldenrod, dock, and Maximilian sunflower. A healthy mix of diverse shoreline vegetation is also important in controlling erosion caused by floods and wave action.

SHALLOW FLATS

A pond with wide, flat benches of various depths provides essential feeding habitat for dabbling ducks and wading shorebirds. When water levels diminish during the dry summer months, mudflats around the edge of

Ring-necked Duck
(Photo by Lora L. Render)

Green-winged Teal
(Photo by Lora L. Render)

Pond with emergent vegetation and shallow feeding area

a pond, tank, or lake become important feeding grounds for resident Killdeer, Spotted Sandpiper, and Wilson's Snipe, which feed on invertebrates such as snails and crayfish plus seeds, frogs, and minnows. These exposed-perimeter mudflats also have ideal growing conditions for vital waterfowl food plants such as pink smartweed, spikerushes, and wild millet. And when mudflats are flooded by the seasonal fall rains, these areas become critical winter feeding habitat for a variety of dabbling ducks. Aquatic invertebrates, such as larval stages of dragonflies and damselflies, and snails that live on and around the plants are also food for waterfowl and shorebirds.

WATER-CONTROL STRUCTURES

A valve on the pond's overflow pipe is a common water-control structure that allows control of water level and production of more consistent shoreline vegetation.

- Lowering water level during the spring and early-summer growing season exposes mudflats for growth of shoreline vegetation.
- Raising water level in the fall and winter floods the mudflats and the plants that have grown there. Flooded mudflats become shallow pond habitat where dabbling ducks and wading birds can feed on the plants and their associated aquatic invertebrates.

EMERGENT PLANTS AND THE WET/DRY CYCLE

Spikerushes and bulrushes are perennial "emergent" plants, sprouting and growing up through the water. Pink smartweed is an annual emergent plant that must first germinate on a mudflat and then, while growing, can tolerate being flooded. These emergent plants and others found in the Hill Country grow with their roots in the water and some parts reaching out of the water. They are in shallow water, often less than 12 inches deep. As water levels drop in the middle of the summer, more and more of the emergent plants are exposed. When the water level falls far enough, an emergent may eventually have no water around it at all.

Emergent plants provide a unique combination of aquatic and terrestrial habitat. The submerged part of the plants is excellent habitat for a wide variety of aquatic invertebrates, including water bugs, aquatic snails, tiny crustaceans, and the more conspicuous larval stages of dragonflies and damselflies. These are vital to most wintering ducks as well as many other birds, from Wilson's Snipe and Green Heron, which eat the aquatic invertebrates, to Barn Swallows and Purple Martins, which snatch the flying adult insects from the air.

CONSTRUCTED TANKS, PONDS, AND LAKES MANAGEMENT GOAL

The goal is to create a body of water that mimics excellent natural pond habitat for fish and attracts many different bird species throughout the year. Suitable bird habitat will have these characteristics:

Killdeer (Photo by Jeff Forman)

Pond with shoreline mudflat

- Shallow and deep water
- Wide, flat benches in pond's bottom contour or shape
- Emergent and submerged aquatic vegetation
- Underwater fish habitat
- An irregular, undulating shoreline with shallow-water shelves and a wide mudflat along the edge
- Diverse woody and herbaceous shoreline vegetation

- When possible, consistent water during both spring and fall bird migration seasons

Tank Building

Because this chapter is about management of constructed habitat, we start with how to do it right the first time. Solutions to problems with existing artificial ponds and lakes will be covered in the specific problems section that follows. The first question is whether or not creating a pond habitat on your property is the best management strategy for birds and other wildlife in your area. This is not necessarily an easy decision. When deciding to build a tank or not, consider the following factors.

TANK BUILDING MAY NOT IMPROVE WILDLIFE HABITAT

Just because you can and want to build a tank does not mean this is what will be most beneficial for birds and other wildlife on your property. You may already have plenty of naturally existing water in a stream, spring, or seep. In other words, water may not be the habitat factor limiting bird diversity, so adding another water source may not be the answer to improving your wildlife habitat. There may be greater habitat management needs where your time and money resources could be better spent. The native plants on your property might be suffering from overbrowsing. Increasing woody and perennial plant diversity might be the one management goal necessary to produce a huge increase in the bird numbers and species on your land. Beginning an annual practice of harvesting many deer rather than building that "dream" tank or lake may be an important but difficult decision to make.

However, if there are no ponds or lakes in your area, water may indeed be a limiting factor for birds and other wildlife, and adding a tank to your property may be a good stewardship decision for birds such as Spotted Sandpiper, Great Egret, Great Blue Heron, and Yellow-crowned Night-Heron. Over the years, changes in land use and groundwater demands have caused some perennial streams that used to flow nearly all the time to become seasonal and flow only during wet months. Also, some historically seasonal streams have now been reduced to flowing only immediately after a heavy rainfall. Thus, the Texas Hill Country now has less water flowing across its landscape than in the past, and constructed pond structures have become more important to birds and other wildlife.

Of course, you may have other reasons besides wildlife habitat for wanting to build a tank or lake on your property. Sometimes many uses can be accomplished in careful tank design and placement. You might be able to provide flood control, a place for the family to fish and picnic, water for livestock, and habitat for birds and wildlife.

Spotted Sandpiper
(Photo by Lora L. Render)

Tank with muddy shoreline and adjacent vegetation

TANK BUILDING CHANGES STREAM HABITAT AND HYDRAULICS
Building a tank changes habitat features already present. You might want a tank so you can attract the charismatic Belted Kingfisher, but by building a tank, you may be giving up important downstream habitat for less charismatic species such as wrens, warblers, and vireos. Consult with your local wildlife biologist to help you determine which birds are in greatest need of conservation through habitat enhancement on your property.

Tanks or lakes built on seasonal or perennial streams change the riparian system: they flatten the stream's hydraulics, alter the vegetation, and narrow the riparian area downstream by drying it out. This depresses seasonal stream-flow variations, but seasonal variation is desirable because floods help build the floodplain and habitat. An example of the dependence of riparian vegetation on these hydrologic variations is that the seeds of some trees (cypress, cottonwood, and sycamore) can germinate only on the moist soils found on the banks of streams as the water level drops.

TANKS INCREASE EVAPORATION AND WATER LOSS
In a dry climate, a pond with its inevitable evaporation may not be the best use of the water, a limited and valuable resource. The pond will make the land downstream more xeric. Water loss through evaporation from the surface of a pond may be more important ecologically than the aquatic habitat created by a pond on your property.

ALTERNATIVES TO BUILDING A TANK ARE AVAILABLE
Tank building is not the only way to provide water for birds. Here are some alternatives:

- Modify an existing livestock trough with a few bird-friendly additions. You can add ramps, rocks, or floating structures so the water is accessible to different kinds of birds. Water should be less than 3 inches deep where the birds will be drinking or bathing—the shallower the better. A wide, flat area 1 inch deep is best for most birds.
- Install water guzzlers as an easy and inexpensive way to collect water for birds. Guzzlers, short for "gallinaceous guzzlers," were first designed for quail and other gallinaceous birds in habitats of the Southwest and are good water resources for Hill Country birds of all kinds, including the Dark-eyed Junco, a popular winter "snowbird." Guzzlers are simple water-harvesting structures that have three basic parts: a catchment area like a roof; a storage tank; and a way to deliver the water to the birds, such as a water trough. One of the most basic guzzler designs has a 10-foot by 10-foot catchment area that drains into two 55-gallon (food grade) drums that store the water. Cut out one corner of the drum and add a ramp for access to the water collected. Many variations on the basic guzzler design exist and are limited only by your ingenuity and/or available materials.
- While we do *not* recommend modifying the habitat of a natural

spring to create a tank or pond, some springs have silted in so that little wet habitat remains. In this case, it may be possible to remove some of the silt material and restore spring flow to re-create a wetland habitat that includes plants for food and shelter as well as a shallow place for birds to drink. Also, a spring may already have been modified so the water is held in a deep pool with edges so steep that the birds cannot reach the water safely. In this case, you can fill the pool with rocks or reshape its edges to make water access shallow and safe for birds and other small creatures.
- Restoring seeps may be a little harder because of their seasonal nature. This is most appropriate for seeps associated with rock ledges or shelves. It is necessary to collect the water on the lower end where the water can be held in a small shallow receptacle appropriate for birds. A narrow, deep container filled with rocks can allow safe access for the birds, and native flat stone that is slightly concave can be used to retain just enough water for drinking.

Do I need a permit to build a pond on my creek?

Probably not. Section 11.142 of the Texas Water Code exempts a pond built for either domestic or agricultural use that is 200 acre-feet or less. (A 50-acre lake that averages 5 feet deep or a 10-acre lake that averages 25 feet deep is 250 acre-feet and requires a permit.) Other regulations are that the water must leave the owner's property where it did before the pond was built, and the pond must be entirely on the owner's property and not back water onto any other property.

Although the Texas Commission on Environmental Quality (TCEQ) is the state regulatory agency that handles permits, your local NRCS conservationist is available for planning and design assistance. The best publication on pond building is USDA agricultural handbook AH590, *Planning, Design, Construction*. The PDF file can be downloaded free at http://naldc.nal.usda.gov/catalog/CAT11133077. There are a number of things to consider when constructing a tank.

Size

Do not rely on the bulldozer operator to determine tank size and how or where to build your tank. Some may have thorough knowledge of the watershed, but it is not in their financial interest to tell you if your soil is not suitable or the watershed is not large enough to provide water to maintain a satisfactory tank in most years. Go to the nearest NRCS office

Guzzler with accessible water for wildlife

Dark-eyed Junco
(Photo by Lora L. Render)

for help when considering a new tank. There is an NRCS office in almost every county, and the personnel can be especially helpful with tank design.

Tank size should be determined by its catchment area. The size of the watershed is extremely important in determining how much water can be funneled to your tank. It is not just the creek, drainage, or draw that carries water into a tank but also the land that slopes toward those areas. The watershed for your tank is all the higher land where rainwater moves downhill across the land to your collection site. Tank size needs to be large enough to hold the water it receives in rain events and small enough so the natural drainage (its catchment area) will produce enough water to maintain an acceptable water level in most years. A tank should be built with an emergency spillway to allow floodwater to bypass the tank in large floods. Many upland sites do not have an adequate watershed to supply enough water for a tank with a decent water level. For more detailed information on the relationship between watershed area and tank size, consult your local NRCS office.

Existing upstream tanks may limit your tank size or even prevent you from building one. Your tank's watershed may not be contained within your property. All upstream tanks must be filled before any water flows into yours. Topographic maps are very useful in determining the size and coverage of a watershed and are readily available on the Internet. Those that are combined with other information such as an aerial photo or your property boundary are most desirable. One of the more useful sources is Web Soil Survey, which allows you to add, or turn on, a topographic map layer. There is also a program you can run with Google Earth called MapFinder by Digital Data Devices, which allows you to overlay a topographic map on Google Earth and even adjust its transparency. This program is somewhat cumbersome, but online map programs are changing rapidly and should improve dramatically in the future.

Location

Occasionally, an upland tank built on a hill for livestock watering might have an adequate watershed and will create an "aquatic island" that serves wildlife and waterfowl even in dry seasons. These are always on a draw of some kind where water moves naturally, and a dam simply holds the water in years when there is adequate rainfall. Never build a pond over obviously fractured rock. The tank will leak and be virtually impossible to seal. Trying to repair a tank built in this kind of poor location is outrageously expensive and frustrating. In the end, it most likely will never hold water. Also, do not build a tank in steep topography. These places are too narrow and steep-sided to provide good bird and wildlife habitat.

Material

The best tanks are built in soils with high clay content so they have clay sides and bottom. Karst limestone of the Hill Country can create big headaches when trying to build a tank where the limestone is exposed. Tanks built in these areas are rarely successful because the rock often has seams or faults that will cause them to leak.

Catchment Area and Tank Location

Property Line

⊗ Proposed tank location
----- Outline catchment area feeding proposed tank

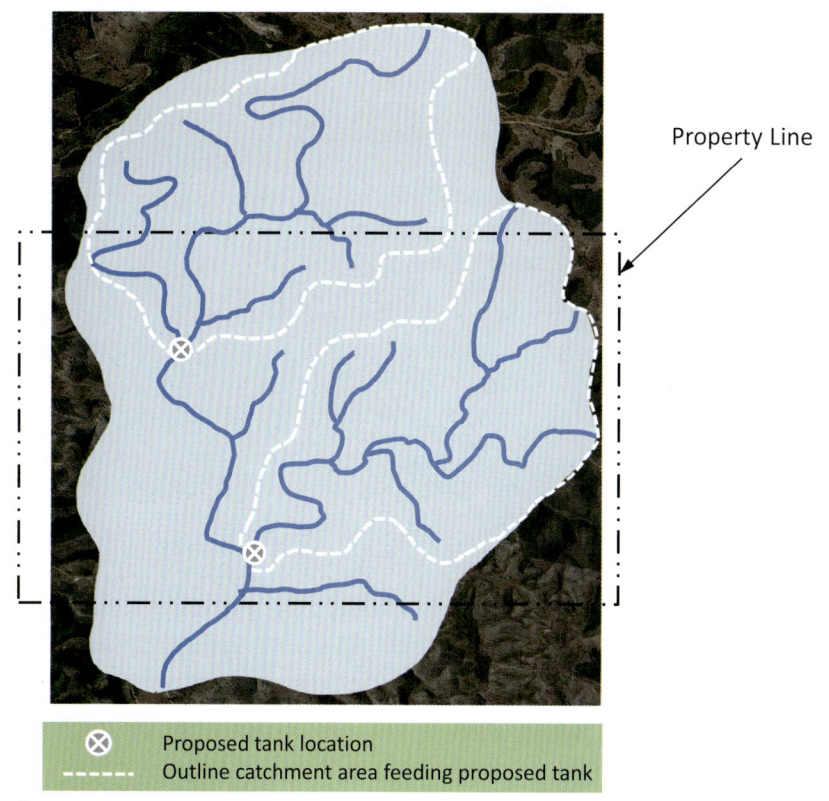

Figure 5.1

Watershed Condition, Erosion, and Water Quality

While we want the tank to collect water draining off the land, we also want the water to be clean and not carry eroded soil. A pond with muddy water is poor fish and wildlife habitat. Quick, heavy runoff after a rainstorm might seem desirable, but it shortens the life of your tank. If the water that fills your tank is cloudy with soil particles, those particles will be deposited in the tank, and it will fill in sooner than it should. To judge the quality of water you might be collecting, look at the creek running into your tank and the land in your watershed.

Healthy watersheds have a diverse plant community, plenty of deep ground cover, little bare ground, and very little or no erosion visible. Unhealthy watersheds have bare ground, extremely short grass, rocks on the ground, cedar and few other plants, and signs of extensive erosion cuts. The most common reason for an unhealthy watershed is overgrazing, but it also can be caused by large-scale brush removal or even extensive stands of regrowth cedar. If the watershed is unhealthy with lots of bare ground, you might want to delay building a tank until the upland

Recently exposed karst limestone with water moving along rock seams

area is restored to deep, diverse ground cover. If you are not sure whether or not the condition of your watershed will create problems with your tank, consult with NRCS.

Construction

Native plants provide valuable bird habitat to a completed tank, and carelessness during construction can destroy a large area around the new pond or lake. Throughout construction, limit heavy equipment to a specific area so you can avoid damaging existing native plants any more than is absolutely necessary. Think of pond construction modifications as a continuum. Greatest change happens exactly where the dam is built; some alteration occurs in the middle where benches are built for habitat; very little change is made on the upper end where water enters the tank. This avoids unnecessary disturbance and maintains the natural shape, stability, and vegetation of the drainage.

Structure

Good design provides good habitat. Like natural systems, constructed ponds and lakes that best serve waterfowl and wading birds have an irregular shoreline and a variety of depths at different water levels. Northern Pintail, Gadwall, American Wigeon, Blue- and Green-winged Teal, and Northern Shoveler are wintering and migrant dabbling ducks that need a place to rest and feed. Because of their feeding behavior and size, the ideal depth for dabbling duck feeding is 1 foot, but they will use areas with water from 0.5 to 3 feet deep.

Pond with shallow edge and shoreline vegetation

Black-bellied Whistling-Duck
(Photo by Jeff Forman)

Weather and Waterfowl

Fluctuations in a pond's annual habitat conditions will have an impact on the presence of food and how many birds will come to your tank in a given year. You may get lots of ducks one year and few or none the next because in extremely dry years even a well-designed pond may provide little or no food for waterfowl. Most plants with seeds eaten by ducks grow in the late spring or early summer. A dry spring and summer mean poor food production. In these years, even though you may have water during the fall and winter, migrant and wintering ducks will go where there is food. This is a natural process. If your pond is well designed, the food and the ducks will return.

Tank Designed for Waterfowl Use

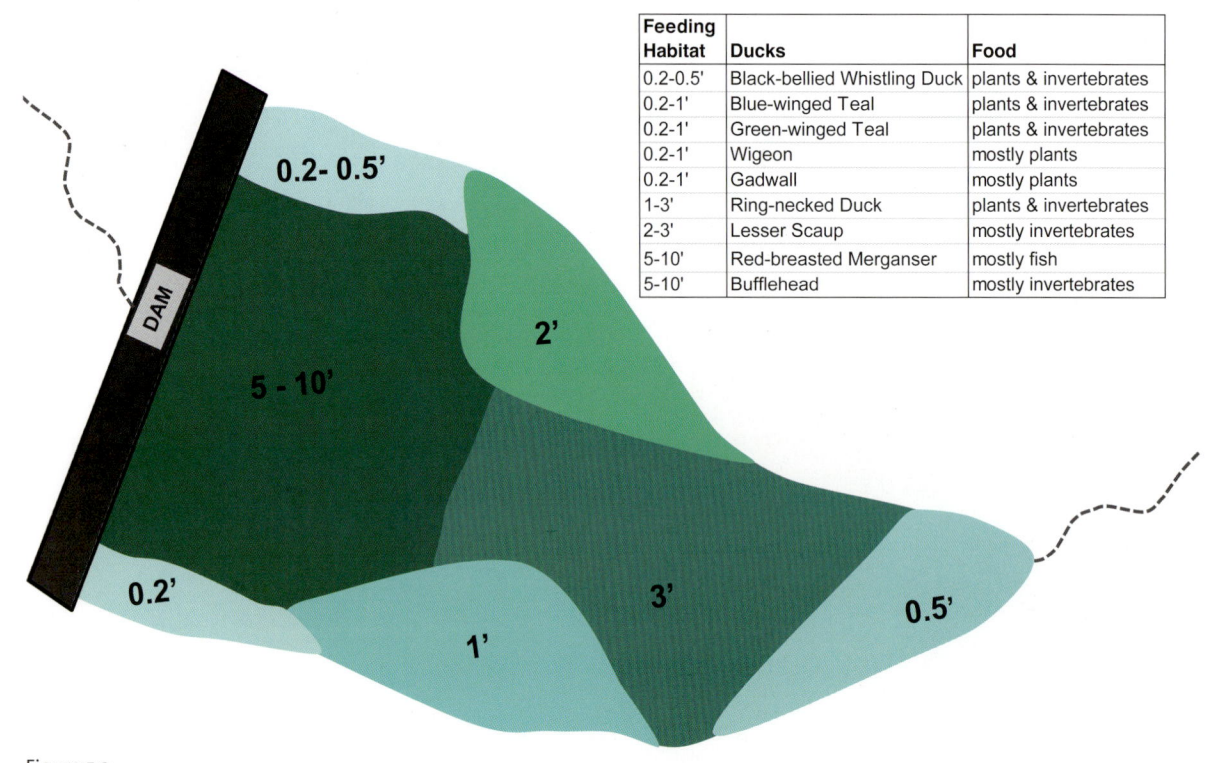

Feeding Habitat	Ducks	Food
0.2-0.5'	Black-bellied Whistling Duck	plants & invertebrates
0.2-1'	Blue-winged Teal	plants & invertebrates
0.2-1'	Green-winged Teal	plants & invertebrates
0.2-1'	Wigeon	mostly plants
0.2-1'	Gadwall	mostly plants
1-3'	Ring-necked Duck	plants & invertebrates
2-3'	Lesser Scaup	mostly invertebrates
5-10'	Red-breasted Merganser	mostly fish
5-10'	Bufflehead	mostly invertebrates

Figure 5.2

Water Fowl Feeding Areas

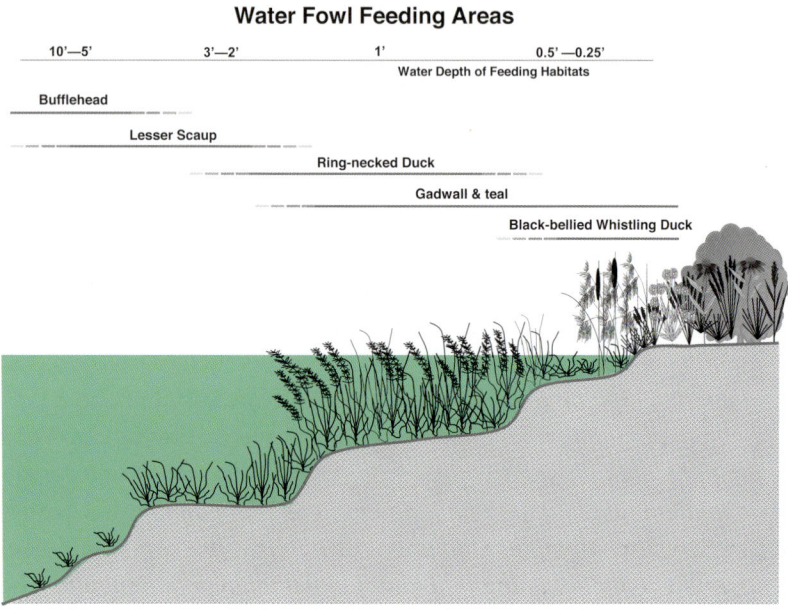

| 10'—5' | 3'—2' | 1' | 0.5'—0.25' |

Water Depth of Feeding Habitats

Bufflehead
Lesser Scaup
Ring-necked Duck
Gadwall & teal
Black-bellied Whistling Duck

Figure 5.2a

Blue-winged Teal
(Photo by Lora L. Render)

Shallow tank with emergent vegetation and adjacent vegetation

Waterfowl Food and Pond Shape

Sculpt your pond to have mudflats at as many different water levels as possible. Ducks, including Blue-winged Teal and Gadwall, like to eat seeds of plants such as dock, pink smartweed, and bulrush, which grow on mudflats or in very shallow water. Since many of these plants grow in late spring and early summer and produce seed in late summer and fall, ponds that provide lots of places for these plants to grow will attract these ducks. The flats can be next to each other in stair-step fashion or on opposite sides of the lake or tank. A small tank may have enough room for only a couple of flats, and a lake may have five or more. As in naturally occurring bodies of water, flats in an artificial tank or lake usually have irregular shapes and sizes.

Fish Habitat

Wading birds, kingfishers, and some diving ducks eat fish. Fish need places to reproduce, hide from predators, and hunt for their food. Shallow water with submerged and emergent aquatic vegetation is good spawning and rearing ground for all types of fish. Fish spend most of their time in underwater structures rather than in featureless, open water. Underwater ridges, rock piles, standing or fallen trees, and stumps provide feeding areas, good escape space, and protective habitat. Leave or add these structures when sculpting your pond.

If you need to add underwater structures to provide the necessary fish habitat, you can add rock piles, stumps, and trees. Where cedar is plentiful, you might wire medium-sized cedar trees together and weight them down with concrete blocks. This is an inexpensive way to create fish habitat in a tank lacking underwater features and an easy way to increase your chances of attracting Belted Kingfisher, Great Blue Heron, and Red-breasted Merganser. There actually are several merganser species that winter in the Hill Country. None of these are common, but a pond or lake with plenty of fish has the best chance of attracting them. If you have a well-designed pond and underwater structures providing a lot of hiding places, none of these birds will deplete the fish populations. And they nearly always eat the smaller fish.

Water-Level Control

Today, all tanks are built with a basic emergency spillway adjacent to the dam where water flows around the dam during major floods. But an emergency spillway does not allow you to consistently control water level in your pond or the amount of water that flows downstream. It does not allow you to take an active part in maintaining wildlife habitat in the pond or protecting downstream habitat from becoming unnecessarily xeric. The basic spillway only prevents too much water from collecting behind the dam, rushing over the top, and damaging or destroying the dam.

Dams built on perennial streams usually have an overflow pipe for consistent release of some water and erosion prevention. For an ideal waterfowl pond, include a valve on the overflow pipe, whenever building or retrofitting a tank. This valve allows true control of pond water

Red-breasted Merganser
(Photo by Lora L. Render)

Deepwater tank with submerged vegetation and fish

level, so for much of the year, you can provide shallow water for dabbling ducks, habitat for wading shorebirds, and a safe place for all birds present.

Bird Safety

Safe water is an essential habitat component for all birds using the aquatic habitat. Safety means not only clean water but also protection from disturbance and predation. Sora, Wilson's Snipe, and Yellow-

Cross–section of Pond with Water Control Structure

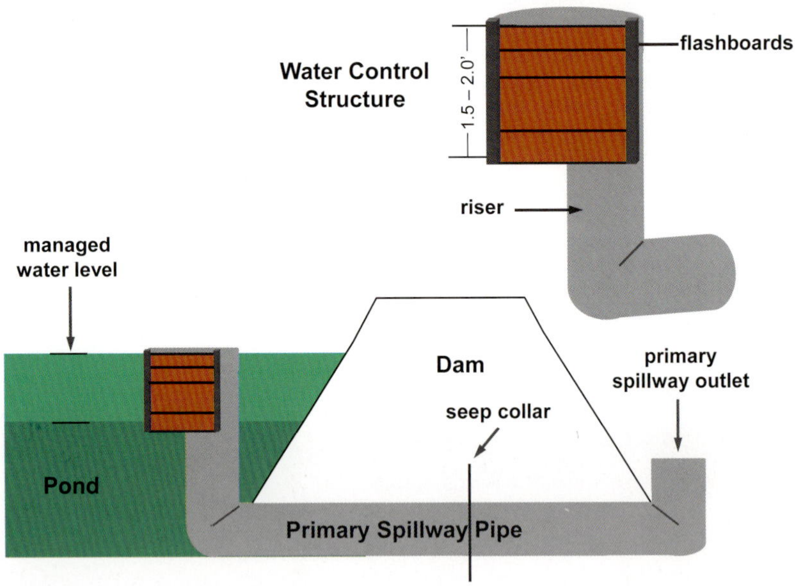

Figure 5.3

crowned Night-Heron are waterbirds that need a lot of plant cover. Water safety is also important and somewhat different for small songbirds such as buntings, vireos, wrens, finches, and warblers. They come to drink and bathe at water's edge, where they need shallow water and vegetation that protects them from predators and bad weather.

To be good habitat for birds, pond or lake water should have these characteristics:

- Be clean and fresh
- Have vegetation where birds can feed and hide from predators
- Provide accessibility for many bird species with different sizes and habits
- Remain available in all seasons in most years

PROBLEMS OF CONSTRUCTED TANKS, PONDS, AND LAKES
Problem 1: Poor Design That Results in Poor Waterfowl, Shorebird, and Songbird Habitat

Most existing tanks, ponds, and lakes in the Hill Country were *not* built for wildlife habitat and do not attract nearly as many waterfowl, shorebirds, and other birds as they might. A regular bowl shape with steep sides and no flat benches on the edges is poor waterfowl habitat. Water-loving birds that migrate, winter, or breed in the Hill Country have specific needs:

Remember: The water level in ponds and lakes will fluctuate radically from season to season. This is normal and can produce valuable bird habitat in a constructed tank.

Sora (Photo by Lora L. Render)

Pond with abundant wetland vegetation

- Diving ducks, such as the Red-breasted Merganser, need deep water with fish.
- Dabbling ducks need shallow water with plenty of emergent plants and aquatic invertebrates.
- Wading birds such as the Yellow-crowned Night-Heron need shallow edges with abundant vegetation.
- Spotted Sandpiper and Killdeer need open mudflats for feeding.

Yellow-crowned Night-Heron

Pond with open shallows and dense trees and shrubs

- Kingfishers need hunting perches, clear water, and fish.
- All birds living in wetlands need vegetation for territorial perches, nest sites, food, and shelter from harsh weather and predators.

What You Can Do: Reshape the Shoreline

1. Consult with a wildlife biologist who knows the habitat needs of wetland birds to help with your location.

2. Carefully select a machine operator who will follow your specific directions for reshaping tank/lake contours.
3. Reshape the shoreline to be irregular and wide, with wide, flat steps in bottom contour.
4. Repair during drought when the water is very low or has dried up.

Problem 2: Water Surrounded by Bare Soil Instead of Herbaceous Plants and Shrubs

Land damaged by livestock, overclearing, feral hogs, and too many deer is common throughout the Hill Country. All of these destroy vegetation and create bare ground around ponds and lakes. Wetland birds need vegetation near the water's edge. Red-winged Blackbirds set up their breeding territories and build their nests in dense wetland vegetation. A Gadwall's diet is about 95% vegetation. Other birds such as the Sora, Wilson's Snipe, and Yellow-crowned Night-Heron hide from predators and escape bad weather in tall wetland grasses and rushes. Many species of migrant, wintering, and resident birds hide and hunt insects in both woody and herbaceous vegetation growing along the shoreline of constructed ponds and lakes.

What You Can Do: Protect and Restore Shoreline

1. Fence to exclude livestock completely, or limit their water access to one place that may already be open and has the least favorable bird habitat.
2. Set up recreational access to protect the most valuable shoreline vegetation areas.
3. Control feral hogs, if you see signs of hog wallows. For more information, see chapter 7.
4. Use the deer browse evaluation chart in appendix 1 to determine if you have a deer browse problem. Practice ongoing deer control in areas damaged by deer browse to reduce deer density and ensure the survival of plants needed by the birds. For more information, see chapter 8.
5. Restore the natural wetland plant community. You can either just wait for the plants to return on their own after excluding livestock or speed the process by planting key species. In any case, be patient because recovery may be slow.

Problem 3: Inadequate Aquatic Vegetation

Ducks, kingfishers, herons, and other waterbirds depend on both emergent and submerged plants growing in the water. Both types of vege-

Red-winged Blackbird
(Photo by Tom Mast)

Tank with dense herbaceous shoreline vegetation

tation are also essential habitat for invertebrates and fish. Underwater roots and stems and even dead trees or stumps can be important places for fish to hide and feed. Kingfishers, herons, and mergansers usually eat the pond fish that live on invertebrates.

In winter, dabbling ducks feed mostly on vegetation in shallow water. The Gadwall is unique because in winter it prefers or specializes in eating

Gadwall (Photo by Lora L. Render)

Pond with plenty of shallows and emergent vegetation

plants that are relatively low in protein. These include leafy portions of pondweeds, southern water nymph, duckweed, watermilfoil, and algae, as well as seeds of pondweed, pink smartweed, bulrushes, and spikerushes. Thus, Gadwalls must feed for longer periods than other dabbling ducks and may not be able to use an otherwise suitable pond or lake when their feeding routine is disturbed regularly.

> **FALL AND WINTER DUCK FOODS IN THE TEXAS HILL COUNTRY**
>
> AQUATIC PLANTS
> Coon's tail
> Duckweed
> Filamentous algae
> Pondweeds
> Southern water nymph
> Stoneworts
> Water primrose
> Watermilfoil
>
> GRASS SEEDS
> Bearded sprangletop
> Panicgrasses
> Paspalum
> Rice cutgrass
> Wild millet or coast cockspur grass
>
> SEDGE SEEDS, RHIZOMES, OR TUBERS
> Arrowhead
> Bulrushes
> Flat sedges
> Jamaican sawgrass
> Sedges
> Spikerushes, spike sedges
>
> FORB SEEDS
> Beggarticks
> Dock
> Pigweeds
> Pink smartweed
>
> WOODY PLANT SEEDS OR FRUITS
> Common buttonbush
> Grapes
> Hawthorn
>
> Source: Adapted from Michael Porter, "Fall and Winter Duck Foods Chart," Samuel Roberts Noble Foundation.

What You Can Do: Restore Wetland Plants

Provide deep and shallow water for different types of vegetation that support an abundance of invertebrates and fish.

1. Build flat, irregular, undulating underwater benches where aquatic plants can grow.
2. Limit "cleaning up" the water as much as possible. It is best to allow aquatic vegetation to grow naturally.
3. Limit or remove livestock access to prevent unnecessary disturbance to aquatic and shoreline vegetation.
4. Foster a diverse emergent-plant community essential to healthy marsh habitat.
5. Regulate water levels so shoreline mudflats are often wet and support benthic (bottom-dwelling) organisms that are food for shorebirds.
6. Retain plenty of underwater vegetation. Excessive aquatic plant control (either chemical or biological) removes too much of the vegetative habitat for aquatic invertebrates. Fewer invertebrates mean fewer small and large fish. Moderation is the key here.

Problem 4: Turbid or Cloudy Water

The Belted Kingfisher is a common and popular bird living around Hill Country ponds and lakes. Kingfishers need perches over clear water so they can see and catch fish. Unnaturally turbid or cloudy water may be due to livestock hooves or waste, feral hog damage, human use, or poor watershed management.

What You Can Do: Prevent Disturbance from Livestock, Feral Hogs, or Humans

1. Control livestock waste.

Water livestock somewhere else, so tank edges will regenerate plant cover. When livestock are present, the water may be cloudy with algal blooms and contain an unusually high bacteria count. Algae affect water clarity and oxygen content; the bacteria affect water quality. When plants return, they reduce shoreline erosion and lengthen the life of the tank, filter the water, and increase water clarity. The newly improved aquatic and shoreline plant communities enhance wildlife habitat. The second choice is to limit livestock access so they can use only the deep end near the dam far away from the best bird habitat at the opposite, shallow end of the pond. In this case, run the fencing far enough into the water so livestock access is controlled even at low water levels.

Pond with high perches and clear water

Belted Kingfisher
(Photo by Ann Mallard)

2. Control feral hogs.

Hog wallows damage and in severe cases eliminate natural vegetation. Without protection from the roots of these plants, the soil is subject to erosion. The turbidity caused by erosion diminishes water quality and reduces the number of aquatic invertebrates and fish that can live there. For information on feral hog control, see chapter 7.

3. Monitor human use.

Some tanks or small lakes may have been cleared for landscape aesthetics or recreational access. Again, this makes the soil open to erosion, and eroded soil causes turbid water that limits aquatic invertebrates and

Pond with Livestock Excluded

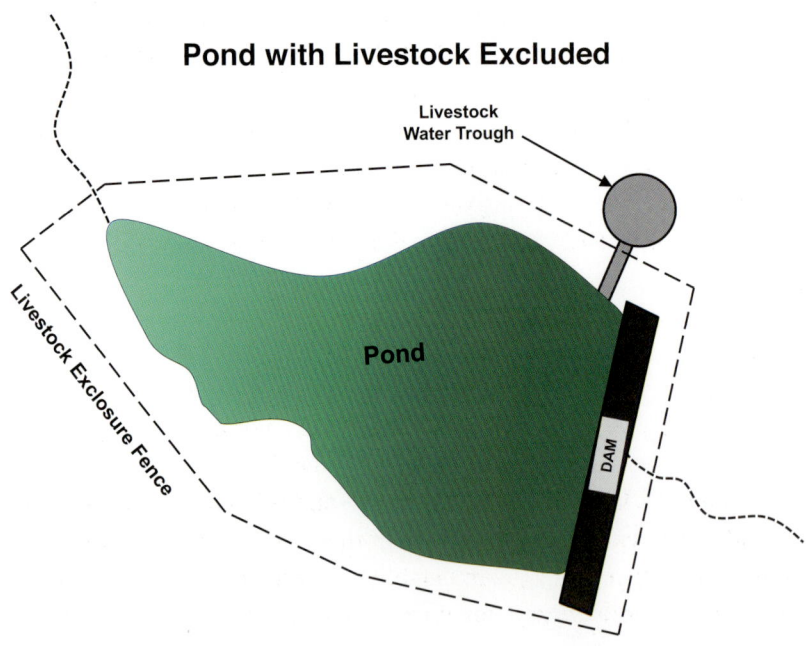

Figure 5.4

Pond with Limited Livestock Access

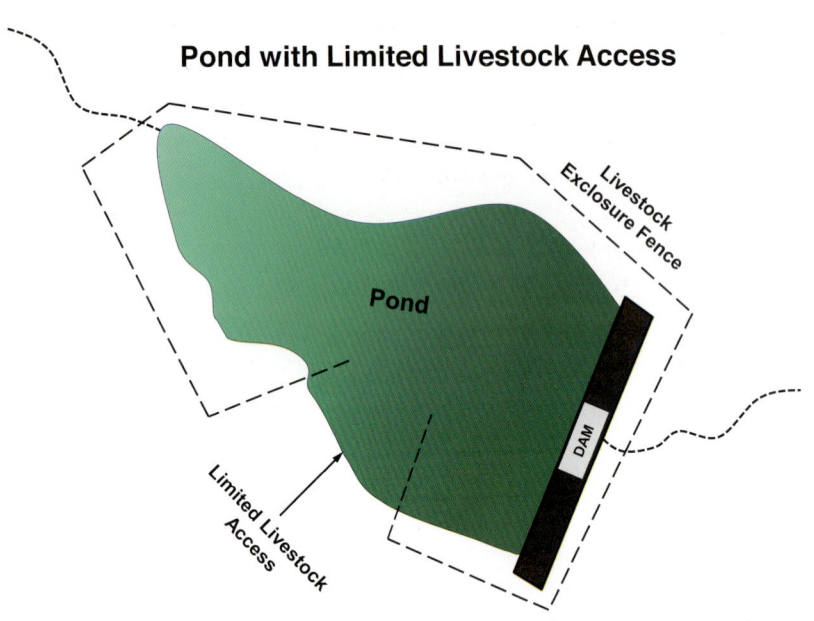

Figure 5.4a

fish eaten by waterfowl and wading birds. Less food means fewer birds. Where you have an area kept open for easy human access and recreation, pay attention to its condition by checking and photographing it at least annually. When the plants in this spot become worn down and you see signs of erosion, move the special human access spot to another place.

Replant the shoreline vegetation or let it return on its own naturally and prevent future erosion. Find a balance between woody and herbaceous wetland plants. Each species has a different root system that contributes to holding soil. Rushes grow on the water's edge. Bushy bluestem, switchgrass, common buttonbush, and black willow nearby are especially useful in buffering wave action and preventing excessive erosion when the water level rises into this slightly higher ground.

4. Improve upstream watershed management.

When a tank is fed by land that is poorly managed and has damaged ground cover, there will be very few or no shrubs, grasses, or forbs to slow runoff during a major rain event. If the watershed is on your property, it is easiest and cheapest to simply rest the land and let it recover on its own. However, it is more likely that your pond is fed by a watershed that extends well beyond your property. In this case it will be necessary to cooperate with your upstream neighbors to accomplish watershed restoration. For more information on watershed restoration, see chapter 3.

Problem 5: Unstable Water Level

Many tanks are built with the expectation of a preferred and consistent water level. In the Hill Country, this is impossible. Because of the hot, dry climate and erratic rainfall, tanks, ponds, and lakes all over the region have radical water-level variations. You may consider a dry tank unattractive, but low water levels must be accepted. Since beauty is in the eye of the beholder, it is wise to learn to appreciate the natural beauty of a lake with variable water levels that support stable plant communities and provide habitat for many waterbirds and other wildlife.

Leaks in a working tank may also contribute to severe water-level fluctuation. We cannot control rainfall, but in many cases it is possible to fix a leaking tank. The goal is a no-leak structure that holds plenty of water as long as possible. This is best for human use and most beneficial for all birds.

BLOWN-OUT DAM

Because of its steep topography, the Hill Country has severe floods, and valuable dams are sometimes blown out by rare high-water events. Floods are dramatic, and a good dam is often hard to replace.

What You Can Do: Repair Dam and Pond Structure
1. Construct water-control structures.

If your dam was blown out and the tank is now empty, this is the time to put in water-control structures, which allow you to manage water

Remember: *Do not pump groundwater to fill a tank. Groundwater is too precious. The Texas Hill Country water table is declining, and abundant groundwater is essential for springs and wells.*

Barn Swallow

Large tank near open fields and nest habitat

levels to benefit birds and other wildlife. These are easiest to install when rebuilding the dam structure. For more information, see the tank building section in this chapter.

2. Do not rely on the "good guy with a bulldozer."

Go to the professionals at agencies such as NRCS for help with engineering the tank layout, size, spillway structure, and so forth and TPWD to consult with a wildlife biologist for design features that benefit water birds and wildlife in general.

LEAKS

A pond may leak and lose water for several reasons: a porous dam, a poorly sealed bottom, or fractures in the bedrock where the pond was built.

What You Can Do: Consult with Professionals

Talk to experts at agencies such as NRCS and TPWD. If they cannot help you, ask for a referral to someone who can. Here is information that will prepare you to question the experts so you get the best possible outcome.

1. Over time, a dam may develop leaks due to poor maintenance.

- Woody plants should be cleared off the dam on a regular basis. If not, their roots may create "pipes" that open up small water channels through the structure. Once water finds a path, it wears away the soil and the opening grows larger and larger. Such a leak may ultimately cause dam failure when it becomes so severe that the dam can no longer hold water.
- Although it is important to keep woody plants off the dam itself, plenty of healthy riparian vegetation on the dam's downstream side is necessary to protect the dam from erosion that can occur when powerful floodwater rushes over the spillway. Erosion of this nature can produce a serious leak that means the tank has a severely unstable water level.
- Ranch vehicles such as trucks, jeeps, gators, four-wheelers, or dirt bikes used indiscriminately on the dam and in its vicinity damage vegetation that holds the soil needed for dam stability. Limit or avoid their use in this area.
- Nutria, an imported exotic rodent, lives in wet habitat with quiet water such as marshes, ponds, and lakes. The burrows they build in the dam soil can cause leaks and even structural failure. Remove them by trapping or shooting as soon as they are detected.

2. Some tanks leak because they are poorly designed in one of the following ways:

- No emergency spillway
- Spillway too small
- Spillway in wrong position
- Dam built in the wrong location

Any of these design problems may require a major overhaul. In the case of a dam built in the wrong spot, it may be necessary to move your pond to a different location altogether—if that is possible.

3. If the bottom of your tank was never sealed properly, it may leak, but do not automatically rely on the bentonite dealer for the best solution. Go to the professionals for advice before attempting a fix. Here are some options to consider:

- Place round bales in the bottom of the tank and feed cattle there so manure will mix with the soil and make it more like clay. This is the least expensive option but will not be effective in areas with many leaky places.
- Add salts, which can serve as an adhesive agent that causes clay particles to align, making them less permeable and creating a seal. If a salt source is not nearby, this will also be too expensive.
- Spread bentonite over the bottom and up the sides. This is the most expensive but usually the most effective fix if the bottom has a high clay content. Unfortunately, many leaky tanks do not have clay bottoms and cannot be fixed even with bentonite. If a tank has been built in a place lacking clay and the bottom is so porous that it leaks everywhere, it may never seal even with a large and very costly amount of bentonite.

4. If the tank is in a location that makes it impossible to repair, you may or may not have a more desirable location on your property.

- Go to a tank design specialist for help making this decision.
- Be sure to consider wildlife needs when planning a new tank in a new location.

Problem 6: Unable to Build a Tank

Water for shorebirds, kingfishers, and waterfowl is quite different from the simpler needs of most wintering, migratory, and breeding birds. Waterbirds need the entire pond or lake habitat for food, water, and shelter. However, in terrestrial habitat birds may find plenty of food and shelter but have a hard time finding a place to drink. You may have discovered that your area of the watershed does not have a spring, wetland, pond, or lake suitable for waterfowl and wetland birds. You may also have discovered that building a structure that provides pond habitat for wintering ducks, shorebirds, and kingfishers requires a substantial investment that you cannot afford right now, for a very long time, or ever. If you cannot build a tank or lake, you will not be able to attract waterfowl, shorebirds, or kingfishers to your property.

Pond overpopulated with domestic waterfowl

Muscovy Duck
(Photo by Lora L. Render)

Egyptian Goose
(Photo by Lora L. Render)

What You Can Do: Choose Alternatives

Water may be a limiting factor for upland birds during most seasons. Small guzzlers work great, are inexpensive to build, and may attract a surprising number of breeding, migrant, or wintering bird species. Modification to an existing livestock tank is an easy, inexpensive, and effective way to supply water to these birds.

Remember: Domestic/exotic ducks and geese compete with native waterfowl and can spread diseases to wild populations. Domestic ducks and geese belong in a farmyard, not in natural settings. Do not release them on a tank or lake. If they arrive uninvited, remove them.

Problem 7: Artificial Overfeeding of Waterfowl

Sometimes we are tempted to put out pellets or grain to attract ducks to a tank with little natural food. Unfortunately, feeding birds at a tank or lake easily turns into an unsustainable practice of overfeeding year after year. Overfeeding can cause a cascade of unintended consequences.

- A consistent artificial food source attracts so many birds that they overuse, trample, and damage your pond's habitat.
- Some of the birds attracted to feeding at a tank or other water source may be domestic/exotic ducks and geese such as the Muscovy Duck and Egyptian Goose. Although they may be attractive and their reliable presence gives a zoolike quality to a pond, these birds can be aggressive and outcompete native species for food and nest territory. Even a single pair may drive away the shyer birds that could be using your pond or lake.
- In cases of severe overuse, the ability of the habitat to recover can be impaired long term.
- Concentrations of any kind of animal in large numbers may lead to increased disease transmission.
- In areas where hunting occurs, federal law protecting important wild game species such as doves and waterfowl prohibits feeding these migratory birds just prior to and during hunting season.

What You Can Do: Do Not Feed Waterfowl with Pellets or Grain

1. To maximize the number of native ducks and shorebirds using your pond or lake, it may be necessary to restructure its shape to increase food production. Do this during a drought year when the water level is very low. For more information, see the tank building section in this chapter.
2. Right after building a tank, when it is still raw, *do not* put out supplemental food to attract migrant or wintering ducks. You must be patient and wait until the water habitat develops on its own—and it will. Some plants will grow quickly, and others take longer. It is a process to watch and enjoy.
3. Do not feed waterfowl. Neither grain nor pellets are an adequate replacement for natural habitat.
4. If domestic/exotic ducks or geese come to your pond or lake, do not feed them or encourage them in any way. In fact, it is best to trap and remove them as soon as possible.

Remember: Native shorebird and waterfowl species are protected by law and must not be handled or harmed in the process of removing outlaw birds.

SUMMARY OF HABITAT PROBLEMS IN TANKS, PONDS, AND LAKES

HABITAT PROBLEMS	STEWARDSHIP SOLUTIONS	AFFECTED BIRDS
Problem 1: Poor design that results in poor waterfowl, shorebird, and songbird habitat	Plan and build with bird needs in mind Make irregular shoreline and wide, flat steps in bottom contour	Black-bellied Whistling-Duck Blue- and Green-winged Teal Bufflehead Gadwall Many songbirds Northern Pintail Red-breasted Merganser Sandpipers Yellow-crowned Night-Heron
Problem 2: Water surrounded by bare soil instead of herbaceous plants and shrubs	Fence all or part to exclude livestock Reduce deer density Develop natural understory and ground cover Control feral hogs	Green Heron Red-breasted Merganser Red-winged Blackbird Sedge Wren Sora Vermilion Flycatcher Wilson's Snipe Yellow-rumped Warbler
Problem 3: Inadequate aquatic vegetation	Redesign tank for improved shoreline and bottom shape Add underwater structures Limit aquatic weed control	All sandpipers, herons, kingfishers, and ducks
Problem 4: Turbid or cloudy water	Limit livestock access Increase herbaceous shoreline vegetation	Belted Kingfisher Great Egret Green Heron Red-breasted Merganser
Problem 5: Unstable water level	Improve management of upstream watershed Plan for normal seasonal fluctuation Repair leaks if practical	Lesser Scaup Pied-billed Grebe Ring-necked Duck
Problem 6: Unable to build a tank	Add guzzler Improve wildlife access to tank	Any bird needing water to drink and bathe
Problem 7: Artificial overfeeding of waterfowl	Improve native food sources with better design and management Stop feeding pellets and grain	All native waterfowl

Bird Summaries: Featured Birds

BARN SWALLOW
FAVORITE BIRD
SPRING AND SUMMER

Known for graceful, quick flight, low and high over land and water, and for building mud nests under the eaves of human-made structures.

Identification: About 6.75 inches head to tail; forked tail longer in males than in females; male purple-blue above, buffy-orange below, rust forehead and throat; female similar but with off-white below.

Habitats: Old field, plateau prairie, river valley prairie, post oak savannah, shin oak savannah, live oak savannah, river, creek, canyon, springs, tank, pond, lake, and backyard.

Cover: Open fields with adjacent, protected caves, crevices, or caves and water nearby.

Breeding Territory: Singular or in colonies; when in colonies, nests as close as 1.75 feet apart.

Nest: Open mud nest often under eave and porch, inside barn or other outbuilding; nest built 1–98 feet above the ground, but most are about 8–11 feet above the ground; cave or cliff probably the most common naturally occurring nest site.

Food: Flies, beetles, bees, wasps, moths, butterflies, and other fairly large flying insects, usually caught in flight.

Stewardship: Maintain a healthy water source in open habitat with low herbaceous vegetation.
- Maintain nest sites on buildings, and leave old nests to be reused and promote reproduction.
- Avoid broad-spectrum herbicides and insecticides to increase forbs and promote insect diversity.
- Control red imported fire ants with an ant-specific insecticide to increase insect diversity and reduce predation.

Barn Swallow

- Keep cats indoors, and control stray and feral cats to reduce predation.
- Use prescribed burns to increase forbs for insects and maintain low herbaceous vegetation.
- Control spread of regrowth cedar using appropriate mechanical or chemical methods to maintain low herbaceous vegetation.

Other Potentially Useful Management:
- Use rotational grazing to encourage a balance of grasses and forbs.
- Mow herbaceous area periodically in a dormant fall or winter season to promote annual forbs and insects.
- Use limited cultivation, such as occasional light disking, to increase forbs.
- Fix leaks in tank, pond, or lake retention structure to maintain drinking water.
- Use late-spring drawdowns in ponds to expose mudflats and provide mud for nest building.

BELTED KINGFISHER

FAVORITE BIRD

ALL YEAR

Large kingfisher that hunts from high perch over clear water; dives to catch prey near the water surface; solitary in winter; easy to locate by following its loud, rattling call.

Identification: About 12.75 inches head to tail; big head with shaggy crest and long, heavy, daggerlike bill; blue-gray above; female has one blue and one rusty breast band; male has only one blue breast band.

Habitats: River, creek, springs, tank, pond, and lake.

Cover: High perches over clear water; exposed dirt banks.

Breeding Territory: Usually determined by water body and ranges from about 1,000 feet/pair to 1.25 mile of stream/pair.

Nest: Burrow 3–5 inches wide and up to 6.5 feet deep in steep bank, often near water.

Food: Fish less than 4 inches long plus tadpoles, crayfish, and frogs caught from a perch but occasionally by hovering.

Stewardship: Maintain clear, high-quality water with shoreline trees for perches and healthy populations of frogs, small fish, and crayfish.

- Exclude livestock grazing in riparian areas to maintain understory, water quality, and bank integrity.
- Limit livestock access to pond shorelines to maintain clean water and healthy aquatic habitat.
- Control invasive exotics and regrowth cedar using appropriate mechanical or chemical methods to maintain diverse riparian woodlands.
- Control feral hogs to improve and maintain water quality and to protect herbaceous ground cover.
- Avoid broad-spectrum pesticides on uplands to protect aquatic insect diversity.

Belted Kingfisher

- Control browsing deer and exotic animals to encourage healthy riparian habitat.
- Manage for a diverse age structure of fish to provide an abundant food supply.

Other Potentially Useful Management:
- Maintain healthy watersheds with deep and sustainable ground cover to retain riparian health and water quality.
- Avoid human disturbance near nest sites to ensure successful nesting.

BLACK-BELLIED WHISTLING-DUCK
UNDER-THE-RADAR BIRD
ALL YEAR (SOUTHERN AND EASTERN HILL COUNTRY)

Odd duck that might make you say, "What the heck is that?"; colorful, distinctive, and gregarious; "twitters" as it flies overhead; many gather at dusk in large groups to roost side by side in trees or on the ground.

Identification: About 19 inches head to tail; long pink legs, striking red bill, long neck, black belly, and white wing patch; male and female similar.

Habitats: Post oak savannah, live oak savannah, river, creek, canyon, seep, springs, tank, pond, lake, and backyard.

Cover: Understory plants adjacent to streams for security while feeding in shallow water; trees up to a mile from water for roosting and nesting.

Breeding Territory: Not territorial; pairs nest in close proximity to each other.

Nest: Has been nesting in the Hill Country since the 1980s. Lays eggs in unlined tree cavity or nest box and sometimes on the ground under vegetation, usually within a half mile of water. When only a few days old, ducklings are led to shallow water to begin feeding.

Food: Dabbles in shallow water for aquatic plants, insects, and mollusks; also eats grain in fields.

Stewardship: Maintain natural vegetation along streams, ponds, tanks, and lakes, and protect marshy habitat near springs and seeps.
- Avoid broad-spectrum insecticides on uplands to protect aquatic insect diversity.
- Retain snags with cavities (including drought-killed trees) to enhance potential nest sites.
- Erect nest boxes to provide nest sites if snags are in short supply.
- Exclude livestock grazing in riparian areas to maintain understory, water quality, and bank integrity.
- Limit livestock access to tank or lake shorelines to maintain water quality and emergent vegetation.
- Minimize or avoid the use of aquatic herbicides to maintain aquatic vegetation and aquatic insect habitat.

Black-bellied Whistling-Duck

- Control regrowth cedar using appropriate mechanical or chemical methods to maintain diverse riparian woodlands.
- Use pond design features, such as shallow benches, that benefit waterfowl when building or maintaining tanks or lakes to promote feeding and brood-rearing habitat.
- Install water-control structures during dam construction or repair to facilitate water-level and vegetation management.

Other Potentially Useful Management:
- Control Muscovy Ducks to reduce nest parasitism.
- Control feral hogs to maintain shoreline vegetation and water quality.
- Maintain healthy watersheds, including deep and sustainable ground cover to protect water quality and quantity.
- Control white-tailed deer and other browsing animals to encourage woody plant recruitment and healthy understory.

BLUE-WINGED TEAL
FAVORITE BIRD

SPRING, SUMMER, AND FALL

A small dabbling duck found across Texas; in greater abundance in the Hill Country during the spring and fall migration; also nests here where habitat is suitable.

Identification: About 15 inches head to tail; both sexes have a powder-blue patch on the upper inside part of the wing; males have a distinctive white slash between the dark head and bill.

Habitats: River, creek, tank, pond, and lake.

Cover: Shallow water with emergent vegetation for feeding and adjacent medium tall grasses for nesting.

Breeding Territory: Less than 1.7 acres or one pair per small tank; male defends shoreline feeding habitat about 170 feet wide.

Nest: A bowl-shaped nest on the ground in fairly dense herbaceous cover.

Food: Eats seeds during migration and a significant amount of invertebrates when nesting.

Blue-winged Teal

Stewardship: Maintain shallow, clean water with diverse emergent vegetation and adjacent grassy uplands.

- Use pond design features, such as shallow benches, that benefit waterfowl when building or maintaining tanks or lakes to promote feeding and brood-rearing habitat.
- Install water-control structures during dam construction or repair to facilitate water-level and vegetation management.
- Manage water levels on shallow benches during fall and winter when birds are present to maintain water depth less than 12 inches deep.
- Expose mudflats during late spring and early summer to encourage the growth of wetland vegetation.
- Minimize or avoid the use of aquatic herbicides to maintain aquatic vegetation and aquatic insect habitat.
- Avoid broad-spectrum insecticides on uplands to protect aquatic insect diversity.
- Keep residual cover from the previous growing season within approximately 500 feet of water to provide nesting cover.
- Exclude livestock grazing in riparian areas to maintain understory, water quality, and bank integrity.
- Limit livestock access to tank or lake shorelines to maintain water quality and emergent vegetation.
- Control invasive exotics and regrowth cedar using appropriate mechanical or chemical methods to maintain diverse riparian areas.

Other Potentially Useful Management:

- Control feral hogs to maintain shoreline vegetation and water quality.
- Maintain healthy watershed with deep and sustainable ground cover to protect water quality and quantity.

DARK-EYED JUNCO

FAVORITE BIRD

WINTER

Juncos and sparrows are in the finch family and have large beaks for breaking and eating seeds. Dark-eyed Juncos classified into groups based on coloration and location; "slate-colored" junco most common in the Texas Hill Country.

Identification: About 6 inches head to tail; dark eye; males have charcoal head and back with white belly and white outer tail feathers; females have grayish-brown head and back but are otherwise similar to males.

Habitats: Old field, river valley prairie, pocket prairie, river, creek, canyon, seep, springs, tank, pond, lake, and backyard.

Cover: Dense understory or herbaceous vegetation; roosts in dense grass or brush piles.

Breeding Territory: N/A

Nest: N/A

Food: Primarily seeds of forbs and grasses usually collected on or very close to the ground; also some insects.

Stewardship: Maintain diverse understory and dense riparian vegetation with adjacent grasslands with a balance of grasses, forbs, and bare ground.
- Exclude or limit livestock access to shorelines and exclude from riparian areas to maintain dense winter vegetation.
- Encourage balance of grasses, forbs, and bare ground to protect food and feeding habitat.
- Build tepee-style brush piles to provide dense cover that may be lacking.
- Control regrowth cedar using appropriate mechanical or chemical methods, and maintain some dense cedar with branches to the ground to maintain diverse riparian woodlands and promote potential roosting cover.
- Control white-tailed deer and other browsing animals to encourage woody plant recruitment, healthy shrubs, and forbs.
- Maintain bird-friendly, natural or artificial water source to supply reliable and safe water.
- Keep cats indoors, and control stray and feral cats to reduce predation.

Other Potentially Useful Management:
- Remove feral hogs to protect low-level habitat for cover and food.
- Avoid broad-spectrum herbicides and insecticides to increase forbs and promote insect diversity.
- Apply ant-specific insecticides to control red imported fire ants and increase insect diversity.
- Maintain healthy watersheds with deep and sustainable ground cover to protect riparian health and water quality.

Dark-eyed Junco

EGYPTIAN GOOSE

OUTLAW BIRD

ALL YEAR

Introduced from Africa; large ducks, not geese. Graceful stature and awkward in flight but good swimmers. Prolific and harmful free-ranging ducks that are spreading from urban introductions into native habitat, where they outcompete native waterfowl.

Identification: About 24 inches head to tail; dusting of cinnamon on light breast and belly; light head with dark rusty eye ring that appears like glasses; wing dark with a distinctive white patch on forewing.

Habitats: River, creek, springs, tank, pond, and lake.

Cover: Wetlands and shorelines with short "parklike" vegetation.

Breeding Territory: Aggressively territorial even against other species, including native.

Nest: On the ground, sometimes in trees, and even in other ducks' nests.

Food: Grazes on shoots of terrestrial grasses and forbs.

Control: Do not tolerate the presence of domestic ducks or geese in waterfowl habitat.

- Remove these birds to prevent competition with native ducks for food and nest sites and to avoid spread of disease to wild ducks.
- Smear vegetable oil on eggs in nest to control reproduction when all geese cannot be removed.
- Manage tank, lake, or stream shoreline to provide healthy emergent and terrestrial vegetation that favors wild waterfowl.

Egyptian Goose

GADWALL
UNDER-THE-RADAR BIRD
WINTER

Dabbling duck fairly common in the Hill Country during winter months; feeds on seeds and vegetation of water-loving plants that grow in water less than 12 inches deep.

Identification: About 20 inches head to tail; male has black rump and cinnamon patch on forewing; both sexes have white patch on inside trailing edge of wing.
Habitats: River, creek, tank, pond, and lake.
Cover: Slow-moving water up to 12 inches deep with emergent vegetation.
Breeding Territory: N/A
Nest: N/A
Food: Seeds and vegetation of aquatic and wetland plants growing in shallow water.

Gadwall

Stewardship: Maintain shallow, clean water with diverse emergent vegetation.
- Use shallow bench pond design when building or maintaining tanks or lakes to promote feeding and brood-rearing habitat.
- Install water-control structures during dam construction or repair to facilitate water-level and vegetation management.
- Manage water levels on shallow benches during fall and winter when birds are present to provide essential feeding water depth of less than 12 inches.
- Expose mudflats during late spring and early summer to encourage the growth of wetland vegetation.
- Minimize or avoid the use of aquatic herbicides to maintain aquatic vegetation habitat.
- Exclude livestock grazing in riparian areas to maintain understory, water quality, and bank integrity.
- Limit livestock access to tank or lake shorelines to maintain water quality and emergent vegetation.
- Control invasive exotics and regrowth cedar using appropriate mechanical or chemical methods to maintain diverse riparian areas.

Other Potentially Useful Management:
- Avoid broad-spectrum insecticides on uplands to protect aquatic insect diversity.
- Control feral hogs to maintain shoreline vegetation and water quality.
- Maintain healthy watersheds with deep and sustainable ground cover to protect water quality and quantity.

GREEN-WINGED TEAL

FAVORITE BIRD

WINTER

Attractive, small dabbling duck; particularly attracted to feeding in very shallow water with recently flooded aquatic vegetation.

Identification: About 14.5 inches head to tail; male has brown head with a wide iridescent green swath from the eye extending toward the back of the head; female plain brown; both the male and female have a green patch on the inside trailing portion of the wing.
Habitats: River, creek, tank, pond, and lake.
Cover: Emergent wetland vegetation in shallow water and mudflats.
Breeding Territory: N/A
Nest: N/A
Food: A variety of aquatic vegetation and insects, crustaceans, and mollusks; typically feeds in water less than 6 inches deep for dabbling.

Stewardship: Maintain good water quality with shallow flats of emergent aquatic vegetation.
- Use pond design features, such as shallow benches, that benefit waterfowl when building or maintaining a tank or lake to promote feeding and brood-rearing habitat.
- Install water-control structures during dam construction or repair for vegetation management to facilitate water-level manipulation.
- Manage water levels on shallow benches during fall and winter when birds are present to maintain water depth less than 6 inches deep.
- Expose mudflats during late spring and early summer to encourage the growth of wetland vegetation.
- Minimize or avoid the use of aquatic herbicides to maintain aquatic vegetation habitat.

Green-winged Teal

- Exclude livestock grazing in riparian areas to maintain understory, water quality, and bank integrity.
- Exclude or limit livestock access to tank or lake shorelines to maintain water quality and emergent vegetation on shallow flats.
- Control invasive exotics and regrowth cedar using appropriate mechanical or chemical methods to maintain diverse riparian areas.

Other Potentially Useful Management:
- Avoid broad-spectrum insecticides on uplands to protect aquatic insect diversity.
- Control feral hogs to maintain shoreline vegetation and water quality.
- Maintain healthy watersheds with deep and sustainable ground cover to sustain riparian health and water quality.

KILLDEER
PRIORITY BIRD
ALL YEAR

Unlike most shorebirds, Killdeer often found on dry land far from water; common and owes at least some of its success to living close to humans; uses broken-wing strategy to lure predators away from young but charges large ungulates approaching its nest; named for its unmistakable *kill-dee* call.

Identification: About 9.5 inches head to tail; slender with long wings and pointed tail; adult has two vivid black-on-white breast bands; immature has one.
Habitats: Old field, plateau prairie, shin oak savannah, live oak savannah, river, creek, tank, pond, and lake.
Cover: Level ground with very short or no vegetation, such as mudflats or ball fields; nest sites often near water.
Breeding Territory: Average distance between nesting pairs about 100–800 feet.
Nest: Shallow place usually 3.0–3.5 inches across, placed on ground in open area with short grass; pale objects sometimes added to nest after eggs laid.
Food: Opportunistic ground forager; feeds on invertebrates of all kinds plus some seeds and even small vertebrates.

Stewardship: Maintain areas of low, mixed grasses with abundant insects and some short vegetation areas near water for nesting.
- Avoid broad-spectrum insecticides to promote insect diversity.
- Apply ant-specific insecticides to control red imported fire ants and increase insect diversity.
- Use shorebird-friendly pond design features, including water-control structures, when building or maintaining a tank or lake to encourage mud flats for feeding.

Killdeer

- Build and maintain tanks and lakes with shallow benches that benefit waterfowl and shorebirds to promote good winter habitat.
- Mow periodically except during May through August nesting season to maintain short vegetation.
- Use moderate to heavy rotational grazing to maintain short vegetation.

Other Potentially Useful Management:
- Conduct prescribed burns to control cedar and maintain relatively short, open grassland, especially during growing season right after the burn.
- Control spread of regrowth cedar using appropriate mechanical or chemical methods to maintain savannah and grasslands.

LESSER SCAUP

PRIORITY BIRD

WINTER

A formal-looking black and white diving duck found on still or slow-moving bodies of water. Arrives late in the fall from its breeding ground far to the north and forms sizable winter flocks.

Identification: About 16.5 inches head to tail with males larger than females; dark head, neck, and breast; back light gray and sides white; black tail and rump.
Habitats: River, creek, tank, pond, and lake.
Cover: Open water.
Breeding Territory: N/A
Nest: N/A
Food: Primarily aquatic invertebrates but also some aquatic plants.

Stewardship: Maintain or encourage good water quality with submerged aquatic vegetation.
- Maintain adequate aquatic vegetation to provide essential aquatic invertebrate habitat when controlling aquatic vegetation either by appropriate chemical or biological (grass carp) methods.
- Exclude livestock grazing in riparian areas to maintain understory, water quality, and bank integrity.
- Exclude or limit livestock access to tank or lake shorelines to maintain water quality and aquatic vegetation.
- Avoid broad-spectrum insecticides on uplands to protect aquatic insect diversity.

Lesser Scaup

Other Potentially Useful Management:
- Install water-control structures during dam construction or repair to facilitate water-level manipulation for vegetation management.
- Use design features that benefit waterfowl when building and maintaining a tank or lake to promote habitat for aquatic invertebrates.
- Control feral hogs to maintain shoreline vegetation and water quality.
- Maintain healthy watersheds with deep and sustainable ground cover to protect water quality and quantity.

MUSCOVY DUCK

OUTLAW BIRD

ALL YEAR

Distant domesticated relative of a protected Mexican species found in far South Texas along the Rio Grande. Large, awkward, domestic duck has hybridized with a variety of domestic ducks and geese, giving it variable plumage. Usually introduced from domestic sources such as feed stores but may move in from adjacent water bodies. Can be the unaffected host of serious waterfowl diseases, such as duck plague, which have spread to wild waterfowl populations, causing large die-offs on coastal wintering grounds.

Muscovy Duck

Identification: Males about 32.5 inches and females 25 inches head to tail; coloration variable with body white to black and often with white patches on the side; head often covered with red and/or black warty skin and waddles.
Habitats: River, creek, springs, tank, pond, lake, and backyard.
Cover: Calm water with shorelines of short "parklike" vegetation.
Breeding Territory: Males polygamous; will breed with many females and defend them against other males.
Nest: In cavities, displacing native cavity-nesting ducks; also may nest on the ground.
Food: White bread and corn chips; fish, aquatic invertebrates, seeds.

Control: Do not tolerate the presence of domestic ducks and geese in waterfowl habitat.
- Remove domestic ducks to prevent competition with native ducks for food and nest sites and to avoid spread of disease to wild ducks.
- Smear vegetable oil on eggs in nest to control reproduction when all domestic ducks cannot be removed.
- Foster tank, lake, and stream shoreline plants to provide healthy, emergent, and terrestrial vegetation that favor wild waterfowl.

RED-BREASTED MERGANSER

UNDER-THE-RADAR BIRD

FALL AND WINTER

Large duck with white patch on upper wing obvious in flight; flies low and fast over water. Most migrate through the Hill Country to winter on the Texas coast, but some stay to feed on fish in deepwater ponds.

Identification: Large diving duck, about 23 inches head to tail; winter plumage of male, female, and immature all similar with large gray body, rusty head, shaggy rusty-brown crest on back of head and long, narrow, orange bill.

Habitats: Tank, pond, and lake.

Cover: Deep water with good water quality and plentiful small fish.

Breeding Territory: N/A

Nest: N/A

Food: Mostly fish and some tadpoles, crayfish, and dragonfly nymphs.

Red-breasted Merganser

Stewardship: Control winter water depth, and conserve complex underwater habitat for fish.
- Use diverse submerged habitat (such as stumps and rock piles) in a tank or lake to provide an abundant food source for various life stages of fishes.
- Maintain adequate aquatic vegetation when controlling aquatic vegetation either by appropriate chemical or biological (grass carp) methods to provide essential fish habitat.
- Exclude or limit livestock access to tank or lake shorelines to maintain water quality and aquatic vegetation.
- Avoid using broad-spectrum insecticides on uplands to protect aquatic diversity.
- Use design features, such as shallow benches, that benefit waterfowl when building and maintaining a tank or lake to promote diverse fish habitat.

Other Potentially Useful Management:
- Install water-control structures during dam construction or repair to facilitate water-level manipulation for vegetation management.
- Control feral hogs to maintain shoreline vegetation and water quality.
- Maintain healthy watersheds with deep and sustainable ground cover to protect water quality and quantity.

RED-WINGED BLACKBIRD

PRIORITY BIRD

ALL YEAR

Easy-to-identify blackbird known for male's striking red and yellow wing marking. During winter, found in large flocks; during summer, disperses to nest in dense vegetation. Breeding bird surveys since 1966 show a disturbing population decline in the Texas Hill Country.

Identification: About 8.75 inches head to tail; long, slender bill; male black with distinctive red and yellow epaulets on wing; female dark brown with white streaks on breast and belly; female can show faintly reddish epaulet on wing.

Habitats: Old field, river valley prairie, river, creek, canyon, seep, springs, tank, pond, and lake.

Cover: Thick herbaceous vegetation; often shoreline or wetland vegetation but also dense grassland.

Breeding Territory: Approximately 0.5 acre and smaller in wetland habitat.

Nest: Well-built, open cup woven onto vertical vegetation; approximately 10–30 inches above the ground in wetlands and up to 22 feet in wooded habitat.

Food: Usually eats seeds, including grain crops, when not breeding; usually eats insects during breeding season. Young are fed various arthropods, especially aquatic species.

Red-winged Blackbird

Stewardship: Maintain and encourage dense native herbaceous shoreline vegetation and high water quality to provide healthy wetland habitat.

- Limit or eliminate livestock access to tank or lake shoreline to maintain dense vegetation for cover, feeding, and nesting.
- Avoid broad-spectrum herbicides and insecticides to maintain dense wetland vegetation and promote insect diversity.
- Use waterfowl-friendly pond design features when building or maintaining tanks or lakes to promote additional wetland habitat.
- Install water-control structures that allow for water-level manipulation during dam construction or repair to enhance wetland plant growth.
- Trap and dispatch cowbirds to reduce nest parasitism and increase nesting success.

Other Potentially Useful Management:
- Remove feral hogs to protect low-level habitat for cover and food.
- Control red imported fire ants with an ant-specific insecticide to promote insect diversity and maintain food supply.
- Control regrowth cedar using appropriate mechanical or chemical methods to maintain riparian woodland and shoreline diversity.
- Maintain healthy watersheds with deep and sustainable ground cover to enhance water quality and quantity.

RING-NECKED DUCK

FAVORITE BIRD

WINTER

Common diving duck found on ponds and lakes throughout the winter; distinctive and formal-looking duck that makes short dives to hunt for its food in shallow water.

Identification: About 16.5 inches head to tail; sharply contrasted black with gray sides and a yellow eye; unique white markings on bill.

Habitats: River, creek, springs, tank, pond, and lake.

Cover: Open water usually less than about 5 feet deep with emergent vegetation along the margins.

Breeding Territory: N/A

Nest: N/A

Food: Mostly seeds and tubers of various sedges, aquatic forbs, and grasses; also eats snails, mollusks, and other invertebrates.

Ring-necked Duck

Stewardship: Maintain diverse emergent vegetation and good water quality.
- Maintain adequate aquatic vegetation when controlling aquatic vegetation by appropriate chemical or biological (grass carp) methods to protect essential food.
- Use pond design features, such as shallow benches, that benefit waterfowl when building or maintaining a tank or lake to promote feeding and brood-rearing habitat.
- Install water-control structures during dam construction or repair for vegetation management to facilitate water-level manipulation.
- Expose mudflats during late spring and early summer to encourage the growth of wetland vegetation.
- Exclude livestock grazing in riparian areas to maintain understory, water quality, and bank integrity.
- Exclude or limit livestock access to tank or lake shorelines to maintain water quality and emergent vegetation on shallow flats.
- Avoid use of broad-spectrum pesticides on shorelines and uplands in the watershed to protect a pond's aquatic vegetation and invertebrates.

Other Potentially Useful Management:
- Control feral hogs to maintain shoreline vegetation and water quality.
- Control invasive exotics and regrowth cedar using appropriate mechanical or chemical methods to maintain diverse riparian areas.
- Maintain healthy watersheds with deep and sustainable ground cover to sustain riparian health and water quality.

SORA

PRIORITY BIRD

FALL, WINTER, AND SPRING (SEPTEMBER–MAY)

Uncommon and secretive rail found in dense, marshy habitat; wanders through thick shoreline vegetation where it probes for food with bill and feet; reluctant to flush or fly.

Identification: About 9 inches head to tail; a plump bird with black between its yellow bill and eye; brown back, gray breast, and long legs.
Habitats: River, creek, canyon, seep, springs, tank, pond, and lake.
Cover: Thick, herbaceous shoreline or wetland vegetation.
Breeding Territory: N/A
Nest: N/A
Food: Seeds and vegetation of wetland plants plus some invertebrates collected while walking or standing.

Stewardship: Maintain and encourage dense, native herbaceous shoreline vegetation and high water quality.
- Limit or eliminate livestock access to tank or lake shoreline to maintain dense vegetation.
- Minimize or avoid the use of pesticides to prevent unnecessary damage to wetland vegetation and aquatic insect habitat.
- Use waterfowl-friendly pond design when building or maintaining a tank or lake.
- Install water-level control structures during dam construction or repair to enhance wetland plant growth.

Sora

Other Potentially Useful Management:
- Remove feral hogs to protect low-level habitat for cover and food.
- Control regrowth cedar using appropriate mechanical or chemical methods to maintain riparian woodland and shoreline diversity.
- Maintain healthy watersheds with deep and sustainable ground cover to enhance water quality and quantity.

SPOTTED SANDPIPER
PRIORITY BIRD
ALL YEAR

Fairly common shorebird with spotted front; found wherever there is a bare shoreline; known for "seesaw" movement whether walking or standing still; low flight with quick wing beats. Both sexes defend the territory and incubate the eggs, but male does most brood feeding; not known for sure if it breeds in the Hill Country.

Identification: About 7.5 inches head to tail; proportionally long neck and legs; grayish body; in breeding birds, breast covered with dark spots.
Habitats: River, creek, tank, pond, and lake.
Cover: Often bare mud shoreline and adjacent herbaceous vegetation for breeding.
Breeding Territory: Averages about 0.2 acre/pair.
Nest: Shallow depression or scrape on the ground; lined with grass and some protective herbaceous vegetation; usually within about 325 feet of water.
Food: Primarily invertebrates, both aquatic and terrestrial.

Stewardship: Maintain or encourage good water quality with adjacent herbaceous vegetation and abundant insect populations.
- Minimize or avoid the use of pesticides, especially aquatic herbicides, to prevent unnecessary damage to aquatic insect habitat.
- Limit or eliminate livestock access to tank or lake shoreline to maintain water quality and herbaceous vegetation for nesting and foraging.
- Use waterfowl-friendly pond design features to supply shallow flats for feeding.

Spotted Sandpiper

- Install water-control structures that allow for water-level manipulation during dam construction or repair to manage mudflats.
- Control red imported fire ants with an ant-specific insecticide to promote insect diversity and maintain food supply.

Other Potentially Useful Management:
- Remove feral hogs to protect low-level habitat for cover and food.
- Control regrowth cedar using appropriate mechanical or chemical methods to maintain riparian woodland and shoreline diversity.
- Maintain healthy watersheds with deep and sustainable ground cover to enhance water quality and quantity.

VERMILION FLYCATCHER

PRIORITY BIRD

ALL YEAR

Brilliant red males are especially eye-catching; breeding bird surveys in the Texas Hill Country show a significant decline for this species. Human water use and habitat destruction have caused drastic population loss in lower Colorado River Valley.

Identification: About 5.25 inches head to tail; unusual for flycatchers, sexes have very different coloring; male is unmistakable scarlet and black; female is gray above with obvious peachy wash on belly and muted streaks on breast.

Habitats: Old field, post oak savannah, shin oak savannah, live oak savannah, river, creek, canyon, seep, springs, tank, pond, lake, and backyard.

Cover: Deciduous trees; especially small trees and shrubs, often with an understory of short herbaceous cover and an open area nearby.

Breeding Territory: Territorial but no specific territory size identified by researchers.

Nest: Loose cup mostly made of twigs and grasses and placed in fork of small horizontal branch averaging about 12 feet above the ground.

Food: Flying insects caught on the wing; males spend about 90% of the day flying off favorite perches to hover and catch food.

Vermilion Flycatcher

Stewardship: Maintain mix of woody plant and grassland diversity near water with understory and perches.
- Avoid broad-spectrum herbicides and insecticides to increase forbs and promote insect diversity.
- Control white-tailed deer and other browsing animals to encourage woody plant recruitment and growth.
- Control cedar that may be encroaching on riparian areas with appropriate mechanical or chemical methods to protect essential habitat of small trees, shrubs, and short herbaceous ground cover.
- Apply ant-specific insecticides to control red imported fire ants, increase insect diversity, and reduce possible predation.
- Use light to moderate rotational grazing to encourage balance of forbs and grasses and woody plant diversity.

Other Potentially Useful Management:
- Use prescribed burns in savannahs to increase forbs and maintain diverse woody plant communities.
- Mow in fall or winter dormant season on no more than 30% of herbaceous area to promote annual forbs.
- Maintain bird-friendly, natural or artificial water source to supply reliable and safe water.
- Trap and dispatch cowbirds to reduce possible parasitism.

YELLOW-CROWNED NIGHT-HERON
PRIORITY BIRD
SPRING–FALL (APRIL–OCTOBER)

Medium-sized heron of eastern Hill Country; eye-catching black, white, and yellow colors on head; makes its living at the interface of the aquatic and terrestrial environment; prefers water surrounded by trees and shrubs.

Identification: About 24 inches head to tail; stout gray bird with long legs and heavy bill; black head with white under eye and on the forehead; has a very long white plume; in flight long legs distinctly trail behind body.

Habitats: River, creek, springs, tank, pond, lake, and backyard.

Cover: Dense woody or grassy cover at the water's edge; also mudflats.

Breeding Territory: Colonial nester often in small groups.

Nest: Platform located on outer portion of low branches often over water; from an average of 7.5 to over 60 feet above the ground but probably on lower part of this range in the Texas Hill Country.

Food: In fresh water, diet is more than 70% crayfish; usually caught while standing still and watching; also (but much less common) eats insects, fish, amphibians, reptiles, birds, and small mammals.

Yellow-crowned Night-Heron

Stewardship: Maintain and encourage dense native herbaceous and woody shoreline vegetation to provide productive aquatic habitat.
- Limit or eliminate livestock access to tank or lake shoreline to maintain dense vegetation.
- Avoid excessive maintenance to protect the canopy of woody plants along and adjacent to water sources.
- Use waterfowl-friendly pond design features when building or maintaining tanks or lakes to promote healthy wetland habitat.
- Install water-control structures that allow for water-level manipulation during dam construction or repair to provide good crayfish habitat.

Other Potentially Useful Management:
- Remove feral hogs to protect low-level habitat for cover and food.
- Control regrowth cedar using appropriate mechanical or chemical methods to maintain riparian woodland and shoreline diversity.
- Avoid broad-spectrum herbicides and insecticides to maintain dense wetland vegetation and promote insect diversity.
- Maintain healthy watersheds with deep and sustainable ground cover to protect riparian health and water quality.

FEATURED BACKYARD BIRDS

PRIORITY
Bewick's Wren
Black-chinned Hummingbird
Black-crested Titmouse
Chimney Swift
Chipping Sparrow
Golden-fronted Woodpecker

UNDER-THE-RADAR
Blue Grosbeak
Eastern Screech-Owl
House Finch
Lincoln's Sparrow
Spotted Towhee

FAVORITE
American Goldfinch
American Robin
Northern Cardinal
Northern Mockingbird
Purple Martin

OUTLAW
European Starling
House Sparrow

Note:
Priority = on at least one of the lists provided by TPWD or Oak and Prairies Joint Venture naming species in need of conservation assistance, or shown by USGS-sponsored Breeding Bird Surveys in the Edwards Plateau to be a species whose population is in decline
Under-the-radar = a regular bird in Texas Hill Country but often unnoticed
Favorite = popular with just about anyone who knows the bird
Outlaw = undesirable because of its behavior and impact on other bird species

6 Backyards

Uniformity is not nature's way; diversity is nature's way.
—Vandana Shiva, Indian philosopher and environmental activist

INTRODUCTION

Birds are dependent on plants for food and shelter. Because they have evolved together, Hill Country birds generally prefer native plants. Thus, if you want to attract birds to your yard, it is best to landscape with native plants. Plants are adapted to specific soil, moisture, sunlight, and topography requirements that set natural limits to the plants you can grow and the birds you can attract to your yard. To have the greatest variety of bird species, you must work with what you have.

The Texas Parks and Wildlife Texas Wildscapes Certification program guides homeowners and recognizes backyards that provide food, water, and shelter for wildlife. Since wildlife thrives on the native plants they need, the Texas Wildscapes program requires at least 50% of the plants in a certified wildscape to be native to Texas. For more information on this program, visit http://www.tpwd.texas.gov/huntwild/wild/wildlife_diversity/wildscapes/wildscape_certification.phtml.

DESCRIPTION: BACKYARD HABITAT

Backyard habitat for birds includes three layers of vegetation: tree canopy, understory, and ground cover.

TREE CANOPY

Tree canopy is high overhead at the upper layer of your tallest shade trees. Including its high-growing vines, the canopy creates a unique aggregation of shelter, nest sites, seeds, and fruit. Most land in the Hill Country is high and dry with scattered oaks and cedar. These areas are often cut by canyons, which have a mixed canopy of oaks and cedar plus escarpment black cherry and hackberries. If you live on or near a creek or drainage, the tree canopy may include American sycamore, elm, pecan, or perhaps red mulberry.

A natural diversity of tall trees provides excellent habitat for many bird species. For example, Golden-fronted Woodpeckers are cavity nesters that use holes in different kinds of large trees and eat insects and acorns in the oak trees. Turkey, quail, and flickers feed on acorns after they fall to the ground. Migrant Cedar Waxwings and American Robins eat hack-

SEASONAL OCCURRENCE OF BIRDS IN WELL-MANAGED BACKYARDS

WINTER	SPRING AND FALL (MIGRANTS)	SUMMER (BREEDING BIRDS)	ALL YEAR (RESIDENTS)
American Goldfinch	Clay-colored Sparrow	Barn Swallow	Bewick's Wren
American Robin	Ruby-throated Hummingbird	Black-chinned Hummingbird	Black-crested Titmouse
Cedar Waxwing		Blue Grosbeak	Cactus Wren
Lincoln's Sparrow		Chimney Swift	Carolina Chickadee
Orange-crowned Warbler		Common Nighthawk	Carolina Wren
Pine Siskin		Orchard Oriole	Chipping Sparrow
Ruby-crowned Kinglet		Painted Bunting	Common Yellowthroat
Rufous Hummingbird		Purple Martin	Eastern Bluebird
Spotted Towhee		Summer Tanager	Eastern Phoebe
White-crowned Sparrow		White-eyed Vireo	Golden-fronted Woodpecker
Yellow-rumped Warbler		Yellow-crowned Night-Heron	House Finch
			Ladder-backed Woodpecker
			Lesser Goldfinch
			Northern Cardinal
			Northern Mockingbird
			Vermilion Flycatcher
			White-winged Dove

berry and escarpment black cherry fruit; finches and Carolina Chickadees feed on sycamore and elm seeds.

UNDERSTORY

Understory is composed of small trees, shrubs, and vines growing beneath the tree canopy. In the Hill Country, there is great diversity in the size and structure of natural understory. And these vegetation levels contain many different kinds of shelter, nest sites, seeds, fruit, and insects. Thus, many bird species depend on this type of vegetation for their livelihood. They include the wintering Yellow-rumped and Orange-crowned Warblers that hide and hunt insects among the understory plants.

A natural variety of understory plants also supports popular birds such as mockingbirds, cardinals, orioles, robins, finches, and catbirds that feed on fruit-bearing plants. If you have rusty blackhaw; hawthorn; evergreen, aromatic, and flameleaf sumacs; elbow bush; American beautyberry; Carolina snailseed; white or coral honeysuckle in your yard, you can attract these and many other birds.

Year-round birds like Bewick's and Carolina Wrens and Black-crested Titmice, as well as migrant warblers and Ruby-crowned Kinglets and nesting White-eyed Vireos, also live in a healthy understory because this is where they find the bugs they like to eat. These birds also supplement their insect diets with elm seeds and poison ivy berries.

Remember: Shelter from bad weather, nest sites secure from predators, fresh water, and clean food are essential bird habitat components. You can provide all of these, even in a small backyard.

Golden-fronted Woodpecker
(Photo by Lora L. Render)

Yard with tall mature trees

GROUND COVER

Some Hill Country bird species are attracted by low-level or ground-covering plants. This habitat layer can grow to various heights and produce many different kinds of seeds and fruit. Some natural understory areas might have flowering plants such as primroses, Texas thistle, purple cone-flower, and sunflowers mixed with native grasses.

Nesting Eastern Bluebirds and Painted Buntings need open areas with a mixture of low plants near midlevel brush. Bluebirds feed on insects living among plants near the ground. Although Painted Buntings are mainly seedeaters, in the summer they also consume grasshoppers,

Spotted Towhee
(Photo by Lora L. Render)

Backyard with understory and litter

caterpillars, beetles, and other insects and need the natural habitat that produces plenty of seeds and insects. Other birds like goldfinches, as well as Lincoln's and Chipping Sparrows, feed throughout the winter on seeds produced by natural ground-cover plants but not a manicured lawn. Native grasses are a great source of unique seed, cover, and nesting sites that will attract birds. Mourning Dove, Blue Grosbeak, Red-winged Blackbird, meadowlarks, and many species of native sparrows feed on grass seed. The habitat surrounding your property will determine if you can attract these birds to a small native grass prairie planted in your yard.

Some native grasses like switchgrass, yellow indiangrass, and eastern gamagrass grow more than 5 feet tall in deep soils with better soil mois-

Remember: The habitat that surrounds your backyard greatly influences the birds you can attract.

ture—and even wetlands—making them an especially good source of cover. On drier sites, little bluestem, hairy grama, sideoats grama, and the bristlegrasses grow 1 to 3 feet tall. In your lawn you can use buffalograss, the Hill Country's shortest native grass that naturally grows to only 6 inches. Bristlegrasses produce excellent food for seed-eating birds and compose as much as 70% of a Painted Bunting's diet.

Yards and other habitats with low-level plants that provide good cover and food will also have plenty of leaf litter that maintains moisture on the ground. Thus, during winter months, in backyards with diverse vegetation, you can expect to see White-crowned Sparrows and the Spotted Towhees hunting insects living in the litter.

COMMON AND POPULAR BACKYARD BIRDS
Northern Cardinal, Black-chinned Hummingbird, and Northern Mockingbird are native species associated with fairly high-density developments because they are well adapted to living near humans. They are especially popular birds because of their beauty, behavior, and abundance. These common backyard bird species do not need special attention to protect their populations. For the most part, they are not easily disturbed by human activity. Being well adapted to urban and suburban habitat conditions has led to their present success.

Northern Cardinals are well known as seedeaters because they have huge seed-eating beaks and come regularly to backyard feeders, but they eat insects in spring and summer and feed insects to their nestlings. They eat seeds all year long, but the percentage of seeds consumed increases from midspring through the winter. By midwinter, seeds compose more than 80% of their diet, so these are the times when they may be most dependent on the seeds in your feeders.

Black-chinned Hummingbirds breed and rear their young in the Hill Country. Adults drink nectar from plant blossoms (or sugar water from hummingbird feeders), and they consume protein in the tiny insects they catch. In addition, hummingbird nestlings and fledglings must have insect protein to grow and survive. Thus, if you want hummingbirds to nest in your yard, be sure to keep hummingbird-friendly, flowering plants as well as the tiny insects that naturally live on these plants.

Also, it is important to know that hummingbird feeders may be used all year. Each winter some Ruby-throated, Rufous, Broad-tailed, and Black-chinned Hummingbirds winter here and can be attracted to your feeders. Other species may be uncommon visitors as well. Put out only what will be consumed in a day or so. In the summer, when hundreds may be feeding, you might have to fill the hummingbird feeders more than once a day. In the winter, when only a few individuals are present,

Yard with mixture of trees and shrubs

Northern Cardinal
(Photo by Lora L. Render)

put out only a very small amount. Hummingbird food is best made with sugar and water in a ratio of 1:4 with *no* food coloring dye. The birds are attracted to the red of the feeder itself, and food coloring may be harmful to some individuals.

Recent research in Arizona found Black-chinned Hummingbird nests to be clustered around active and inactive accipiter hawk (Northern Goshawk and Cooper's Hawk) nests. Also, the hummingbirds are five times

Black-chinned Hummingbird
(Photo by Lora L. Render)

Yard with nectar-producing blossoms

more successful at producing fledglings when their nest is within approximately 1,000 feet of an active hawk nest. The hawk does not eat the hummingbirds and probably protects the tiny birds from predation by other smaller predators that it does eat. We also surmise that proximity to human habitation could limit nest predation and have a similar affect on fledgling success.

Northern Mockingbird diet varies with the seasons, from mostly insects in spring and summer to mostly seeds and fruit in fall and winter.

Northern Mockingbird
(Photo by Lora L. Render)

Yard with berry-producing shrub (Photo by Tyra Cox Kane)

Some of the mockingbird's favorite native fruits are hackberry, Virginia creeper, and poison ivy. Unmated male mockingbirds sing more than mated ones, and only unmated males sing at night. Both sexes sing in the fall to claim winter feeding territories. Mockingbirds will attack humans, and they have the ability to distinguish between individual humans and may attack one person while ignoring others. Mockingbirds also have been known to attack predatory birds, even bald eagles, when their territory is invaded.

Chipping Sparrow
(Photo by Lora L. Render)

Yard with diverse mix of native trees, shrubs, and perennials

PROBLEMS IN BACKYARD HABITAT
Problem 1: Lack of Native Plant Species Diversity

In the Hill Country, many natural native plant communities are severely degraded because they have been cleared or consumed by livestock and/or deer. Native plants attract a variety of different bird species, so when the plants are reduced or lost, the birds no longer have food, shelter, or nest space essential to survival. A diverse mix of plants attracts an

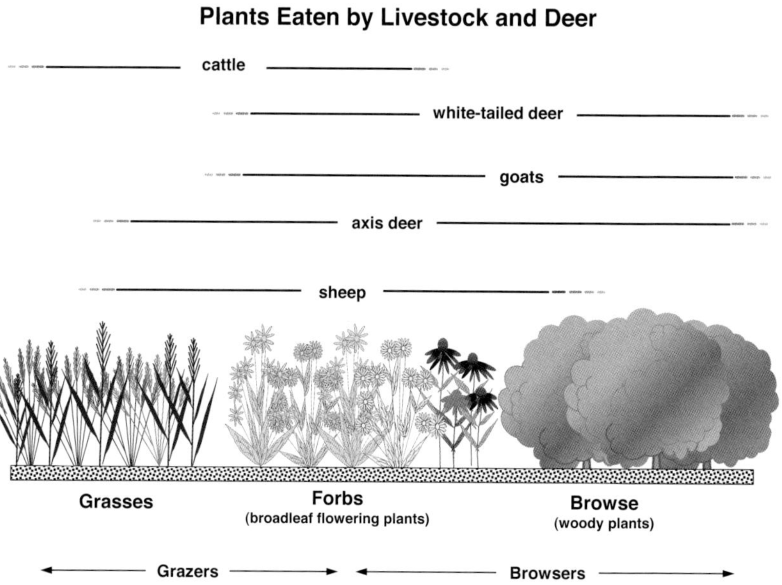

Plants Eaten by Livestock and Deer

```
--------   cattle   --------
        --------   white-tailed deer   --------
              --------   goats   --------
        --------   axis deer   --------
    --------   sheep   --------
```

| Grasses | Forbs (broadleaf flowering plants) | Browse (woody plants) |

←—— Grazers ——→ ←—— Browsers ——→

Figure 6.1

amazing number of birds, including wrens, woodpeckers, vireos, warblers, finches, chickadees, titmice, buntings, and sparrows such as the Chipping Sparrow. Restoring your land with native plant communities can make a huge impact on the quality of habitat for birds.

What You Can Do: Add Native Plants

Plant a diverse mix of native plants in your yard. As the basis of good bird habitat, native plants provide food, shelter from weather and predators, and space for breeding behavior and nesting. A healthy backyard "birdscape" is made of a wide variety of plant species suited to the physical limitations in your yard. If you have them, the birds will come. This can be especially important to the Bewick's Wren, which used to live as far east as New York. By 1980 it had disappeared from most of its eastern territory probably because of habitat loss, increased pesticide use, and nest space competition with other more aggressive birds. It is now listed as endangered, threatened, or Species of Concern in 13 eastern and midwestern states.

Problem 2: Little or No Understory or Ground Cover

Birds are adapted to feeding, nesting, and escaping predators and bad weather in different vegetation layers. Consequently, the greater the variety of plants and layers in your yard, the more species of birds you can expect to attract.

Remember: Diversity breeds diversity. A plant monoculture, even of a major native species such as live oak, cannot support the variety of bird species expected in a healthy backyard habitat.

> **THE VALUE OF NATIVE PLANTS FOR BIRDS**
>
> Because most of the native plants in the Preserve, a suburban San Antonio development, are protected and pesticide use is limited, it was certified as a Texas Wildscape. This designation requires provision of the basic habitat components: food, water, shelter, and space. A research project that compared numbers of birds in the Preserve to nearby traditionally landscaped developments showed that insect-eating birds, such as White-eyed Vireos, cuckoos, phoebes, and Bewick's Wrens, were more common in the Preserve than in the traditional developments.

What You Can Do: Restore Understory and Ground Cover
1. Think mosaic.

A healthy plant mosaic has layers. The ideal is landscaping with native and adapted plants that supply the necessary food and shelter for birds with different habitat needs.

2. Deal with the deer.

Throughout the Hill Country, deer present a problem for gardening and landscaping. They seem to eat just about everything and make selecting plants for our property a challenge. Often we give in and use only plants the deer do not like and rarely eat. In most cases, this does not help the birds. There is an alternative called conservation gardening in which priority native plants are introduced by selecting, planting, and protecting them from deer. Most are also valuable for healthy natural bird habitat. Use the deer browse evaluation chart in appendix 1 to determine the extent of deer damage to your woody plants. For help dealing with a deer problem, see the conservation gardening section in chapter 8.

3. Add brush piles.

If your property lacks adequate understory shelter for birds, brush piles create excellent cover for resident, migrant, and wintering species. Well-constructed shelters will last many years before they sink to the ground and need renovation. Carolina Chickadee, Chipping Sparrow, Black-crested Titmouse, Carolina Wren, Bewick's Wren, Northern Mockingbird, Northern Cardinal, House Finch, and in some areas the Rufous-crowned Sparrow are resident Hill Country birds that may use brush piles year-round. Sturdy brush piles appear to be useful to mi-

Bewick's Wren (Photo by Lora L. Render)

Side yard with diverse native shrubs

grant and winter birds when they encounter cold fronts with high wind and cold temperatures; small flocks disappear into the shelters where they are out of the wind and in a space heated by the warmth of the earth. House Wren, Hermit Thrush, Spotted Towhee, Dark-eyed Junco, and Chipping, Clay-colored, Field, Song, Lincoln's, and White-crowned

Tepee brush piles

Sparrows are migrant and wintering birds seen using brush piles in the Hill Country.

Problem 3: Pitfalls to Feeding Birds

For many of us, attracting birds to the backyard is an important first step to understanding and becoming an active steward of bird habitat. Watching birds in your backyard lifts the spirits and is satisfying. Whether you

Lincoln's Sparrow
(Photo by Lora L. Render)

Backyard with brush pile

live in a suburban or a rural setting, providing good seed in well-placed feeders can benefit many desirable native birds. This is most important during years or seasons when there are seed shortages in their natural environment.

Our feeders supply seeds only for seed-eating birds. They will attract and feed resident species such as Northern Cardinal, Carolina Chickadee, Black-crested Titmouse, Golden-fronted Woodpecker, Lesser and

American Goldfinch
(Photo by Lora L. Render)

Backyard with active bird feeding station

American Goldfinches, and Bewick's and Carolina Wrens. Feeders also attract wintering species, including Chipping and Lincoln's Sparrows, and occasionally colorful migrants such as Indigo Bunting and Rose-breasted Grosbeak. But feeding birds has some pitfalls: wrong seed, too much seed, poor feeder placement, predation, and feeder thieves.

What You Can Do: Select Proper Food in Proper Amounts and Protect Feeders

1. Select the proper types of seeds.

The vast majority of seed-eating birds prefer black-oil sunflower seed. Many people who feed birds regularly provide only this one kind of seed and are very happy with the many species that come to this food. It attracts just about everything. In most cases, it is fine to buy black-oil sunflower seed wherever you find the best price. This is not true with other kinds of seed. Do not buy an inexpensive birdseed mix from grocery or discount stores. These include red millet, rapeseed, and/or radish seed that wild birds will throw on the ground and not eat. Stay away from the large striped sunflower seeds. For small birds, they are hard to crack, and their fat content is inferior to that of the smaller black-oil sunflower seed.

The second most popular birdseed is white millet, especially with the native winter sparrows and towhees. It can be spread directly on the ground, placed in shallow trays on the ground, or fed from a traditional feeder. When feeding large numbers of birds through the winter, you will probably find it cheapest to buy white millet in bulk from a feed store if possible. The disadvantage of millet is that it attracts House Sparrows.

The third most popular seed is thistle. It is consumed voraciously by the House Finch and Lesser Goldfinch, as well as American Goldfinch and Pine Siskin in the winters when they are present. You could find yourself feeding over 25 pounds of thistle a week in the winter months when the Hill Country is occupied by these species. Pine Siskins and American Goldfinches periodically winter here in abundance, and this seems to be determined by the severity of the weather up north. It is fine to feed thistle in black nylon socks. These cost only about five dollars, and the birds prefer them to the more expensive thistle tube feeders. In cold months when the wintering finches are here by the hundreds, you may have to fill these socks more than once a day. In the summer, when only year-round residents are present, cut back to a partly filled sock so you do not waste food or have it out at night to be raided by a raccoon.

Which birds will be attracted to suet?

A steady supply of suet in the winter and early spring will attract Golden-fronted and Ladder-backed Woodpeckers, Orange-crowned and Yellow-rumped Warblers, and Western Scrub-Jays, as well as other wintering birds that are also attracted to your seed feeders. If you live where there are many raccoons, suet can become "an attractive nuisance" and will need to be brought in at night. You can buy ready-made suet squares at your favorite specialty store or, even better, make your own.

Remember: Bird feeders do not provide a diverse, natural food supply. They are not a substitute for a healthy native plant community in your yard.

Mix together 1 cup lard and 1 cup crunchy peanut butter.
Add 2 cups oatmeal flakes, 2 cups plain cornmeal, 1 cup wheat flour, and ⅓ cup sugar.
Stir and pat into a low baking dish to form 2-inch-thick "bird butter."
Freeze and then cut into squares that can be stored frozen in plastic bags.

2. Feed the proper amount of seed.

"Feed only what will be eaten in one day" may be the most important advice on feeding wild birds. Old seed spoils. If you do not follow this rule, at least half of what you put out can be wasted.

3. Place feeders to avoid predation and unwanted birds.

- Set bird feeders in safe places. Feeders should not be either overexposed or surrounded by too much vegetation.
- Where squirrels and/or raccoons are present, feeders probably should be hung on a slender metal pole at least 10 feet from the nearest tree.
- A heavy stovepipe-type baffle installed properly on the pole can be used to exclude squirrels and probably raccoons.
- Sometimes feeders that are caged in wire mesh with 1-inch openings are effective deterrents to both raccoons and squirrels.
- These feeders and poles are available at specialty bird and nature stores as well as some discount outlets.
- If your bird feeders are not emptied during the day and your area has many hungry and destructive raccoons, you will have to bring in feeders every night.

Do feeder halos eliminate House Sparrows?

House Sparrow is an exotic species that occurs only near human habitation and sometimes is so abundant that it keeps other birds away from bird feeders. The feeder halo is an experimental device designed to exclude House Sparrows from feeders. It is a sturdy ring fastened around and slightly above the feeder and has wires that hang at 1- to 2-foot intervals around the ring that are held down with fishing weights or attached to the ground. It can be purchased at wild bird supply stores or made with found objects like a coat hanger bent into a circle and shiny hobby wire. Do not use fishing line because birds become entangled.

Research at the University of Nebraska found that feeder halos did not deter breeding female or juvenile House Sparrows but significantly reduced males at feeders. Others report initial avoidance and later re-

Remember: Do not feed an exotic invasive species. If your bird feeder attracts only House Sparrows, stop feeding.

Feeder with House Sparrows

duced numbers plus fewer House Sparrows at nearby feeders. Halos also are reported to reduce visits by Blue Jays and Common Grackles without affecting use by other birds. How it works is uncertain and may have something to do with the different angle at which House Sparrows approach a feeder. For more information, see "Do Feeder Halos Keep House Sparrows at Bay" on the Project FeederWatch Blog at http://projectfeederwatch.wordpress.com/2012/10/09/do-feeder-halos-keep-house-sparrows-at-bay.

Problem 4: No Reliable, Safe Place for Birds to Drink and Bathe

A reliable and safe water supply for drinking is essential to most birds. Water can be a limiting factor for resident, migratory, and nesting birds in the Hill Country because of the regular periods of extended drought.

What You Can Do: Maintain a Reliable, Auxiliary Water Supply

Providing water in your backyard will become more and more important to the birds for both drinking and bathing as drought conditions become severe. In addition to helping the birds, keeping a continuous water supply in your yard will provide many happy hours of backyard bird-watching.

Which birds will be attracted to water in my backyard?

A safe, dependable, appropriate supply of water attracts both seedeaters that come to standard bird feeders and insect eaters that are not usually attracted to feeders. Here is just a partial list of resident and nesting species that commonly come to human-made water sources in the Hill Country: Wild Turkey, Lesser and American Goldfinches, Golden-

Blue Grosbeak (Photo by Lora L. Render)

Backyard with shallow running water adjacent to live oak savannah

fronted and Ladder-backed Woodpeckers, Eastern Phoebe, American Robin, Western Scrub-Jay, Carolina and Bewick's Wrens, Eastern Bluebird, Summer Tanager, Blue Grosbeak, Painted Bunting, and adjacent to appropriate rural habitat, perhaps Scott's Oriole.

What are the best ways to supply water in a backyard habitat?

The traditional birdbath on a pedestal is not the best choice. It usually has water that is too deep for most birds and is hard to keep clean. But your traditional birdbath or any container with deep water can be modified to be safe for birds by adding stones for easy access. A good birdbath has a shallow basin with water less than 1 inch deep. This allows

Traditional birdbath

smaller birds such as wrens, bluebirds, and finches to wade and still keep their bodies above the water. A large, shallow, potted-plant saucer made of plastic, ceramic, or terra-cotta works well. These materials are easy to clean and are durable. Probably because this is the way water is available to them in nature, birds tend to prefer water provided at ground level.

How do I protect birds from predators at a backyard water feature?

Unless there are cats in the neighborhood, you can imitate a natural pool and put the birdbath right on the ground. If cats are present, place the birdbath 2 or 3 feet above the ground and away from anything like a brush pile that might hide a predator from the birds' view. Birds are vulnerable to predation when their feathers are wet and flight is more awkward than usual. Thus, immediately after bathing, they preen or clean individual feathers and add a protective coating of oil from the gland at the base of the tail. Most birds prefer to bathe near good preening perches, where they are safe from a potential predator. To observe preening, you can add perches in the form of loose vegetation or small brush piles in the vicinity of the birdbath. Birds also will preen when perched on wire cages placed to protect landscape plants growing near the water source. To prevent the easy predation that can happen when too many birds are attracted to one location, make sure that the birdbath and preening perches are not immediately adjacent to a bird feeder.

What is the advantage of running or dripping water?

In a yard providing or surrounded by appropriate habitat, the pleasing sound of running or dripping water has an almost magical at-

American Robin
(Photo by Lora L. Render)

Yard with water dripper

traction to birds from the ever-popular American Robin to the colorful Painted Bunting. You usually can find a commercially made dripper in a feed store or birding supply store. These are attractive to birds and humans and work well in most situations. If you choose to install a permanent water feature, offer running water and plenty of shallow places for birds to drink and bathe. One design is patterned after a livestock tank where birds line up along the rim and in the shallow water that runs over the edge. In the backyard version, you can save water by recycling it within the water feature. Place a smaller tank or container on top of a larger tank, and let the water run over the lip of the small tank into the larger one. Use a small submersible pump to lift the water back into the higher tank.

How important are cleaning and refilling a water bath?

The most attractive birdbaths are maintained regularly. Keep them filled with water at all times. Place them in the shade to reduce the evaporation rate. Regular cleaning will prevent disease. Replace water and remove buildup of organic debris by scrubbing with a mixture of one part household bleach to nine parts water. Rinse thoroughly with fresh water.

What is the best way to prevent the water in my birdbath from freezing?

Central Texas winters are usually so warm that we do not need special water heaters to prevent birdbaths from freezing. On the few very cold days that occur in some winters, it is best to simply keep the birdbath water open by hand: break or remove the ice and refill the birdbath with warm water as necessary. *Never* place any kind of antifreeze in the water. Antifreeze is lethal to birds and other wildlife when ingested.

Problem 5: Spread of Disease due to High Concentration of Birds at Feeders

Although diseased birds are rarely seen at feeders, sublethal cases, which are difficult to recognize, may be present. Since our feeding stations bring unusually large numbers of birds in close proximity, it is a place where infectious diseases can be transmitted easily. Thus, it is particularly important to keep our feeding stations clean so we are not contributing to the spread of avian disease.

What You Can Do: Practice Good Hygiene

1. Space your feeders.

Allow plenty of room between and around feeders so birds will not be crowded together.

2. Use safe feeders.

Use only feeders that are free of sharp edges and points. Even a small scratch can be an entry point for infectious bacteria or viruses.

3. Keep feeders clean.

Wash and disinfect feeders once or twice a month. After washing with hot water, disinfect feeder with a two- to three-minute immersion in a solution of one part bleach to nine parts warm water. If there are sick birds in the area, disinfect feeders weekly.

4. Use fresh food.

Discard seed that smells musty, is wet, looks moldy, or has fungus growing on it. Disinfect storage containers and scoops that have been used with spoiled food.

5. Avoid food contamination.

Keep rodents out of the food storage area. Mice can carry and spread the bird disease salmonellosis without being affected themselves.

6. Clean up waste.

Clean the area under your feeders with a leaf blower or Shop-Vac set to blow; even better, vacuum up waste food and droppings.

7. Maintain a healthy feeding station routine.

Do not wait to "clean up your act" until disease is reported in your area. With good prevention, you can avoid having sick or dead birds at your feeders.

8. Spread the word.

Encourage your friends and neighbors who feed the birds to follow these precautions. Birds move among feeders and can spread disease from one yard to another.

What are the common avian diseases seen at bird feeders?

Salmonellosis is the most common bird-feeder disease. It is caused by a salmonella bacteria, and the symptoms are not always obvious. Sick birds may appear thin, fluffed up, and depressed and may have swollen eyelids. They are often lethargic and easy to approach. Some have no symptoms but are carriers and can transmit the disease to healthy birds. Salmonellosis is usually spread by fecal contamination of food and water and can be transmitted by ingestion of contaminated feed or bird-to-bird contact.

Mycoplasmosis or House Finch disease is a bacterial disease that causes red, swollen, runny, or crusty eyes. In severe cases, the eyes become swollen shut or crusted over. An infected bird might sit quietly or clumsily scratch an eye against its foot or a tree. Some recover, but many diseased birds die from starvation, exposure, or predation. This disease is also found in American Goldfinches and is spread by direct contact, airborne droplets, or dust. This is the most recently discovered bird disease, first noticed in 1994. For more information, visit Project FeederWatch at http://birds.cornell.edu/hofi/index.html or call 800-843-2473.

Trichomoniasis most commonly affects pigeons, doves, and the raptors that eat them. Sick birds have raised lesions in the mouth, esophagus, and crop and may appear to have trouble closing their mouth. The disease is caused by a protozoan often present in the mouth secretions of birds that appear healthy but are carriers of the disease. It is spread when birds consume contaminated food or water. Mortality from this disease varies and can be high.

Aspergillosis is caused by a fungus that affects the respiratory system. Infected birds may have difficulty breathing, appear thin, drink a lot, and appear to have difficulty walking. When the eyes are infected, they may be covered with a white film and have discharge. Healthy birds normally resist the disease, but birds with depressed immune systems are vulnerable. The infection causes lesions in the lungs and air sacs and has been reported in many species of birds. Aspergillosis fungus or mold grows on damp feed or debris beneath feeders and is spread when birds ingest or inhale the spores. Outbreaks of the disease can cause significant mortality in certain species.

Avian pox has two forms. The most common produces wartlike growths on featherless surfaces of face, legs, and feet. The second form affects mouth, throat, trachea, and lungs. Infected birds have impaired breathing and difficulty in feeding. Secondary infections that lead to a bird's death are common. Avian pox is caused by avipox virus and has been reported in 60 species of birds, including turkeys, hawks, owls, and sparrows. The virus is usually spread by infected mosquitoes but also by direct contact with infected birds or contaminated feeders or ingestion of contaminated food or water.

To learn more about these diseases, visit the USGS National Wildlife Health Center website at http://www.nwhc.usgs.gov.

How do I recognize a sick or dying bird, and what should I do?

Perhaps you will be lucky and never see a diseased or dead bird in your yard. Just in case your luck runs out, it is good to be prepared and know how to recognize a sick bird and what do to do if you see one. Sick birds in a flock are less alert, less active, feed less, have feathers that look unkempt, and often are reluctant to fly away. (Molting or very young birds such as a juvenile cardinal may look ratty as they replace feathers.

House Finch (Photo by Lora L. Render)

Clean backyard feeding station

At this time they can be mistaken for sick birds, but they differ by being active and alert and eating normally.)

If you see more than one dead bird at a time, call the Texas Parks and Wildlife 24-Hour Communication Center at 512–389–4848. Immediate response is key to helping a successful investigation. Because chemicals in plastic interfere with toxicology tests for pesticides or antifreeze, do not put a dead bird into a plastic bag. Instead, wrap it in aluminum foil. If necessary, it can be put in a freezer for storage.

What do I do if avian disease is reported in my area?

Some birds move among food sources over a large area and can spread disease as they move from feeder to feeder. Disinfect your feeders weekly if diseased birds are seen at your feeders or in your area.

Problem 6: Effects of Pesticides and Herbicides

Nearly all birds eat insects at least sometime during the year. Chimney Swifts, Purple Martins, swallows, and flycatchers all feed on flying insects and cannot live where they are absent. Recent research shows that preserving their food supply is important to Chimney Swift conservation as well as protection of all birds that feed on the wing. Six- and eight-legged invertebrates are also indispensable food for woodpeckers, warblers, wrens, vireos, and bluebirds. And nearly all birds, even those we know as seedeaters, must feed invertebrates that have high amounts of protein to their nestlings.

What You Can Do: Stop or Minimize the Use of Chemicals

Welcome insects and spiders to your backyard. Warblers, wrens, vireos, and bluebirds thrive on invertebrates. There are very few birds that do not eat insects and spiders at least part of the year. Even a bird like the Northern Cardinal that we know as a seedeater has a diet that is about 60% insects in the spring.

Limit even the use of organic pesticides because they are not always harmless. They can be detrimental to birds by killing the insects that birds eat. For example, *Bacillus thuringiensis* (BT) kills all caterpillars, an important food for wrens and cuckoos. And BT may be washed off into ponds or wetlands where it can kill invertebrates needed by waterfowl and shorebirds.

Problem 7: Inadequate Nest Sites for Cavity-Nesting Birds

When nest cavities are not available in otherwise suitable habitat, they are a limiting factor for cavity-nesting birds, both passerines and non-passerines. Eight Hill Country passerine species nest in cavities: Eastern Bluebird, Black-crested Titmouse, Bewick's and Carolina Wrens, Carolina Chickadee, Ash-throated Flycatcher, Great Crested Flycatcher, and rarely, House Finch, which is known to use nest boxes in other parts of the country. Purple Martins are a special case since these passerine birds are colonial cavity nesters. They are covered separately in another section.

Cavity-nesting birds normally make their nests in dead trees (snags) or large branches with cavity holes. If your property or neighborhood has appropriate habitat for the cavity-nesting passerines known to occur

Chimney Swift (Photo by Lora L. Render)

House chimney and open yard (untreated with chemicals)

in the Texas Hill Country but lacks natural cavities, nest sites will be a limiting factor. Three Hill Country cavity-nesting birds—Black-crested Titmouse, Carolina Chickadee, and Bewick's Wren—are priority species. Thus, by adding nest boxes, we are promoting nest success for birds that need conservation. Black-crested Titmouse, Carolina Chickadee, Carolina Wren, and Bewick's Wren prefer somewhat wooded habitat. Eastern Bluebirds frequent open grasslands; Ash-throated Flycatchers prefer a higher, drier open prairie landscape. However, individual birds have different preferences, and the habitat they use will vary.

> **PASSERINES**
>
> Passerines are birds in the order Passeriformes, which includes 60% of all bird species. Some of these birds are often called songbirds (they make sounds pleasant to human ears as part of their courtship) or perching birds (based on the structure of the foot). The other 40% of birds as a group are nonpasserines, such as ducks, woodpeckers, and birds of prey.

What You Can Do: Add, Maintain, and Monitor Nest Boxes

1. Add nest boxes to increase cavity-nesting bird populations.
2. Maintain nest boxes regularly to maximize production. This includes annual inspection and repair as well as removal of old nests after each nesting to promote renesting.
3. Monitor weekly to document success and remove nests of exotic species.

What are the best nest boxes for passerines in the Texas Hill Country?

In the Texas Hill Country, Eastern Bluebirds prefer the Gilwood design over other types. Years of field tests at the Cibolo Nature Center have shown that the Gilwood box attracts three to six times more bluebirds than standard oval- or round-holed boxes. Developed by bluebird expert Steve Gilbertson, the Gilwood box features a hinged, recessed front door that keeps out rain and allows easy cleaning. Eggs and young stay secure, even when the nest box is opened for monitoring. Field tests at the Cibolo Nature Center also show that other Hill Country cavity-nesting passerine birds use Gilwood boxes as frequently as they use traditional boxes. Thus, we recommend this design for all of these birds.

What is the best way to mount nest boxes for passerine birds?

For all passerine birds, mount your boxes the same way:

- Set them at eye level or about 4–6 feet above the ground for easy monitoring.
- Face the entrance east, away from prevailing storms and hot afternoon sun.
- If possible, locate the nest box where it will have afternoon shade.
- Never hang a nest box on a tree or fence because this allows easy access for predators.

Remember: For all cavity-nesting birds, it is important to leave dead trees (snags) standing so their preferred, natural nest sites are available. This reduces the expense and maintenance of bird boxes.

Black-crested Titmouse

Open wooded yard with nest box and plant diversity

- Use PVC pipe, metal electrical conduit, or something similar for the mounting post.
- With this light post material, it is best to use lightweight wood such as cedar for the box itself.
- When adding more than one box in suitable habitat, place the boxes about 100 yards apart or one box per 2 acres.
- When adding nest boxes, always be prepared to monitor them weekly.

EASTERN BLUEBIRDS

The Eastern Bluebird, probably the most well-known cavity-nesting passerine in the Hill Country, is a perfect example of a bird species helped by nest boxes. Eastern Bluebird populations declined by 70%–80% in the late 1960s and 1970s. Increasingly scarce nest sites and fierce competition from two aggressive introduced species, House Sparrow and European Starling, were the primary causes of this decline. In recent years Eastern Bluebird populations have rebounded through widespread habitat enhancement in the form of well-designed and maintained bird boxes. This success is due to organized human intervention addressing the habitat's limiting factor: insufficient cavities in otherwise good habitat with heavy competition for the few available nest sites.

Eastern Bluebird eggs in nest box

Should I provide nest boxes for birds other than passerines?

The Hill Country has several non-passerine cavity-nesting birds that require special boxes, including two woodpeckers, two ducks, and three owl species, such as a popular backyard owl, the Eastern Screech-Owl. Because of other habitat limitations, attracting these species to a nest box is less likely than with the cavity-nesting passerines. Nonetheless, it may be useful to know that all of these birds need medium to larger boxes mounted higher off the ground than the passerine boxes as shown in the table.

Gilwood-style nest box

1. Build nest boxes for Wood Ducks.

For nest box designs for Wood Ducks, see http://www.npwrc.usgs.gov/resource/birds/woodduck/wdnbox.htm or http://www.woodducksociety.com/duckhouse.htm.

2. Build nest boxes for screech-owls and kestrels.

For nest box designs for screech-owls and kestrels, see www.instructables.com/id/American_Kestrel_nest_box/ or http://mn.audubon.org/sites/default/files/documents/american_kestrelnest_planaudubon.pdf. Set American Kestrel nest boxes at least 10 feet high. Place them in an open area of 50 acres or more. They need an opening about 2¾ inches in

NEST BOX INFORMATION FOR CAVITY-NESTING BIRDS NATIVE TO THE HILL COUNTRY

BIRD	BOX SIZE	ACRES/BOX	BOX HEIGHT	HABITAT TYPES	SECTION OF HILL COUNTRY
American Kestrel	Medium	10	At least 10 feet	Savannah, woodland margins	Northern half
Ash-throated Flycatcher	Small	12.5	4–6 feet	Woodlands, savannah, riparian areas	All
Barn Owl	Large	Not territorial	At least 12 feet	All except extensive open grasslands	All
Bewick's Wren	Small	1	4–6 feet	Backyards, woodlands, savannah, riparian areas	All
Black-bellied Whistling-Duck	Large	Not territorial	At least 10 feet	Within 0.75 mile of perennial water	Southern half
Black-crested Titmouse	Small	10.5	4–6 feet	Backyards, woodlands, savannah, riparian areas	All
Carolina Wren	Small	2.5	4–6 feet	Backyards, woodlands, savannah, riparian areas	All
Carolina Chickadee	Small	4–6	4–6 feet	Backyards, woodlands, savannah, riparian areas	All but western edge
Downy Woodpecker	Small	11	At least 12 feet	Backyards, woodlands, savannah, riparian areas	Eastern edge
Eastern Bluebird	Small	5	4–6 feet	Backyards, prairies, savannah, riparian areas	All but western edge
Eastern Screech-Owl	Medium	15–75	At least 10 feet	Backyards, woodlands, savannah, riparian areas	All
Golden-fronted Woodpecker	Medium	40	12–20 feet or more	Backyards, woodlands, riparian areas	All
Western Screech-Owl	Medium	247	At least 10 feet	Backyards, woodlands, savannah, riparian areas	Western edge
Wood Duck	Large	Not territorial	At least 10 feet	Near perennial water	Eastern edge

diameter. The kestrel prefers a box with a couple of inches of wood shavings in the bottom. Face the entrance east, away from prevailing storms and hot afternoon sun. Avoid the use of most pesticides throughout the breeding territory because they may affect food availability and reproduction.

3. Build nest boxes for woodpeckers.

For nest box designs for woodpeckers, see www.birdwatchersdigest.com/bwdsite/solve/howto/woodpeckerbox.php. Bird box building plans on this website should work for both Golden-fronted and Downy Woodpeckers. The Golden-fronted Woodpecker requires an opening of 2⅛ inches, but the smaller Downy Woodpecker will use a 1¼-inch entry

Eastern Screech-Owl
(Photo by Lora L. Render)

Open yard with large trees and limited understory

hole. Place a couple of handfuls of wood shavings in the box so woodpeckers have something to excavate. Always use a design that allows easy opening for regular monitoring.

Will monitoring nest boxes disturb the birds?

Contrary to popular opinion, weekly visits do not deter birds from nest building or feeding their young. Except when nestlings are about to fledge, you can put your hand in the nest, count the eggs, and look at the nestlings with no harm to the young birds. Birds have almost no sense of smell and are so strongly "programmed" to nest and attend to their young that brief, regular monitoring visits do not impact their nesting success. It is best to avoid any contact with the bird box during the day or two before the young are expected to fledge.

You can predict fledge date by knowing that nesting passerines usually lay one egg on successive mornings until they complete their clutch. They begin incubating their eggs on the day the last egg is laid. With ideal weather, these eggs will hatch about 14 days later. If cold or stormy weather occurs, the incubation time will be longer. After hatching, the young birds remain in the nest about 14 days before they are ready to fledge. Inclement weather can also interfere with regular feeding and increase the time nestlings need before they are able to fly.

How do we avoid disturbing woodpeckers when monitoring their nest boxes?

Avoid contact after woodpecker nestlings are 12 days old because they are easily flushed from their nest prematurely. Do annual, general maintenance on woodpecker boxes in the fall or winter because Golden-fronted Woodpeckers are known to conduct their own maintenance during the time between broods.

How do we avoid disturbing American Kestrels when monitoring their nest boxes?

Limit maintenance of raptor boxes to the fall or early winter to be completed by January 1 since they may begin nesting early in some years. Avoid any disturbance of raptor nest boxes during the breeding season. Eastern Screech-Owls and American Kestrels will use boxes and usually raise a single brood per breeding season. Their young require extended care and should not be disturbed.

How do we avoid disturbing Wood Ducks or Black-bellied Whistling-Ducks when monitoring their nest boxes?

Conduct annual nest box maintenance for these ducks from September through December. Wood Ducks and Black-bellied Whistling-Ducks produce one or two broods per season, and Wood Ducks often produce two broods in Texas' warm climate. Nesting may begin from January through late summer depending on weather conditions. Dry years with midsummer rains may lead to late nesting. Wood Duck young leave the nest about 24 hours after hatching. Do not monitor the nest box near this time.

What is the best way to monitor my nest boxes?

It is best to use a routine when monitoring your nest boxes. Make weekly afternoon visits to check each nest box throughout the nesting season.

- Keep written records on a data sheet or in a journal used especially for this purpose.
- Note date, time, nest description, egg color and numbers, nestling numbers, unhatched eggs, and number of birds fledged for each visit,.
- Wear binoculars and watch for the adults so you can positively identify the species in each box.

How to I avoid wasps and ants that may use the bird boxes?

Especially early in the season, wasps might build nests in a bird box. When monitoring, to prevent being stung as they fly out, stand to the side of the box, tap it, and then open the box. The wasps fly straight out and will not bother you if you are out of their flight line. You can also use a stick to make sure all wasps are out of the box before looking inside. To prevent wasps from building nests in boxes, rub a bar of mild soap on all interior surfaces. If wasps are already established in an old box, you may have to remove wasp nests repeatedly and resoap several times to discourage them. Ants can also become established in a bird box untended for many years. In this case, it is probably best to move the box to another location.

How do I protect the birds in my nest boxes from predators?

Always consider predation when adding any kind of bird box. In most locations, the most common predators on cavity-nesting passerine birds are raccoons, snakes, and perhaps cats. Mount the box on a structure least likely to allow access by predators. A well-designed nest box with a predator guard is the best prevention. In the deep Gilwood box design, the bird nest is far below the opening, making it hard for most raccoons and cats to reach in and catch a helpless nestling. Snakes can still enter these boxes with ease, however.

A predator guard made from galvanized dryer vent pipe hung on the poles gives necessary protection from snakes and "long-armed" raccoons and cats. An excellent and easy-to-make stovepipe predator baffle design is available at https://tpwd.texas.gov/publications/pwdpubs/media/pwd_bk_w7000_0512.pdf. Predator guards are also effective on non-passerine nest boxes. At the Welder Wildlife Foundation in Sinton, Texas, a 1980s survey of their whistling-duck box monitoring data showed that none of the boxes with guards had predation, but 55% of boxes with no guards had predation. Predator guards were of two types: galvanized

Manufactured gourd assembly

Purple Martin

sheet-metal cones or razor ribbon wire. The major predators at the Welder Preserve were raccoons and snakes.

PURPLE MARTIN COLONIES

Purple Martins naturally nest in hollow trees and holes made by woodpeckers, and they are colonizers that benefit from close proximity to their own kind as well as human habitation. Early European settlers learned about attracting martins to dwellings from the Choctaw and Chickasaw, who used to hang hollow gourds near their lodges. Today, although most martins farther west nest in natural cavities, homeowners from Texas to the Dakotas and east attract Purple Martins by erecting large multistoried birdhouses and gourd assemblies.

Which kind of housing is best for my martin colony?

No matter what type of housing you provide, the best colors are white or very light pastel; trim can be any color. White housing seems to be most attractive to martins because it reflects the heat of the sun, keeping nestlings cooler. For over 10 years, the Cibolo Nature Center martin colony in Boerne, Texas, had two sets of hanging gourds and two apartment houses. All were mounted on separate metal poles and were easy to lower for cleaning and monitoring. They provided a total of 24 gourds and 36 apartments. Based on information from many years of weekly monitoring throughout the breeding season, it is clear that martins preferred gourds to the houses. For example, in 2007, birds built nests in 73% of the gourds and in only 47% of the apartments.

When do martins arrive, and when should I open the houses?

In the Texas Hill Country, male martins first appear sometime in February, and their arrival time for your area is predictable. Open the chambers soon after the middle of February, so the early birds can roost in them and get protection from late winter wind and cold. At a well-established colony, martins typically are present for weeks before they actually begin building their nests. When the females show up, they apparently select the male and location of choice. Then, both members of a pair build their nest out of grass litter, leaves, or tiny twigs, and sometimes mud at the front. Most nests are made in late April and early May. However, young nesting birds may find and use your martin house as late as June.

The female lays one egg each morning on successive days and begins incubation right after the last egg is laid. She usually lays three to five eggs and incubates them for 16 days before they hatch. Some female martins will leave the nest at night for the "group roost" preferred by all Purple Martins. If this happens, the incubation period will be longer. Most martin young fledge in late June, the time of greatest mortality.

*I have had a good house up for several years
but no martins. What should I do?*

If you live where there are no nearby martin colonies, it may take two to four years for them to find your new house. If you live in a neighborhood where colonies already exist, you may get martins nest-

LOCATION, LOCATION, LOCATION

The major reason people fail to attract martins is that they place their martin housing incorrectly or their property is inappropriate martin habitat. Martins have very specific aerial space requirements.

- Housing should be in the center of the most open spot available, about 30–120 feet from human housing.
- In general, there should be no trees taller than the martin housing within 40 feet and preferably 60 feet. However, the Hill Country is in the southern half of the Purple Martin breeding range, where they are somewhat less particular about house placement. Southern landlords can sometimes place housing within 15–20 feet of trees and still attract martins. Rule of thumb is the farther the housing is placed from trees, the better.
- Place martin housing 10–20 feet above the ground.
- Keep tall bushes, shrubs, and vines away from the pole.
- Do not attach wires to a martin house, especially if they lead to trees, buildings, or to the ground.

ing in a new house almost immediately. Martins prefer to nest in houses in or near open spaces, where they can swoop and soar unhindered by tall trees. If there are trees near the martin house that has not attracted birds, relocate the house to a more open area, mount the house higher, and/or prune or remove trees to create a more open site. If present, remove any wire to trees, building, or ground. To help the birds find a new location, many experts recommend playing a tape of the Purple Martin's "dawn song" for an hour early each morning until they begin to nest in your martin house.

What do I need to do to maintain a martin colony?

Being a martin colony landlord requires more than just putting up a proper martin house. When hosting Purple Martins, you become the maintenance person, record keeper, and security guard. To cultivate a successful colony, you must take your duties seriously.

Each spring before the first birds arrive is the time for vital martin housecleaning:

- Remove old nesting material, and scrape the mess from each nest compartment.

- Wipe with a cleaning solution of one part bleach to nine parts warm water. It should kill any residual bacteria and/or mites. Rinse thoroughly with water.
- Dry in sun for a day or two.
- Repair any damage.

Regular monitoring will also allow you to remove dead birds and stop a parasite infestation or predation before you lose too many eggs, nestlings, or adult birds. A martin monitoring data sheet is provided in appendix 5.

- Lower the house or gourd assembly to monitor your colony weekly.
- Label each cavity with a letter and number that represents both the house and the individual chamber for the most accurate records.
- Record martin nests, eggs, nestlings, and fledges and removal of sparrow nests for each chamber weekly.
- Continue monitoring until all young birds have fledged.
- Judge the success of your colony by the number of birds fledged.

What is the best way to deal with mites and other nest parasites?
Some weather conditions may contribute to a population explosion of common, external nest parasites, including fleas, mites, and blowfly larvae. An outbreak of mites can occur very quickly and will kill many nestlings. These sudden parasite population explosions usually happen in the summer after most eggs have hatched. They seem to be most prevalent in houses in which the nest compartments are very close together rather than in gourds, in which the chambers are separate. Never use pesticides in nests or boxes.

If parasite infestation is severe, the recommended method to deal with this problem is "nest replacement." Before you do nest replacement, visit www.purplemartin.org and read "Why and How to Do Nest Replacements for Purple Martins" by James R. Hill III of the Purple Martin Conservation Association.

How do I protect my martin colony from predators?
Prey attracts predators. In the Texas Hill Country, martins most often are preyed on by rat snakes, raccoons, Great Horned Owls, and accipiter hawks (Cooper's and Sharp-shinned). Martin colony landlords who lose an entire colony from one year to the next often suspect that their flock died in a storm during migration or was poisoned by pesti-

Martin monitoring

cides on their wintering grounds. These scenarios are unlikely; the martins that share a breeding site do not migrate or overwinter as a group. Loss of an entire colony is most often caused by something that happened in the landlord's own backyard during the nesting season.

- Keep the colony size modest to lower the chances of attracting a serious predator and losing the entire colony. Purple Martins most commonly abandon a colony site because predators have raided their nests. A large martin colony is more attractive to predators. It may take only one foray up a martin colony pole by a snake, raccoon, or squirrel or a few visits by an owl, hawk, or crow to cause all the surviving birds to abandon the site.
- Be consistent with weekly nest checks so you know when martins, eggs, or nestlings are disappearing.
- Place guards on all martin colony poles (wooden or metal) to eliminate easy climbing by predators.
- Equip the house with an owl guard if it has become a regular target for hawks, owls, or crows. Hawks are especially difficult to deter.
- Federal law protects all birds of prey, and harming them is both unethical and illegal. Aside from spending a lot of time outside to discourage hawks from hanging around your yard, not much can be done about hawk predation.

How do House Sparrows and European Starlings affect a martin colony?

Martin colonies can also be abandoned because of interference from other cavity-nesting birds. One of the secrets to success with martins is controlling two exotic bird species that are also cavity nesters: House Sparrows and European Starlings. House Sparrows were introduced to Brooklyn, New York, in 1851, and there were two other introductions in San Francisco and Salt Lake City during the early 1870s. They are now common throughout most of the United States and Canada. This highly successful "city" bird is unrelated to other sparrows in North America. European Starlings were first introduced to New York City's Central Park in the early 1890s by a group who wanted America to have all the birds mentioned by Shakespeare. It worked. Today, over 200 million starlings live from Alaska to Mexico.

Never allow House Sparrows or European Starlings to nest in your martin colony! At established colonies, these exotic species often break martin eggs, kill nestlings, and/or fight with adults. You may not witness any of these aggressive activities, but you can assume that they do occur. Allowing House Sparrows and European Starlings to nest in martin housing at the least will significantly reduce productivity and in severe cases will drive the martin colony away from your site.

House Sparrows and European Starlings can build a nest in a day or less, and they are quick to choose a martin house. If you live in a place where House Sparrows are present, putting up a martin house without regular monitoring along with sparrow and starling removal is worse than no martin house at all. It just provides a place for these aggressive, nonnative birds to flourish and outcompete the martins.

How can I control exotic cavity-nesting birds in my martin colony?

Because House Sparrows and European Starlings are nonnative, it is okay to remove them at any time. House Sparrows are most common, and if you have a martin house, you will probably have to deal with them. Although less common, European Starlings are larger and more aggressive. When present, starlings are a serious threat to the sustainability of a martin colony and can quickly take over martin nests and displace the entire colony.

To stop House Sparrows from using your house, check and remove their nests weekly. By weekly monitoring, you might have to remove a nest or a nest with eggs but will not have to remove their young. If you see European Starlings in the vicinity of your martin house, do whatever you can to scare them away. You may have to modify your martin

Monitor cleaning martin house

House Sparrow
(Photo by Lora L. Render)

European Starling
(Photo by Lora L. Render)

house entrance holes to exclude the starlings. When measured precisely, starling-resistant entrance holes can be effective. A height of 1³⁄₁₆ inches for the entrance opening is critical. It will exclude most starlings (and screech-owls). If made a hair too big, starlings can get in; if made a hair too small, martins will not be able to enter.

How are martins affected by the weather?

Since martins feed solely on flying insects, they are extremely vulnerable to weather that impairs or kills insects. Prolonged bad weather—rain, snow, cool temperatures, and/or heavy winds—reduces or eliminates insect flight. If bad weather persists for more than two or three days, martins begin to die of starvation. Heat waves and droughts can also be a problem. When air temperatures go above 100°F for many days, nestlings may perish from overheating. Prolonged drought can also adversely affect insect numbers and produce parasite problems.

Problem 8: Invasive Native Urban Birds That Have Become Pests

Great-tailed Grackle, Blue Jay, and White-winged Dove have become three of our most unpopular common urban birds. They are native species that thrive in the habitat changes made by humans in urban and suburban developments. In some areas, their numbers are so high that they have become pests.

GREAT-TAILED GRACKLE

Today, the Great-tailed Grackle is widespread in most of Texas and considered a pest in many places. It has rapidly expanded its range since the 1940s when it was found only on the Gulf Coast. In the Hill Country, these birds may move south during especially cold winters. Great-tailed grackles are highly successful urban birds. They are omnivorous and flourish by scavenging food from fast-food restaurants and gas stations and by drinking from sprinkler system water or the ubiquitous air-conditioning condensate released near urban buildings.

What You Can Do: Discourage Grackles

Grackles often gather to roost in large numbers at commercial properties that have tall signs, utility lines, and trees. There appears to be a tipping point at which open parking spaces with widely scattered trees become ideal habitat for roosting flocks. Limiting big-box development is usually not practical. An alternative is landscaping urban development with blocks of small trees and shrubs under larger trees to create thick understory that is less desirable to birds that roost in large colonies. If roosting by Great-tailed Grackles is not caught early enough, scare techniques such as noisemakers can be useful. If grackles have become a nuisance at your backyard bird feeders, you might try some of these methods to discourage them. However, remember that nothing is guaranteed.

1. Avoid feeding on platforms and ground preferred by grackles.
2. Put black-oil sunflower seeds in a feeder without perches, one with short perches, or a caged feeder that excludes large birds.
3. Regularly clean up seeds spilled onto the ground where grackles feed.
4. Add a thistle feeder for finches and other small birds.

Great-tailed Grackle
(Photo by Lora L. Render)

Urban commercial property with roost trees and no understory

5. Use a suet feeder for birds that feed upside down.
6. Try feeding safflower seeds, which are unattractive to most grackles.

BLUE JAY

Blue Jays are intelligent and known for their confident, boisterous behavior and their ability to imitate. If Blue Jays live in their territory, they can do an impressive imitation of a crow, Red-shouldered Hawk, Cooper's Hawk, or Sharp-shinned Hawk. As omnivores, Blue Jays are well adapted to living in suburban areas and small towns. Thus, Blue Jays are expanding their range in association with the rapid growth of suburban areas throughout the Hill Country.

Blue Jay (Photo by Lora L. Render)

Suburban yard with trees

Blue Jays can be aggressive and usually are dominant birds at a feeder. They also prey on other birds' eggs and young and may be associated with the decrease of some bird populations. If you do not see Blue Jays regularly in your yard or neighborhood, this probably means that the natural habitat is not overly disturbed and, for the most part, has a normal mix of plants and birds. This is good because Blue Jays tend to disturb an area when they move in. They often outcompete other less aggressive species, and you may see a decline in some local birds.

What You Can Do: Discourage Blue Jays

If you do have Blue Jays and do not live in a town or urban area, you may be able to reduce their numbers by restoring the natural plant community. In general, go for less lawn and a mixture of plants naturally common in your area. Problem birds often are attracted to land impacted by domestic animals. Where domestic animals are present, do your best to limit the area they occupy and keep them as far away as possible from the rest of your more natural areas.

WHITE-WINGED DOVE

White-winged Doves were once limited to the Rio Grande Valley but are now moving north. Since the early 1990s they have expanded their population into the Texas Hill Country and beyond. Our towns and suburban settings are particularly good habitat because they have plenty of trees needed for nesting and a seemingly endless supply of seeds at backyard feeders. White-winged Doves become a nuisance at backyard feeders when large flocks eat up all the birdseed, leaving none for the smaller and less abundant birds.

What You Can Do: Discourage White-winged Doves

When a large flock becomes a nuisance in your yard, it is best to stop feeding for a few weeks and force them to move elsewhere. You might also try the following:

1. Avoid feeding on platforms and the ground.
2. Put black-oil sunflower seeds in a feeder without perches, one with short perches, or a caged feeder that excludes large birds.
3. Regularly clean up seeds spilled onto the ground where they feed.

Problem 9: Birds Flying into Windows

Window strikes are a serious threat to birds. No one knows precisely how many birds are killed by window collisions, but researchers estimate it to be 100 million to 1 billion per year in North America. That is a lot of birds and includes many that hit the windows of our homes and outbuildings. They may die immediately after a severe impact or be stunned and incapacitated. Of those that do fly away from a window strike, about 50% die of internal injuries.

Almost everyone has either seen or heard a bird fly into a window. It is very common, and we may think of it as a suicide, but a bird does not think or see like a human. The bird that strikes a window does not see the window and intends to move safely through its habitat—not kill itself. For birds, the reflective and transparent character of window glass is dangerous because they cannot see the glass. To them the window looks like sky, clouds, or trees, and it is not perceived as part of the building that we see clearly.

White-winged Dove
(Photo by Tom Rust)

Large feeder with many White-winged Doves

What You Can Do: New Kinds of Windows and Window Treatments

Ideally, window glass must be visible to the birds but invisible to humans to prevent birds hitting windows. Researchers and architects are working on this "miracle glass." One type, now available in North America, has a pattern painted on it in a substance that reflects UV light that birds can see but humans cannot (see http://www.ornilux.com/).

Until our windows are made of "miracle glass," there are some ways to reduce bird collisions. First, pay attention to which windows the birds are hitting. Perhaps, just one or two need treatment to deter most bird strikes. Also, check the location of your feeder or birdbath. Might the birds be flying into the window as they leave them? If so, moving them to within 2–3 feet of the window will help. If a bird hits a window from so close, they are flying more slowly and have a much better chance of surviving.

Here are a few window treatments that may be helpful:

1. On the outside of the window, place strips of tape or a collection of feathers, bark, or other lightweight objects within at least 10 inches of each other. They will give the birds something they can see and avoid. If the problem window is large, this may require many objects, and it will limit your view.
2. On the outside of windows, place screens that reduce reflection and transparency. The birds that do hit the window will be cushioned on collision.
3. The Bird Crash Preventer is a product that has monofilament strings stretched vertically on the outside of a window. Most birds will see the strings and avoid the window. Since the strings are set a few inches from the window, those that strike them can bounce off and have a better chance of survival.
4. Window decals are not very effective unless they are mounted very close together. The decals must be placed on the outside of the window and removed to wash the windows.

Preventing bird strikes is an active area of research, and many articles are available online. For a useful place to start learning about this field, visit http://www.birdwatchingdaily.com/featured-stories/15-products-that-prevent-windows-strikes/.

What should I do when a bird is injured by flying into a window?
The bird may be temporarily incapacitated. In cold or hot weather or if there are any potential predators nearby, place it in a dark box until the bird becomes alert and then release it. Handle the bird as little as possible, and do not offer food or water. If you want the assistance of a permitted wildlife rehabilitator, names are available at http://tpwd.texas.gov/huntwild/wild/rehab/.

SUMMARY OF HABITAT PROBLEMS IN BACKYARDS

HABITAT PROBLEMS	STEWARDSHIP SOLUTIONS	AFFECTED BIRDS
Problem 1: Lack of native plant species diversity	Landscape with native plants Do not feed the deer	Bewick's Wren Blue Grosbeak Chipping Sparrow Finches Lincoln's Sparrow Painted Bunting Ruby-crowned Kinglet Warblers Woodpeckers
Problem 2: Little or no understory or ground cover	Think mosaic Use deer exclosures Practice conservation gardening Add brush piles	Black-crested Titmouse Blue Grosbeak Carolina Chickadee Carolina Wren Spotted Towhee White-eyed Vireo White-crowned Sparrow
Problem 3: Pitfalls to feeding birds	Feed everything black-oil sunflower seed Feed finches thistle seed Make your own suet Feed amount eaten in one day Place feeders in safe place	American Goldfinch Bewick's Wren Black-crested Titmouse Carolina Chickadee Carolina Wren Golden-fronted Woodpecker Lesser Goldfinch Pine Siskin
Problem 4: No reliable, safe place for birds to drink and bathe	Maintain constant water supply Use shallow basin and dripping water Keep water clean Protect water area from predators	Bewick's Wren Eastern Bluebird Orange-crowned Warbler Painted Bunting Western Scrub-Jay
Problem 5: Spread of disease due to high concentration of birds at feeders	Space feeders Remove waste Use safe feeders Keep feeders clean Use fresh food and avoid contamination	All birds eating at or below feeders
Problem 6: Effects of pesticides and herbicides	Welcome bugs and spiders Stop or minimize chemical use in yard Even limit organic pesticides	Barn Swallows Blue Grosbeak Chimney Swift Purple Martin Vermillion Flycatcher

Problem 7: Inadequate nest sites for cavity-nesting birds	Keep mature snags and large, dead branches Add, maintain, and monitor nest boxes Use Gilwood nest boxes for most species Install on metal posts with predator guards Use special boxes for larger cavity nesters Limit size of martin colony	Bewick's Wren Black-crested Titmouse Carolina Chickadee Carolina Wren Eastern Bluebird Eastern Screech-Owl House Finch Purple Martin
Problem 8: Exotic urban birds that have become pests	Remove nest material weekly Use boxes with holes too small for starlings Remove nest box if you cannot get rid of starlings	European Starling House Sparrow
Problem 8: Invasive native urban birds that have become pests	Restore yard to native plants Keep domestic animals in limited area that is cleaned regularly Do not use feeding platforms preferred by larger birds Put black-oil sunflower seeds in a feeder without perches Limit feeding to thistle Try safflower seeds Stop feeding for a month Use only small feeders and small amount of seed Clean area beneath feeder	Blue Jay Great-tailed Grackle White-winged Dove
Problem 9: Birds flying into windows	Cover windows with screen Hang moving obstacles outside window Mount the Bird Crash Preventer on windows Place decals close together on outside of windows	Migrant, breeding, and inexperienced fledgling birds most vulnerable

Bird Summaries: Featured Birds

AMERICAN GOLDFINCH
FAVORITE BIRD
WINTER

Lively bird with distinctive undulating flight; winters in flocks; common backyard visitor that eats seeds on and under bird feeders.

Identification: About 4.75 inches head to tail; thick beak; forked tail; in winter males and females have nondescript, olive-green plumage, dull black wings, and two light wing bars; in early spring males turn bright yellow with shiny black patch on head and wings.

Habitats: Old field, river valley prairie, pocket prairie, post oak savannah, shin oak savannah, live oak savannah, river, creek, springs, seep, and backyard.

Cover: Foliage of trees and shrubs near openings, weeds and grasses, and evergreen trees and shrubs.

Breeding Territory: N/A

Nest: N/A

Food: Seeds of annuals, perennials, and some trees, especially those of annual and perennial composites such as sunflowers or Mexican hat; collects seeds by clinging to upright heads of dried wildflowers.

Stewardship: Maintain evergreen cover and a diversity of seed-producing plants through the winter.
- Keep enough dense structure and branches all the way to the ground to maintain winter cover.
- Keep cats indoors, and control stray and feral cats to reduce predation.
- Use traditional feeders with black-oil sunflower or thistle feeders with thistle to provide supplemental food.
- Use fresh seed, and keep feeders clean to prevent disease.

American Goldfinch

- Maintain bird-friendly, natural or artificial water source to supply reliable and safe water.
- Avoid broad-spectrum herbicides to increase a diversity of forbs.
- Plant a mixture of different seeds to increase forbs.

Other Potentially Useful Management:
- Use rotational grazing at moderate levels to encourage a balance of grasses and forbs.
- Use prescribed burns to increase seed-producing forbs.
- Mow periodically in fall or winter dormant season on no more than 30% of herbaceous area to promote forbs.
- Use limited cultivation such as occasional light disking to increase forbs.
- Control white-tailed deer and other browsing animals to improve forb diversity.

AMERICAN ROBIN

FAVORITE BIRD

ALL YEAR

Considered a generalist species that can thrive in many parts of North America; cheerful, well-known songbird commonly seen in the Hill Country during wet winter months in large flocks feeding on short-grass habitat such as found on golf courses and old fields; most of the Hill Country too dry for nesting but does breed successfully in a few moist Hill Country habitats and irrigated suburban/urban areas.

Identification: About 9.75 inches head to tail; dark head, long tail, gray-brown back, and orange breast.

Habitats: Old field, pocket prairie, mixed wooded slope, post oak savannah, shin oak savannah, live oak savannah, river, creek, canyon, springs, seep, and backyard.

Cover: Understory of low shrubs and vines for nesting and protection from predators and bad weather; near water for frequent drinking and bathing; short-grass prairies, old fields, moist wooded slopes, savannahs, backyards, parks, and golf courses.

Breeding Territory: Usually less than 0.5 acre and territories can overlap. Male's territorial aggression strongest in vicinity of nest.

Nest: Built by female from inside out with grass, twigs, also feathers, rootlets, moss; often placed under dense leaves.

Food: Mostly soft-bodied invertebrates (bugs, spiders, and worms), especially in spring; fruit, including honeysuckle, sumac, and cedar berries, in fall and winter (90%+), often collected on the ground.

American Robin

Stewardship: Maintain diverse woody understory with a variety of native fruit-producing plants, including a diversity of short grasses and bare ground.

- Avoid use of insecticides to protect invertebrate food.
- Control red imported fire ants with ant-specific insecticides to increase insect diversity.
- Protect or plant berry-producing trees, shrubs, and vines to provide winter food.
- Provide dependable clean water through both natural and artificial sources to provide water and a safe place to bathe.
- Keep cats indoors, and control stray and feral cat populations to reduce predation.
- Control white-tailed deer and other browsing animals to improve forb diversity and maintain a healthy shrub community.
- Control regrowth cedar using appropriate mechanical or chemical methods to maintain shrubs and woody plant diversity.

Other Potentially Useful Management:
- Use prescribed burns in savannahs to increase forbs and maintain diverse woody plant communities.
- Use moderate to heavy rotational grazing in savannahs and grasslands to encourage balance of forbs, short grasses, bare ground, and woody plants.

BLACK-CHINNED HUMMINGBIRD

PRIORITY BIRD

SPRING, SUMMER, AND FALL

Only hummingbird species to breed in the Texas Hill Country; a few individuals stay here in winter; male repeats a stunning 66–100-foot U-shaped courtship/breeding territory dive with unique whirring or buzzing sound.

Identification: About 3.5 inches head to tail; slender body; long slender bill; green back, head, and sides with gray below; male has black throat with a purple band seen only in bright light.

Habitats: Old field, pocket prairie, mixed wooded slope, post oak savannah, live oak savannah, river, creek, canyon, springs, seep, and backyard.

Cover: Understory to midlevel woody vegetation.

Breeding Territory: About 1.4–7.1 acres/pair; defends about 16–50-foot radius around nest.

Nest: Well-camouflaged, tiny cup made of lichen, spider webbing, and plant down that stretches as nestlings grow; often placed near end of slender downward-bending branch usually 6.5–13 feet above the ground.

Food: Flower nectar, tiny insects, and spiders; sugar water at feeders.

Stewardship: Encourage nectar-producing native plants, insect diversity, and a healthy understory and overstory.
- Plant shrubs and perennials to supply nectar preferred by hummingbirds.
- Avoid broad-spectrum insecticides to preserve essential insect food.
- Do not dye the sugar water in hummingbird feeders, keep them free of mold, and immediately replace cloudy liquid to provide a safe alternative to nectar.
- It is acceptable but not necessary to hang hummingbird feeders throughout the year to provide food for wintering birds.
- Maintain very shallow water to supply reliable and safe water for drinking and bathing.

Black-chinned Hummingbird

- Control white-tailed deer and other browsing animals to improve forb diversity and maintain a healthy shrub community.

Other Potentially Useful Management:
- Avoid clearing riparian areas and moist canyons to protect native breeding habitat.
- Maintain springs or other very shallow water to provide an appropriate place for drinking and bathing.
- Mow periodically in fall or winter dormant season on no more than 30% of herbaceous area to promote forbs.
- Use light disking to increase forb production.
- Use prescribed burns in savannahs to increase forbs and maintain diverse woody plant communities.
- Use light to moderate rotational grazing to encourage balance of forbs and grasses and woody plant diversity.
- Manage for nesting Cooper's Hawks to increase potential nesting success.

BLACK-CRESTED TITMOUSE

PRIORITY BIRD

ALL YEAR

Locally common with small range limited to Central Texas and northeastern Mexico. Known by its loud, demanding voice, neat appearance, and active nature. Lives in almost all Hill Country woodland habitat types as well as in wooded yards. In winter, regularly moves in small mixed feeding flocks. Breeding bird surveys available from 1967 through 2005 show a disturbing population decline in the Texas Hill Country.

Identification: About 6.5 inches head to tail; male and female similar with big dark eyes; distinguishing black crest; gray above, light below with peach-colored sides; immature has no black crest and less color on sides.

Habitats: Mixed wooded slope, post oak savannah, shin oak savannah, live oak savannah, river, creek, canyon, springs, seep, and backyard.

Cover: Middle to lower canopy.

Nest: Made of leaves, moss, bark strips, and dry grass; lined with fine hair or fur; placed in cavity 3–23 feet above the ground; frequently uses bird boxes.

Breeding Territory: Estimated at 10.5 acres/pair.

Food: Insects and seeds; usually feeds in small trees and shrubs and occasionally on the ground; uses feet to hold food when pecking at it on a branch.

Black-crested Titmouse

Stewardship: Promote a diverse understory and overstory and seed-producing perennials with an abundance of insects.
- Avoid herbicides that kill broadleaf plants to protect seed-producing forbs.
- Avoid broad-spectrum insecticides to protect insects.
- Use traditional feeders with black-oil sunflower to provide supplemental food.
- Use fresh seed, and keep feeders clean to prevent disease.
- Add bird boxes where few or no snags are present to provide auxiliary nest sites and aid successful reproduction.
- Keep snags, including trees killed by drought or oak wilt, for nest sites and important feeding sites.
- Keep cats indoors, and control stray and feral cat populations to reduce predation.
- Control browsing animals (white-tailed deer and exotics) to encourage woody plant recruitment and healthy shrubs.

Other Potentially Useful Management:
- Maintain diverse natural understory to provide essential insects, seeds, and cover.
- Control regrowth cedar using appropriate mechanical or chemical methods to maintain diverse woodlands.
- Use prescribed burns in savannah to increase forbs, create snags, and maintain woody plant communities.
- Use light to moderate rotational grazing to encourage balance of forbs and grasses and woody plant diversity.

BEWICK'S WREN

PRIORITY BIRD

ALL YEAR

Active grayish wren with a variable, loud, and cheerful song that a male learns from neighbor males before its first winter and uses to defend its breeding territory. Texas Hill Country species in need of special attention to prevent its demise as happened in the Midwest and eastern states after 1975.

Identification: About 5 inches head to tail; male and female look the same; clear white eye line; slender; gray-brown back, light gray below; very long tail often held tipped up at an angle.

Habitats: Pocket prairie, mixed wooded slope, post oak savannah, shin oak savannah, live oak savannah, river, creek, canyon, springs, seep, and backyard.

Cover: Open understory and shrubby thickets in open areas.

Breeding Territory: Averages 0.8–9.4 acres/pair.

Nest: Small cup made of sticks and lined with soft material; in cavity or on ledge; usually 2.0–10.7 feet above the ground; readily uses nest boxes.

Food: Mostly insects and spiders gleaned from low-growing twigs and foliage; in winter occasionally seeds, fruit, and other plant material.

Stewardship: Maintain a diverse native plant community with dense understory and open spaces with sufficient snags for nesting.
- Retain snags, including drought-killed trees with cavities, to enhance nest sites and provide feeding sites.
- Provide and maintain nest boxes where snags are in short supply to enhance reproductive success.
- Keep cats indoors, and remove feral and stray cats to reduce predation.
- Avoid broad-spectrum insecticides and herbicides to protect insect food.

Bewick's Wren

- Control red imported fire ants with ant-specific insecticides to increase insect diversity.
- Control white-tailed deer and other browsing animals to improve forb diversity, encourage woody plant recruitment, and maintain a healthy shrub community.

Other Potentially Useful Management:
- Keep overgrown fence lines and brushy spaces to preserve natural habitat.
- Build tepee-style brush piles to help offset lack of shrub-level habitat.
- Use prescribed burns in savannahs to increase forbs, create snags, and maintain shrubby plant communities.
- Control spread of regrowth cedar using appropriate mechanical or chemical methods to maintain savannah and shrubby habitat.
- Use light to moderate rotational grazing to encourage balance of forbs and grasses and woody plant diversity.
- Mow periodically in fall or winter dormant season on no more than 30% of herbaceous area to promote forbs.

BLUE GROSBEAK
UNDER-THE-RADAR BIRD
SPRING AND SUMMER

Beautiful dark blue songbird found at edges of woods; male sings from treetops to defend its territory; often raises two broods per season; commonly parasitized by Brown-headed Cowbirds.

Identification: About 6 inches head to tail; male deep blue with rust wing bars; female light brown with darker wing bars and pale blue on rump; immature similar to female; large bill and rounded tail.

Habitats: Old field, river valley prairie, pocket prairie, post oak savannah, shin oak savannah, live oak savannah, river, creek, springs, seep, and backyard.

Cover: Some trees and shrubs and open ground cover with a forb and grass mixture.

Breeding Territory: About 15 acres/pair.

Nest: Tight cup woven with plant strips, twigs, and bark; in low tree or shrub in or near open area; nest height from 6 inches to over 25 feet above the ground.

Food: Insects, especially grasshoppers, snails, and grass seeds.

Blue Grosbeak

Stewardship: Maintain open savannah and recently disturbed woodland habitat with adjacent grasses.
- Control red imported fire ants with ant-specific insecticides to increase insect diversity.
- Avoid broad-spectrum herbicides and insecticides to increase forbs and promote insect diversity.
- Keep cats indoors, and control stray and feral cat populations to reduce predation.
- Control white-tailed deer and other browsing animals to improve forb diversity and maintain a healthy shrub community.
- Control spread of regrowth cedar using appropriate mechanical or chemical methods to maintain open savannah with shrubby habitat.

Other Potentially Useful Management:
- Trap and dispatch cowbirds to prevent nest parasitism and enhance reproductive success.
- Use prescribed burns to maintain grasses, increase forbs, and maintain shrubby plant communities.
- Use light to moderate rotational grazing to encourage balance of forbs and grasses and woody plant diversity.
- Use periodic mowing in fall or winter dormant season on no more than 30% of herbaceous area to promote forbs for insects production.

BLUE JAY

ALL YEAR

Expansion of their range west corresponds to urban and suburban sprawl. Noisy with many vocalizations, intelligent, close family bonds; nestlings may wander up to 15 feet from nest a few days before they can fly but often not fed until they return to the nest; known to cache, or store, food for later use.

Identification: About 11 inches head to tail; bright blue with white and black markings; crest used in aggressive behavior.
Habitats: Mixed wooded slope, post oak savannah, shin oak savannah, live oak savannah, river, creek, springs, seep, and backyard.
Cover: Woodland edges and large scattered trees with understory and midcanopy of small trees and shrubs.
Breeding Territory: Does not hold a traditional territory but is aggressive toward other Blue Jays during breeding season.
Nest: Usually large, open cup of twigs, grass, sometimes mud and lined with rootlets; 10–25 feet above the ground in fork of tree branch.
Food: Mostly insects, acorns, other nuts, fruits, and grains; occasionally bird eggs, dead or injured birds, and other small creatures; can store up to 5,000 food items in caches.

Stewardship: Maintain large mast-producing trees such as live oaks and diverse woodland edges with well-defined understory.

Blue Jay

- Avoid broad-spectrum pesticides, and control red imported fire ants with an ant-specific pesticide to promote insect diversity and maintain food supply.
- Maintain bird-friendly, natural or artificial water source to supply reliable and safe water.
- Control white-tailed deer and other browsing animals to encourage woody plant recruitment and diversity.

Other Potentially Useful Management:
- Use prescribed burns in savannahs to maintain diverse woody plant communities and encourage forbs.
- Control spread of regrowth cedar using appropriate mechanical or chemical methods to maintain diverse and relatively open woodlands, including some cedar.
- Use light to moderate rotational grazing to encourage balance of forbs and woody plant diversity.

CHIMNEY SWIFT

PRIORITY BIRD

SUMMER

Small, dark, gregarious birds whose acrobatic flight and charming twitter are welcome summer entertainment. Do not perch but cling to vertical surface with their clawed feet.

Identification: About 5.5 inches head to tail; gray all over; cigar-shaped body; almost always seen flying with quick, tight wing beats.
Habitats: River valley prairie, mixed wooded slope, post oak savannah, shin oak savannah, live oak savannah, river, creek, tank, pond, lake, and backyard.
Cover: Roost in hollow trees, caves, or dark shafts with surface irregularities such as chimneys and smokestacks made of clay tile or masonry where they can cling to sides.
Breeding Territory: Usually only one pair per suitable nesting site (chimney or hollow tree).
Nest: Shallow shelf made of small twigs stuck together and to the wall with saliva; only one nest per chimney, but others also roost there.
Food: Flying insects; can consume more than 1,000 insects per day.

Stewardship: Encourage terrestrial and aquatic insect diversity, and maintain large snags and chimneys for nesting.
- Use interior masonry chimney construction to provide safe nest sites.
- Remove chimney covers that exclude chimney swifts to make suitable nest sites available.
- Add Chimney Swift tower or extensions on existing wood chimney to expand nest sites.

Chimney Swift

- Preserve hollow trees or snags, including drought-killed trees, to provide natural nest sites.
- Maintain leaf litter, especially in shady places, to protect sources of essential insect food.
- Avoid broad-spectrum insecticides and herbicides to protect flying insect food.
- Control red imported fire ants with ant-specific insecticides to increase insect diversity.

Other Potentially Useful Management:
- Provide aquatic insect habitat in streams, ponds, and lakes to increase food supply.
- Use prescribed burns in savannahs to increase forbs, create snags, and maintain woody plant communities.
- Control regrowth cedar using appropriate mechanical or chemical methods to maintain healthy diverse habitats.

CHIPPING SPARROW

PRIORITY BIRD

ALL YEAR

Active and distinctive sparrow, most common in winter throughout open, wooded areas of the Hill Country; often in flocks; regular visitors to backyards, where it forages on the ground under bird feeders. Breeds primarily in western portion of Hill Country and considered rare during the breeding season; Texas Hill Country breeding bird surveys report an alarming decline in this species.

Identification: About 5 inches head to tail; bright rusty cap, black eye line, and gray breast.
Habitats: Mixed wooded slope, post oak savannah, shin oak savannah, live oak savannah, river, creek, canyon, springs, seep, tank, pond, lake, and backyard.
Cover: Open woodlands with shrubby understory and grassy areas.
Breeding Territory: About 0.5–2.5 acres/pair.
Nest: Well-hidden, delicate cup lined with fine hair or plant material, near end of branch, 3–10 feet above the ground; sometimes parasitized by Brown-headed Cowbird.
Food: Mostly seeds of grass and forbs; feeding often close to or at ground level; also insects and sometimes fruit.

Stewardship: Encourage a diverse understory, including evergreen shrubs and healthy herbaceous area with native grasses and seed-producing annuals and perennials.
- Keep cats indoors, and remove feral and stray cats to reduce predation.
- Set up a clean water source protected from predators to provide a safe place to drink and bathe.
- Construct tepee-style brush piles to increase shrub-level habitat.
- Control spread of regrowth cedar using appropriate mechanical or chemical methods to maintain savannahs and woodlands.
- Control white-tailed deer and other browsing animals to improve forb diversity and maintain a healthy shrub community.

Chipping Sparrow

Other Potentially Useful Management:
- Trap and dispatch cowbirds to prevent parasitism and enhance reproduction.
- Use rotational grazing at moderate levels to encourage a balance of grasses, forbs, and bare ground.
- Use prescribed burns in savannahs to increase forbs and maintain diverse woody plant communities.
- Mow periodically in fall or winter dormant season on no more than 30% of herbaceous area to promote forbs.
- Control feral hogs to protect food, cover, and nest sites in natural habitat.

EASTERN SCREECH-OWL
UNDER-THE-RADAR BIRD
ALL YEAR

Common in large trees but camouflaged and almost invisible because its small size, color, and mottled appearance blend in with the bark. Active at night; usually hunts at dawn and dusk but occasionally in daylight. Male hunts food fed to brooding mate and nestlings.

Identification: About 8 inches head to tail; male and female similar but male smaller; usually mottled gray but occasionally red; ear tufts may be held up or down.

Habitats: Mixed wooded slope, post oak savannah, shin oak savannah, live oak savannah, river, creek, canyon, and backyard.

Cover: Open woodlands with large trees and limited understory.

Breeding Territory: About 15–75 acres/male; larger territories in rural areas.

Nest: Does not build a nest; uses existing cavity often made by woodpecker in tree or snag and in human-made box, usually 13–20 feet above the ground.

Food: Insects and any small animal, including rats, birds, mice, and squirrels; also crayfish, earthworms, lizards, and frogs.

Stewardship: Encourage healthy mature woodlands with well-developed overstory, open understory, and abundant insects and small mammals, with sufficient snags for nesting.

- Retain snags, including drought-killed trees, to provide nest cavities and important feeding sites.
- Install nest boxes where natural cavities are limited to add nest sites and increase reproduction.
- Control red imported fire ants with an ant-specific pesticide to promote insect diversity and maintain food supply.
- Avoid broad-spectrum herbicides to increase forbs and promote insect diversity.

Eastern Screech-Owl

- Avoid use of d-Con or any other anticoagulant rat/mouse poison to prevent poisoning small mammals and killing owls.
- Control white-tailed deer and other browsing animals to encourage deciduous woody plant recruitment and diversity.

Other Potentially Useful Management:

- Construct low brush piles to increase shrub-level habitat for prey.
- Control regrowth cedar using appropriate mechanical or chemical methods to maintain savannahs and open woodlands.
- Use prescribed burns in savannahs to increase forbs, create snags, and maintain woody plant communities.

EUROPEAN STARLING

OUTLAW BIRD

ALL YEAR

A marvel of survival; intelligent, aggressive, and gregarious; mobs aerial predators in flight; an excellent mimic; gathers in large flocks during winter. Unlike native birds, this exotic bird is not protected by the Migratory Bird Treaty Act of 1918.

Identification: About 8.5 inches head to tail; male and female similar; short wings and tail; breeding adult body black with iridescent spots and long yellow bill; winter adults have black bill and obvious pale spots; juvenile is brown.

Habitats: Old field, post oak savannah, shin oak savannah, live oak savannah, river, creek, springs, seep, and backyard.

Cover: Open, grassy ground cover; trees and shrubs; human-made buildings; usually in urban, suburban, or agricultural habitat altered by humans.

Breeding Territory: Colonial; defends 10–20-inch radius around its nest.

Nest: Mess of grass, trash, string, feathers, and so forth with a depression lined with fine material at back of nest box or other cavity; usually 10–25 feet above the ground and sometimes much higher.

Food: Walks or runs on ground and eats nearly anything, including insects, snails, spiders, fruit, berries, seeds, grain, and garbage.

European Starling

Control: Discourage nesting, and eliminate its favorite food.
- Use specific dimensions and location when installing next boxes for native birds.
- Monitor nest boxes weekly during breeding season to limit use by European Starlings.
- Use martin houses and gourds with entrance opening that excludes European Starlings.
- Monitor martin colony chambers weekly, and remove all European Starling eggs and nest material.
- Use bird food (i.e., thistle, suet, and black-oil sunflower seeds) and feeders that target desired species and are less desirable to European Starlings.
- Remove bird feeders where European Starlings feed regularly.
- Trap and dispatch European Starlings in agricultural sites with large populations.

GOLDEN-FRONTED WOODPECKER
PRIORITY BIRD
ALL YEAR

Most common and popular woodpecker throughout the Hill Country; not shy at all and a frequent visitor to backyard suet feeders; in the United States found only in Texas and Oklahoma.

Identification: Almost 10 inches head to tail; spot of yellow above bill; golden-orange nape; black and white barred back; white rump.
Habitats: Mixed wooded slope, post oak savannah, shin oak savannah, live oak savannah, river, creek, canyon, and backyard.
Cover: Lives on big trees.
Breeding Territory: Averages 43 acres/pair.
Nest: Tree cavity 3–25 feet above the ground.
Food: Gleans insects from trunks and large branches; also catches flying insects and eats nuts and fruit, including pricklypear cactus.

Stewardship: Maintain mixed woodlands and savannahs of mature trees (including oaks) and snags, including open areas with some bare ground.
- Keep snags, including trees killed by drought or oak wilt, to supply nest sites and important feeding sites.
- Use suet feeders to provide supplemental food.
- Avoid broad-spectrum insecticides to protect insects.
- Maintain bird-friendly, natural or artificial water source to supply reliable and safe water.
- Control regrowth cedar using appropriate mechanical or chemical methods to maintain diverse woodlands.
- Control white-tailed deer and other browsing animals to encourage woody plant recruitment and healthy shrubs.

Golden-fronted Woodpecker

Other Potentially Useful Management:
- Use prescribed burns in savannah to increase forbs, create snags, and maintain woody plant communities.
- Use light to moderate rotational grazing to encourage balance of forbs and grasses and woody plant diversity.

GREAT-TAILED GRACKLE

ALL YEAR

Bold, raucous, and gregarious; birds roost in huge flocks; fly off to feeding sites each morning and return to roost at dusk; most common in human-modified habitat.

Identification: About 15–18 inches head to tail; male shiny black with long flared tail; female smaller, dark brown, and with shorter tail; both have yellow eyes.

Habitats: Old field, plateau prairie, river valley prairie, pocket prairie, mixed wooded slope, post oak savannah, shin oak savannah, live oak savannah, river, creek, canyon, springs, seep, tank, pond, lake, and backyard.

Cover: Areas with dispersed large and small trees, human landscapes, usually near water.

Breeding Territory: Colonial or nearly so; nearly 100% in human-altered locations.

Nest: Crude cup of grass, forbs, other plants, and often with human-made materials; usually lined with mud and soft inner layer; up to 60 feet off the ground.

Food: Eat just about anything; seeds, fruit, invertebrates, and small vertebrates.

Control: Open prairies or dense woodlands with limited water.

- Encourage dense understory to reduce nesting and roosting habitat.

Great-tailed Grackle

- Reduce regular watering with irrigation systems on timers or any unnecessary water source to limit consistent water.
- Use food such as suet and tube feeders that exclude grackles.

HOUSE FINCH

UNDER-THE-RADAR BIRD

ALL YEAR

This is not the invasive European House Sparrow! Common, widespread, popular songbird with long, cheerful song sung by male to defend territory. Females prefer males with brightest red color. Parents often bring their young to bird feeders when fledged and then switch to wild food if it is available. All are subject to mycoplasmosis.

Identification: About 5.75 inches head to tail; brownish male has rosy-red to orange-yellow cap, chest, and rump; color depends on diet with orange-yellow in dry habitat; female brown and heavily streaked.

Habitats: Old field, river valley prairie, pocket prairie, post oak savannah, shin oak savannah, live oak savannah, river, creek, canyon, springs, seep, and backyard.

Cover: Diverse understory with shrubs and trees and ground cover with mix of grass and forbs, often adjacent to woody borders.

Breeding Territory: Very small territory; about 624 square feet/pair and nests as close as 3 feet possible.

Nest: Cup made of plant strips, leaves, string, wool, and feathers and lined with soft down; 4–14 feet above the ground; placed on tree branch, cactus, rock ledge, building, hanging plant, or nest box.

Food: About 95% seeds, buds, and fruit—even fed to nestlings; prefer black-oil sunflower seeds at feeding stations.

Stewardship: Maintain mixture of woody and annual and perennial plants with healthy understory to provide cover near safe water supply.
- Avoid broad-spectrum herbicides to increase forbs.
- Plant and conserve selected trees, shrubs, vines, and forbs to provide plenty of seeds and berries.
- Use traditional feeders with black-oil sunflower or thistle feeders with thistle to provide supplemental food.
- Use fresh seed.
- Keep feeding stations clean, and supply fresh seed to prevent spread of contagious disease.
- Maintain bird-friendly, natural or artificial water source to supply reliable and safe water.
- Keep cats indoors, and control stray and feral cat populations to reduce predation.
- Control regrowth cedar using appropriate mechanical or chemical methods to maintain shrubs and woody plant diversity.
- Control white-tailed deer and other browsing animals to encourage woody plant recruitment, healthy understory, and forb diversity.

Other Potentially Useful Management:
- Mow periodically in fall or winter dormant season on no more than 30% of herbaceous area to promote forbs.
- Use limited cultivation such as occasional light disking to increase forbs and bare ground.
- Use moderate rotational grazing to encourage balance of forbs and grasses and woody plant diversity.
- Use prescribed burns to control woody plant invasion, encourage forbs, and increase bare ground.

House Finch

HOUSE SPARROW

OUTLAW BIRD

ALL YEAR

Invasive and exceptionally aggressive toward other birds.

Identification: About 6.25 inches head to tail; plump body, short tail, and stout bill; male brown back, gray chest and cheeks, distinct black bib; female streaky brown with gray chest.

Habitat: Backyard.

Cover: Human habitation; urban, suburban, and rural backyards; barns where domestic animals are fed grain.

Breeding Territory: Aggressively defends area immediately around nest, known to attack 70 other species of birds, usually at their nest, including cavity nesters such as Eastern Bluebird and Purple Martin.

Nest: Messy collection of dry plant material; also plastic, feathers, and string; usually placed in or on human-made structures, including nest boxes.

Food: Mostly seeds; also insects and discarded human food.

Control: Discourage nesting and eliminate their favorite food.
- Remove sparrow nests and eggs from all nest boxes weekly and without fail to prevent reproduction.
- Remove bird boxes if House Sparrows are using them; do not allow House Sparrows to reproduce in them.
- Remove feeders used by a large number of House Sparrows to curtail food supply.
- Do not use cracked corn, millet, or bread in bird feeders to avoid attracting House Sparrows.
- Trap and remove House Sparrows in agricultural sites with large populations to rid area of this invasive, exotic species.
- Eliminate potential nest sites in eves and overhangs on houses and barns to discourage nesting.

House Sparrow

LINCOLN'S SPARROW

UNDER-THE-RADAR BIRD

WINTER

Common and nondescript sparrow in winter seen under bird feeders and throughout open wooded habitat of the Hill Country.

Identification: About 5.5 inches head to tail; overall gray appearance; broad gray eyebrow, buffy whisker or malar line; chest light tan with thin streaks; male and female similar.

Habitats: Old field, river valley prairie, pocket prairie, post oak savannah, shin oak savannah, live oak savannah, river, creek, canyon, springs, seep, and backyard.

Cover: Low understory foliage in or near a clearing, shrubs, and weedy areas.

Breeding Territory: N/A

Nest: N/A

Food: In winter, feeds on seeds and invertebrates picked off the ground; feeds by itself or with a few others.

Stewardship: Maintain healthy understory and low shrubs with mixed ground cover, including some bare ground for feeding.
- Keep cats indoors, and remove feral and stray cats to reduce predation.
- Conserve brushy areas (including shrubby fences) with mixed vegetation to protect cover and food.
- Construct tepee-style brush piles to increase shrub-level habitat.
- Control regrowth cedar using appropriate mechanical or chemical methods to maintain shrubs and woody plant diversity.

Lincoln's Sparrow

- Control white-tailed deer and other browsing mammals to encourage woody plant recruitment, healthy shrubs, and forbs.

Other Potentially Useful Management:
- Mow periodically in fall or winter dormant season on no more than 30% of herbaceous area to promote forbs.
- Use light to moderate rotational grazing to encourage balance of forbs and grasses and woody plant diversity.
- Use prescribed burns to increase forbs and maintain diverse woody plant communities.
- Remove feral hogs to protect low-level habitat for cover and food.
- Use limited cultivation such as occasional light disking to increase forbs and bare ground.

NORTHERN CARDINAL

FAVORITE BIRD

ALL YEAR

This "redbird" may be the Hill Country's most popular and well-known backyard bird; male defends his breeding territory; pairs stay together all year; males and females sing loud, clear whistled phrases in different combinations; alarm call is a metallic chip that gets louder and faster as danger approaches.

Identification: About 8.75 inches head to tail; males larger than females; thick red bill and long tail; male bright red with crest and black mask; female tan with red highlights on tail, wings, and tip of crest; immature similar to female.

Habitats: Old field, river valley prairie, pocket prairie, mixed wooded slope, post oak savannah, shin oak savannah, live oak savannah, river, creek, canyon, springs, seep, and backyard.

Cover: Dense, low vegetation of shrubs and small trees.

Breeding Territory: About 0.5–6.5 acres/pair.

Nest: Well-built cup hidden 3.0–6.5 feet above the ground in dense foliage of shrub or small tree.

Food: A variety of seeds, fruit, insects, and occasional buds collected on the ground or from branches.

Northern Cardinal

Stewardship: Maintain diverse shrubby understory with a mixture of woody plants and forbs to provide seed and attract insects.
- Avoid broad-spectrum insecticides and herbicides to protect insects and natural seed sources.
- Keep cats indoors, and remove feral or stray cats to reduce predation.
- Maintain bird-friendly, natural or artificial water source to supply reliable and safe water.
- Control regrowth cedar using mechanical or chemical methods to maintain shrubs and woody plant diversity.
- Control white-tailed deer and other browsing animals to improve forb diversity and maintain a healthy shrub community.

Other Potentially Useful Management:
- Build tepee-style brush piles to help offset lack of shrub-level habitat.
- Trap and remove cowbirds to reduce nest parasitism and increase nesting success.
- Use prescribed burns in savannahs to increase forbs and maintain diverse woody plant communities.
- Use light to moderate rotational grazing to encourage balance of forbs and grasses and woody plant diversity.

NORTHERN MOCKINGBIRD
FAVORITE BIRD

ALL YEAR

Active, vocal, and talented mimic; highly territorial in winter and summer and aggressive defender of territory; more common in Texas than any other state.

Identification: About 9.25 inches head to tail; small head, long tail, and gray body; two white wing bars; flashy white spot on wings and white outer tail feathers in flight.

Habitats: Old field, river valley prairie, pocket prairie, post oak savannah, shin oak savannah, live oak savannah, river, creek, canyon, springs, seep, and backyard.

Cover: Open areas of short grass with shrubs and trees.

Breeding Territory: About 1.5–6 acres/breeding pair.

Nest: Well hidden in dense vegetation, usually 3–10 feet above the ground.

Food: Mostly fruit in fall and winter and insects in summer; also some crustaceans and small vertebrates often collected on the ground.

Stewardship: Encourage diverse trees and shrubs with open areas that include a balance of short grasses, forbs, and bare ground.
- Avoid using insecticides to protect the invertebrate food needed by nestlings.
- Keep or plant fruit-bearing vines, shrubs, and trees, including mulberries and possumhaw to provide winter forage.
- Keep woody plants with dense vegetation to supply nest sites and prominent perches for territory defense.
- Keep cats indoors, and control stray and feral cats to reduce predation.

Northern Mockingbird

- Control white-tailed deer and other browsing animals to encourage woody plant recruitment, healthy shrubs, and forbs.

Other Potentially Useful Management:
- Use limited cultivation such as occasional light disking to increase forbs and bare ground.
- Mow periodically in fall or winter dormant season on no more than 30% of herbaceous area to promote forbs.
- Use moderate to heavy short-term grazing to help maintain short grasses and bare ground and encourage forbs.
- Use prescribed burns to increase forbs and maintain woody plant communities.
- Control regrowth cedar using appropriate mechanical or chemical methods to maintain shrubs and woody plant diversity.

PURPLE MARTIN
FAVORITE BIRD
SPRING AND SUMMER

Common and well-loved, colonial bird around human habitation; needs open air space for feeding; may abandon otherwise ideal habitat when predation occurs; first arrivals in mid-February do not stay and nest. After young fledge, birds gather in large flocks. Although very common, breeding bird surveys show their numbers slowly declining since the late 1980s—especially in the upper Midwest but also in parts of East and North Texas.

Identification: About 7.75 inches head to tail; male blue-black all over, female and immature dark above and lighter below; strong, agile flyers.
Habitats: Old field, river valley prairie, post oak savannah, shin oak savannah, live oak savannah, river, creek, springs, seep, tank, pond, lake, and backyard.
Cover: High perches and open sky near colony site, usually near human-made structures.
Breeding Territory: Colonial; small to medium-sized colonies more stable than supercolonies.
Nest: Small twigs, grass, mud, and leaves built in existing chamber; hanging gourds preferred over apartment house.
Food: Flying insects caught up to 150 feet or so above the ground; gathers small pebbles for digestion in crop.

Purple Martin

Stewardship: Provide, monitor, and protect a healthy martin colony to support a stable population in the Hill Country.
- Place martin house or gourds near open sky to provide essential space for aerial feeding.
- Place martin house or gourds near human habitation to give protection from some predators.
- Maintain small to medium-sized colony to provide greatest stability.
- Control House Sparrows and European Starlings to reduce competition and depredation.
- Keep cats indoors, and control stray and feral cat populations to reduce predation.

SPOTTED TOWHEE

UNDER-THE-RADAR BIRD

WINTER

Big, beautiful, and distinctive but shy winter songbird; typically seen solitary or in pairs on or near the ground in wooded areas with plenty of leaf litter and ground-level food.

Identification: About 7.5 inches head to tail; red eyes; head, throat, chest, upper back, and tail are black in males and dark brown or gray in females; all have rufous sides and white on outer tail feathers.

Habitats: Mixed wooded slope, post oak savannah, shin oak savannah, live oak savannah, river, creek, canyon, springs, seep, and backyard.

Cover: Dense vegetation near the ground; shrub thickets, fencerows, and edges.

Breeding Territory: N/A

Nest: N/A

Food: Scratches through leaf litter like a chicken looking for seeds, insects, and spiders; also eats some berries and occasional small lizards.

Spotted Towhee

Stewardship: Maintain healthy thicket-forming understory vegetation.

- Keep or plant low understory plants that produce seeds to supply food.
- Maintain mulch and leaf litter to support insects and other invertebrates for food.
- Avoid using broad-spectrum pesticides to protect forbs and insects for food.
- Control red imported fire ants with an ant-specific insecticide to promote insect diversity and maintain food supply.
- Keep cats indoors, and control stray and feral cats to reduce predation.
- Maintain bird-friendly, natural or artificial water source to supply reliable and safe water.
- Control white-tailed deer and other browsing animals to encourage woody plant recruitment and healthy shrubs.

Other Potentially Useful Management:

- Limit grazing by livestock in riparian areas to maintain understory.
- Construct tepee-style brush piles to increase shrub-level habitat.
- Control regrowth cedar using appropriate mechanical or chemical methods to maintain diverse woodland.
- Use prescribed burns in savannahs to maintain diverse woody plant communities.

WHITE-WINGED DOVE

ALL YEAR

Long-lived dove whose call may sound like *who cooks for you*; also known to "sunbathe" in cold weather; as with other doves, young are fed crop milk.

Identification: About 11.5 inches head to tail; male and female similar; pink feet and legs; plump, tan body; small head; long tail; distinct white edge on folded wing flashy in flight.

Habitats: Old field, post oak savannah, shin oak savannah, live oak savannah, river, creek, springs, seep, and backyard.

Cover: Large and small trees, shrubs, and agricultural fields with annual forbs.

Breeding Territory: Prefers to nest near food in colonies; nests may be only 10 inches apart.

Nest: Loosely built bowl mostly of sticks and often in shrub or forked branch of small tree.

Food: Seeds, grain, berries, and seeds; mostly on the ground; also cactus fruit.

Stewardship: Maintain mature, open woodlands with nearby fields of annual forbs.
- Keep cats indoors, and control stray and feral cat populations to reduce predation.
- Maintain bird-friendly, natural or artificial water source to supply reliable and safe water.
- Control white-tailed deer and other browsing animals to encourage woody plant recruitment and a diversity of forbs.

White-winged Dove

Other Potentially Useful Management:
- Avoid feeding corn that may be contaminated with aflatoxin at deer feeders regularly visited by White-winged Doves to reduce exposure to disease.
- Use limited cultivation such as occasional light disking to increase forbs and bare ground.
- Plant annual food plots followed by yearly disturbance in late winter to provide food.
- Mow periodically in fall or winter dormant season on no more than 30% of herbaceous area to promote annual forbs.
- Use rotational grazing at moderate levels to encourage a balance of grasses and forbs.
- Use prescribed burns to maintain healthy deciduous plant community and encourage forbs.

7 Predators and Other "Dangerous" Animals

An alternative model is to propose that predation
is necessary for cycles, but not sufficient by itself.
—Charles Krebs, vertebrate ecologist

INTRODUCTION

This chapter focuses on the environmental impact of the animals most dangerous to the survival of Texas Hill Country birds and on the management practices that minimize the influence of those animals. These threats are not necessarily hawks, owls, eagles, snakes, coyotes, foxes, or even raccoons, which are native predators that usually contribute to the dynamic equilibrium of their prey populations. The animals most dangerous to Hill Country birds include one native bird (Brown-headed Cowbird), two exotic mammals (feral hog and domestic cat), an insect (red imported fire ant), and a native prey mammal (white-tailed deer). Before discussing these "most dangerous" species and how to control them, we present five ecological concepts important to understanding predation.

1. Population conservation: Wildlife stewardship supports sustainable populations through habitat management. If enough birds survive and reproduce, it is okay that some individuals are eaten or driven away by native predators. The main concern is providing adequate, healthy habitat so bird populations survive from year to year.
2. Critical areas: There is a critical area or smallest number of habitat acres needed to sustain any bird species in your area. In other words, each species has a minimum amount of suitable habitat necessary to support enough individuals to reproduce through both good and bad years and makes it sustainable. The size of this critical area depends on contributing environmental factors:
   - Predation pressure or predation severity
   - Variation in food and water supply from year to year
   - Size of breeding territory needed by the species
   - Basic habitat quality
   - Population size
3. Predator/prey relationship: Native predators are usually not the problem. The natural up-and-down cycling of prey populations

is caused by a combination of environmental factors, including food, weather, and predation. In the Hill Country, skunks, snakes, ring-tailed cats, raccoons, coyotes, gray foxes, loggerhead shrikes, hawks, American Kestrels, owls, eagles, and jays all sometimes prey on bird eggs, nestlings, fledglings, or adults, but all of these predators and their prey have important ecological roles.

Prey animals and their natural predators have survived for a very long time without human intervention. Before major environmental changes were initiated by European settlement, the predator and prey populations remained sustainable through natural controls. Today, on well-managed land, both native predator and prey populations continue to be sustainable in a similar manner.

4. Predator/prey imbalance: If you suspect substantial predation from a native species on your property, the issue is an unnatural situation in the habitat. You must find the root cause of especially severe predation so the problem can be addressed holistically through habitat stewardship. First, look at possible limiting factors in the habitat such as availability of nest sites, food, water, protective cover, or any other environmental need not being fulfilled.

5. Ecological traps: An ecological trap puts predators at an advantage, so at least a segment of their prey population is at extra risk. Wildlife habitat enhancements by a well-meaning person may result in an ecological trap. For example, migratory birds, tired, thirsty, and hungry after a long flight, are vulnerable to predation at backyard birdbaths surrounded by thick vegetation; even a well-placed bird feeder where outdoor cats are present puts the birds at risk; and a bird feeder not kept clean can increase the spread of contagious disease.

Brown-headed Cowbirds

Today, Brown-headed Cowbirds, causing ecological stress and imbalance on a landscape scale, have become a critical problem. They damage breeding-bird populations throughout the Hill Country and all across the southern Great Plains. The Brown-headed Cowbird is a nest parasite that once traveled with herds of migratory bison. It is adapted to eat insects disturbed by grazing animals and to lay its eggs in the nests of other birds. Sometimes a cowbird female lays her eggs alongside the host bird's eggs, and she often removes eggs or nestlings from another bird's

Brown-headed Cowbird

nest before laying her own eggs. The parasitized bird then incubates the cowbird eggs and rears the cowbird offspring. Thus, the reproductive success of parasitized birds is impaired.

When Brown-headed Cowbirds followed migratory bison herds, a local breeding-bird population would be parasitized only occasionally. Today, cowbirds have become sedentary because they stay around confined domestic livestock during breeding season, so they impact the same bird populations year after year. Thus, although it is a native species, the Brown-headed Cowbird creates severe ecological stress wherever it exists.

Cowbird parasitism creates ecological imbalance. There are plenty of native predators (snakes and raccoons) that eat bird eggs and nestlings and limit nesting success. But breeding-bird populations have lived with these predators for a very long time. Their long-standing predator/prey relationship contributes to a dynamic equilibrium that allows both predatory and prey bird populations to survive. Add sedentary cowbird populations to this once sustainable predator/prey equation, and the situation is pushed over a breaking point, making some of the songbird populations that are parasitized by cowbirds unsustainable. Today, wildlife biologists consider nest parasitism by the Brown-headed Cowbird an important contributor to the downward trend of many declining songbird populations of the Hill Country and many other parts of Texas.

Brown-headed Cowbirds parasitize many bird species. Nest parasites impact bird populations of all types from tiny vireos and gnatcatchers to much larger blackbirds and meadowlarks. Usually undetected, cowbirds impair reproduction of the endangered Black-capped Vireo,

Remember: The most dangerous predator for nesting birds in the Texas Hill Country is a *bird*— a parasitic bird.

Orchard Oriole

PARASITISM RATES ON HILL COUNTRY BIRDS COMMONLY AFFECTED BY COWBIRDS

HIGHLY PARASITIZED
Bell's Vireo
Black-capped Vireo
Blue Grosbeak
Dickcissel
Indigo Bunting
Lark Sparrow
Northern Cardinal
Orchard Oriole
Painted Bunting
Summer Tanager

MODERATELY PARASITIZED
Blue-gray Gnatcatcher
Eastern Meadowlark
Field Sparrow
Golden-cheeked Warbler
Grasshopper Sparrow
House Finch
Louisiana Waterthrush
Red-eyed Vireo
Red-winged Blackbird
White-eyed Vireo

Dry live oak savannah with mesquite and cactus understory

common House Finch, popular Painted Bunting, and the little-known Orchard Oriole, among many other species.

Cowbird trapping controls Brown-headed Cowbirds. Trapping cowbirds is a dramatically effective method to increase songbird reproductive success for species, including the endangered Black-capped Vireo. The required training and certification are available online at the Texas

A PERSONAL STORY: COWBIRD TRAPPING

DUSTY BRUNS

To get started, my wife and I took a certification course where we saw film of a female cowbird grabbing a baby bird and dropping it over the side of its nest. After thinking about the little baby bird, we decided we should do some cowbird trapping. Our trap stays in a pasture near our house where I can check it easily. We are thinking about mounting it on wheels so I can move it to different pastures along with the cows. I check the trap daily during the trapping season and release anything that isn't a cowbird.

Our neighbor, who is a serious birder, also traps cowbirds. She helps us identify birds that we don't know. We've never seen a nest parasitized by cowbirds, but then we haven't looked either. We are into our fifth year, and as required by Texas Parks & Wildlife, we trap from March 1 to May 31. This is how many we've caught so far:

First Year—73
Second Year—64
Third Year—210
Fourth Year—179
Fifth Year—146

We haven't noticed an increase in songbirds because of cowbird trapping. But we never surveyed the birds before we started. At least, for every female cowbird we trap, there is one less nest out there being parasitized and probably more. One spring hundreds of cowbirds showed up in one afternoon! I ran down and added some grain near the entrance on top of the trap and caught about 75 of them. There were so many wings batting around that the cows got scared and ran away! Looked like something out of Hitchcock's *The Birds*.

Parks and Wildlife website. Fort Hood (near Killeen, Texas) is a large military base with a long history of land management for birds. Continuing research on Fort Hood shows that in 1987—before cowbird trapping—over 90% of Black-capped Vireo nests were parasitized by the Brown-headed Cowbird. By 1997—after intensive cowbird trapping—the parasitism rate fell to about 10% and continues at this level. During this time, Fort Hood researchers also observed an increase in the number of Black-capped Vireo breeding pairs and nest success.

Portable, walk-in cowbird trap

Feral Hogs

In 2003, Richard Taylor, then of the Texas Parks and Wildlife Department, wrote this: "Early explorers and missionaries brought the first swine into Texas, but the feral populations originated during colonization. In the twentieth century, introductions of domestic hogs and European boars into the wild by landowners and sportsman further enhanced the population. Changing land use practices, improved animal husbandry, and eradication of diseases have enabled the feral hog to adapt and disperse throughout most of Texas. With an estimated population of 1 million animals, feral hog numbers rank second behind the white-tailed deer as a large-mammal population and are an integral part of Texas fauna." A more recent AgriLife Extension estimate of feral hogs in Texas is around 3 million.

- Where feral hogs are present, the greatest harm may be to ground-nesting birds, including Northern Bobwhite and Rio Grande Turkey.
- Across the Hill Country, feral hogs have more direct impact on ground-nesting birds than do red imported fire ants.
- Hogs are attracted to deer feeders, and the feeders concentrate hog damage to the surrounding area.
- Since feral hogs are primarily nocturnal, you may have many of them on your property but rarely see them.

Feral hog wallow habitat disturbance (Photo by Kory Perlichek)

Feral hog (Photo by Grady Allen)

- Feral hog signs are wallows in wet places and rooting disturbance that can be all over the landscape.

Hogs eat almost anything they come across. Hogs are opportunistic foragers, and their diet varies by season. Most of what a hog eats is plant leaves, roots, tubers, bulbs, fruit, and acorns. They also eat a wide variety of animals, including insects, amphibians, reptiles, and birds. Eggs of birds that nest on the ground, including Northern Bobwhite and

Rio Grande Turkey eggs in nest on ground

Rio Grande Turkey, and Common Nighthawk are especially susceptible to feral hog predation.

Results of a short-term study on turkeys relocated to a wildlife management area in East Texas suggest "intensive feral hog control may have increased nest success and poult survival." Prior to control they found 0% nest success for nesting turkeys; after intensive hog removal, it increased to 25%.

Rooting and wallowing cause severe damage to habitat. Hogs forage at ground level and often use their strong snouts to root up food from the ground. Hogs also love to wallow in wet places. The combination of rooting and wallowing in wet areas is a threat to water quality and increases the spread of exotic plants. In this way hogs damage seeps, springs, and riparian areas—critical bird habitat.

What is the best way to control or get rid of feral hogs?

Because feral hogs are prolific and well adapted to life in the Hill Country, they have become a serious problem. Controlling hogs requires consistent and intense removal. A good feral hog control program includes rigorous trapping in combination with shooting them whenever encountered. Thus, if you have hunters on your property, ask them to shoot hogs on sight. Continuing research and advances in hog control techniques are improving the effectiveness of hog control methods. Before beginning a control program, check on the current methods and

Common Nighthawk
(Photo by Lora L. Render)

Dry plateau prairie with bare ground

the regulations associated with trapping, shooting, or disposing of feral hogs. Agencies such as Texas Parks and Wildlife and the Animal Health Commission regulate various aspects of hog removal.

BOX TRAPS AND CORRAL TRAPS

AgriLife Extension has several excellent publications giving details on identifying and controlling feral hogs. Two types of traps in common use are box traps and corral traps. Both have advantages and disadvantages.

The box trap may be the most popular because it is relatively small and easy to move around. Usually made of hog panels welded together to make a fully enclosed cage, the box trap has a single door at one end held

Feral hog corral trap

A PERSONAL STORY: FERAL HOGS

DUSTY BRUNS

I am pretty sure hogs killed and ate one of our newborn calves a few years ago. The cow was upset and appeared to have just had her calf. Saw three very large hogs running from the area but never found any sign of the calf. The hogs tear up our fences. Our hunters had their feeders turned over and destroyed. The hogs also root up plants in our riparian areas, making erosion control difficult. Hogs visited our neighbor's front yard one night, and you would have thought they rototilled their yard!

One even came after me while I was walking the fence making repairs. Apparently, I surprised a sow with pigs. She jumped up and came after me all bristled up. Shot her four times with my small revolver. After the fourth shot, she decided I was too much trouble, which was good because there were only four bullets in the gun. After that I started carrying a larger revolver working fence lines!

I trap them and shoot them on sight. One year, between me and the neighbors, we took out over 100 of them. But apparently, that amounts to emptying a wheelbarrow with a spoon. As they say, "You kill one and six more show up for the funeral!" My son-in-law had 35 hogs at a deer blind one season. As he was relatively new to hunting, he wasn't sure what the survivors would do to him should he shoot some of their number, so he held his fire. I sat in that blind the following day with plenty of bullets, but none showed up. So I plug away, catching them one or two at a time with my traps. We eat the nice ones, feed the rest to the vultures—they've got to eat too!

> **CAUTIONS FOR PROCESSING AND CONSUMING FERAL HOG MEAT**
>
> Feral hog meat is lean, good-quality protein, but there are a number of diseases that can be spread from feral hogs to humans. The most common are brucellosis and tularemia. Use the following basic field and cooking precautions recommended by the Centers for Disease Control (CDC) to avoid these diseases:
>
> - Wear gloves (and eye protection) for field dressing.
> - Use utensils disinfected with bleach for field dressing.
> - Cook meat to an internal temperature of 160°F.

open by a latch attached to a trigger mechanism. When the hog enters and trips the trigger, the door closes, capturing the hog alive. The major disadvantage of a box trap is that it usually catches only a single animal and probably teaches trap avoidance to those that escape.

A corral trap is made of hog panels joined to make a large corral and anchored to the ground. It usually is equipped with a door and a latch-trigger mechanism similar to that in the box trap. In a corral trap, the trigger is located at the far end of the trap, so groups of hogs can be trapped together. Corral traps are becoming more popular because if a deer gets in the corral, it can jump out. These traps are also becoming more technically sophisticated with cameras and an electronic trigger that can be remotely activated by such things as a smartphone. The disadvantages are that they are more expensive and not easily moved from place to place.

SNARES: DO NOT USE

Snares are inhumane. Also, deer and other wildlife can be caught in snares, and it is illegal to catch white-tailed deer in snares.

Cats

The only way to prevent domestic cat predation on wildlife is to keep cats indoors all the time. This is safest for both wild creatures and cats. You do not protect birds by keeping cats inside during the day and letting them out at night. Cats often hunt in the late evening, at night, and during early-morning hours.

Cat predation on wild birds is a serious problem. A recent research review shows the kill rate by domestic cats to be two to four times higher

Outdoor cat

than previous projections. A model based on current statistics predicts that cats stalk and kill around 2.4 billion birds each year in the United States. Nestlings that cannot escape, weak fledglings with little experience, and ground-nesting birds are most vulnerable to cat predation. Cats also kill adult birds; cats are even fast enough to catch a hummingbird! If you have bird feeders or a garden with plants that attract birds feeding on nectar or insects, you may be helping the cats in your yard more than the birds. If you live in a suburban area, habitat fragmentation in your neighborhood provides cats and other predators with easy access to birds and other wildlife living on small tracts of land. Rather than havens for wildlife, your neighborhood may be an especially dangerous place for birds.

Well-fed cats kill birds. Because their hunting instinct is so strong, even well-fed cats prefer live food. In one study, six cats were presented with a small, live creature while eating their preferred food. All six cats stopped eating the food, killed the animal, and then resumed eating the food.

Cats wearing bells kill birds. Studies have shown that bells on collars do not prevent cats from killing birds or other wildlife.

- Bells offer no protection for helpless nestlings and fledglings.
- Bells are not natural warning sounds for birds, and neither fledglings nor adults necessarily associate the sound of a bell with danger.
- Cats with bells usually learn to stalk their prey silently.
- Even if the cat's bell makes a sound, it may ring too late and be the last sound the bird hears.

Most birds that seem to escape a cat do not survive. Most small wild creatures injured by cats do not survive. After being caught by a cat, a bird that looks perfectly healthy may die from shock or internal injuries. Even if treatment is administered immediately, only about 20% of the creatures rescued from cat predation survive. Cats carry bacteria and viruses in their mouths, some of which can be transmitted to their prey. Many animals treated at wildlife rehabilitation centers are cat attack victims. In 1994, Wildlife Rescue, Inc., in Palo Alto, California, found that during nesting season approximately 25% of their patients were cat-caught birds, and almost half of these were young fledglings. Another wildlife rehabilitation center reports that cats caught 30% of the injured birds treated.

Most cat owners change gradually. It is easier to comprehend the terrible impact cats have on wildlife than it is to act in a logical, consistent manner on this information. However, most cat owners gradually change the way they care for their cats. For some, it is easiest to simply keep a cat indoors and supply kitty litter. Others, who have many cats, choose to build special indoor/outdoor buildings or convert storage sheds into special homes for their animals. While this may not be a perfect solution, it protects both the cats and birds. When cats are given a healthy and safe indoor home, pet owners and neighbors can enjoy many more birds on their property.

> *Why not use Trap/Neuter/Release (TNR) programs to save cats and wildlife?*

- Cats living outdoors, even when well fed, kill lizards, small mammals, birds, and insects.
- In order for feral cat TNR colonies to slowly shrink through time, 71%–94% of the cats must be neutered and there must be no immigration.
- Feral cats freely move into TNR colonies from adjacent areas.
- Feral cat populations are generally 10–100 times denser than populations of similarly sized native predators.
- Feral cats in TNR colonies exhibit a form of hyperpredation. (They prey on species whose populations have declined so much that they cannot support even native predators.)
- Cats in TNR programs have high disease and parasite infection rates and carry diseases, including rabies.

We can all agree on humane treatment. Whether you are a proponent of cats living outdoors or cats being kept inside to protect wildlife, we can all agree on the wisdom of spay and neuter for all our feline pets.

A PERSONAL STORY: OUTDOOR CATS BECOME INDOOR CATS

BARBARA KILPPER

My experience with cats has evolved greatly over the years. About 16 years ago a charming cat "adopted" me. I was too clueless to even know she was pregnant until her kittens appeared on a blanket in my bedroom! I was so taken with them that I kept them all. The mother soon disappeared and took up with a neighbor. Throughout the years more cats have adopted me, and I now have eight! They have all been neutered. I actually tried to give the most recent one away, but no one wants an older cat (sad).

It took me quite a long time to understand how many birds my cats were getting. One day I watched one kill a hummingbird feeding on red salvia blossoms in our yard. After that, my husband and I decided to build indoor/outdoor enclosures to protect the birds. Food, water, and litter boxes are inside. I use leaves as "litter" in all outside areas, and the cats usually use this unless it's raining. Indoor kitty furniture enables them to jump up and down to the shelves and move about the building.

We started with two cats living in our home and six in one large enclosure that has "kitty doors" for indoor/outdoor access. Soon, we discovered that some of the cats did not get along and had to be separated. Now, the large house holds only four cats, and we converted two small storage sheds into enclosures that each holds one cat. These have a window seat bed at their windows. This provides a way for the cats to go in and out easily. If we had known at the start that we were going to have to separate some of the cats, we could have just built kitty doors into those too.

One day a cat slipped out of her enclosure when I opened the door. I wasn't too worried because I thought that she was not a hunter. Was I wrong! In less than two minutes, she had killed a female cardinal right in front of me! I also have to admit that at one time I was letting my cats out after dark because

Cat in screened outdoor area (Photo by Barbara Kilpper)

I wanted them to have as much freedom as possible. I would get up before daylight and put them back in the indoor part of their "kitty condos." I always took the same path because every morning they were on my back deck waiting for me. They would follow me from the deck to their enclosures, where I would feed and leave them until after dark and then start the process all over again. I was comfortable doing this because I believed that the cats did not hunt at night and I was averting any possible killing of birds. One morning, a cardinal nest that had had baby birds was empty—on the ground—in the middle of my path! At this point, I decided the cats were hunting in the dark and should not go out at night anymore.

Finally, I'll say that keeping our cats in their kitty condos is not an absolutely perfect situation, but they are happy enough, and now my husband and I enjoy a birdsong symphony every morning. We have many beautiful birds that can safely eat from our bird feeders. In our yard and on our walks, we see cardinals, titmice, Carolina Wrens, goldfinches, scrub jays, hummingbirds, House Finches, phoebes, and even an occasional Painted Bunting and Eastern Bluebird. We have enjoyed some wonderful migrating birds, too. Just a couple of weeks ago a huge flock of Cedar Waxwings spent some time here. They were thrilling to watch; I had never seen them before.

Everyone supports the humane treatment of animals, and no one wishes to contribute to feral cat populations. We hope that you can take the next step and keep your cats indoors.

Red Imported Fire Ant

The red imported fire ant (RIFA) is small and varies in size from about ⅛ inch to ¼ inch long. This variation in size is a distinguishing characteristic; most other ant species are uniform in size. The RIFA bites and stings. Its sting produces a hot, burning sensation. A day or so later, the ant's venom causes a white, fluid-filled blister. Only fire ant venom causes these blisters.

How do you know that you have red imported fire ants?

You have red imported fire ants if you see one or more of their mounds—fluffy "worked" soil with *no* visible entrance hole. Mounds are most common after rain saturates the soil and may be from a few inches to 18 inches high. If you disturb the mound, you will see adult worker ants carrying small white eggs, larvae, and pupae. When agitated, the workers swarm up any nearby vertical surface—like your leg. Native fire ants do not do this.

How do red imported fire ants kill wildlife?

- They can kill fawns, lizards, birds, and horned lizards. These ants are sensitive to vibrations. When they crawl onto a nestling bird (or up your leg) and feel movement, all of them are stimulated at the same time and they sting in unison. Too many stings can

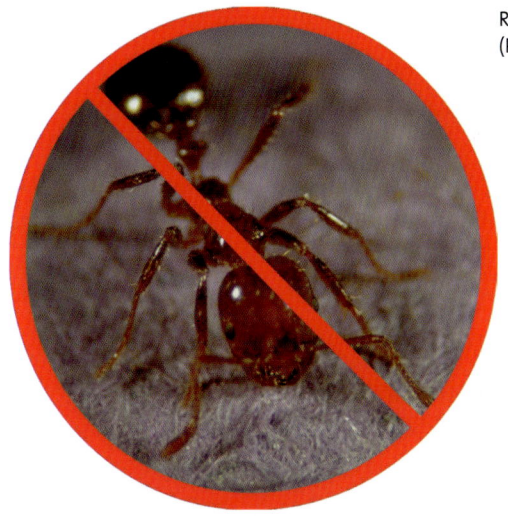

Red imported fire ant
(Photo by B. Drees)

Red imported fire ant mounds in overgrazed pasture

incapacitate young or small animals so they cannot escape. When more and more ants swarm over them biting and stinging, the ant venom finally kills the wild creature.
- Red imported fire ants even can kill fish. Swarms contain thousands of ants, and when many ants fall into water, fish may eat so many that they are poisoned.
- One of the greatest impacts of the RIFA on birds may not be the direct killing of young birds but the impact to the bird's food supply. The ants feed on large numbers of other insects and at high densities can reduce insect diversity. Since insects are a major food source for the young of most Hill Country breeding birds, this reduction in insect diversity can impact breeding success.

RED IMPORTED FIRE ANT CONTROL
There is no single, easy answer for every fire ant situation. If you have fewer than 20 mounds per acre (5 per quarter acre), it may be best to use bait or the Two-Step Method. The Two-Step Method is proven to reduce these ants where there are heavy infestations, and it is appropriate for large areas. It is cost-effective and more environmentally friendly than other chemical approaches, and organic options are available.

1. In late August through October, broadcast a slow-acting ant bait (Extinguish Plus, Amdro, Siege Pro, Logic, Award, Ascend, or Raid Double Control Ant Bait), which reduces ant mounds by 80%–90% when done properly.

Red imported fire ants spread by floodwater (Photo by B. Drees)

2. The following spring when air temperature is above 65°F, treat any remaining live mounds with a fast-acting, individual mound treatment such as a dust, granule, bait, drench insecticide, or home remedy (for example, very hot water poured on the mound).

For more information on the Two-Step Method and other RIFA questions, see the Texas A&M University website on RIFA control: http://fire

Spreading bait for red imported fire ant control

ant.tamu.edu/controlmethods/twostep/. Also, consult with your local AgriLife Extension agent for special instructions in your particular location. Be sure to ask about the best way to protect native ant species while treating for RIFA control.

PROTECT NATIVE ANTS

When controlling the RIFA, take care to protect native ants. Some native ants actually assist with the control of red imported fire ants by attacking and killing them, and we can use all the help we can get. Other ants, especially harvester ants, are a primary food source for the Texas horned lizard. When treating for RIFA, identify other ant colonies and avoid them; time applications to avoid foraging of native species; and use individual mound treatment when possible. The goal here is to control the RIFA and to maintain the natural balance of native ant species on your property.

Why control and not eradicate?

RIFA biology and distribution make it economically, technically, and ecologically impossible to eradicate the ants with methods available today. The best we can do is long-term control. After treatment, the ants can reinvade with mounds sometimes appearing after the next rain and most often within a year. Do not waste your time and money on methods that do not work. There are many ideas for RIFA control not supported by scientific evidence. Some of these are club soda, boric acid, corn grit, diatomaceous earth, Sweet'N Low, or dumping one mound on another.

Scientists have been studying biological control for the RIFA. Phorid flies that parasitize the ants are one of several biological control agents that could help control (but not eliminate) them. Phorid flies have been released and established in parts of Texas and are spreading into adjacent areas. Their effectiveness in Texas and the southern United States is not yet fully known. Researchers predict that they will help reduce RIFA numbers at least to more tolerable levels. Other natural RIFA enemies include predaceous mites, parasitic nematodes, and a fungus. Scientific evaluation continues on the effectiveness for some of these. Consult the Texas A&M fire ant web page (http://www.extension.org/fire+ants) and your local AgriLife Extension agent for the most up-to-date information on fire ant research and control.

Deer

Because this problem is so widespread, we often overlook the fact that deer overpopulation creates ecological instability affecting wild birds. In the Hill Country and throughout much of the nation, "too many deer" harm birds through habitat destruction.

What do we do to protect birds from "dangerous" deer?

- Do not feed deer.
- Harvest more deer.
- Manage for 1:2 buck/doe ratio.
- Plant and protect native plants that are loved by deer.

Habitat destruction causes loss of bird food indirectly and directly. In places where deer overpopulation exists, deer reduce the number of plants on which insects live. Many bird species are primarily insectivorous, and nearly all bird species must feed insects to their young for survival. Thus, by damaging too many plants, too many deer harm insect-eating birds. Deer are in direct competition for the same food that seed-eating birds consume. A huge deer competing with a tiny bird may be hard to imagine, but it is easy to see that those big deer can easily outcompete small birds for a limited supply of forbs, seeds, and berries.

Deer overpopulation limits plant diversity. Deer feed or browse continuously on the plants they prefer. Heavy browsing on the deer's favorite plants—along with the ones they eat only out of necessity—dramatically limits plant reproduction and reduces plant diversity. Thus, the broad environmental impact of deer overpopulation is a change in Hill Country plant communities. Overpopulation of deer causes a shift from plant communities that include plant species deer prefer to ones with only plants that deer cannot eat at all and those they eat only when they are

Deer browsing tree leaves (Photo from TPWD)

starving. If you are uncertain of the plant community condition on your land, use the deer browse evaluation chart in appendix 1.

Habitat destruction removes understory structure needed by wild birds. Deer browse on understory plants that they can reach. Too many deer eliminate or reduce many low-growing woody plants. They also limit the low-level growth on large woody species that normally have branches growing near the ground. This understory structure is critical habitat for many Hill Country resident, migrant, and breeding birds. The Bewick's Wren, Black-crested Titmouse, Carolina Chickadee, Chipping

Sparrow, White-eyed Vireo, Greater Roadrunner, and many others rely on low-growing trees and shrubs for a place to hide from predators, escape bad weather, build their nests, and rear their young.

> *How do we know if a bird population is being limited by predation on our property?*

If they exist on your property, you can assume that deer overpopulation, Brown-headed Cowbirds, fire ants, feral hogs, and outdoor cats are limiting birds on your land. If you are interested in a bird that could be affected by any of these animals, see appendix 2 to find out if the bird is expected to use habitat you have on your property. For a bird that breeds in the Hill Country, you can get the approximate size of its breeding territory from the descriptions at the end of each chapter. Use this to estimate how many pairs could be in healthy habitat. Then, do a survey for the bird on your property during its breeding season. If your survey shows it to be absent or present in numbers lower than expected, cowbirds, deer, fire ants, feral hogs, and domestic cats are suspects. For information on surveys and monitoring in general, see the introduction to this book. Unless you have enough experience to identify the species in question, you will have to find someone qualified to do the bird census during the season the species is expected to be present. Survey in spring or early summer for breeding birds and in January or February for wintering species.

Some species are especially vulnerable to Brown-headed Cowbird parasitism. Some species depend on understory, so they are especially vulnerable to deer overpopulation, and deer overpopulation also is harmful to any bird that depends on deciduous trees or forbs. You can find these birds in appendix 2.

If you see domestic cats, have bird boxes near open grassland on your property, and bluebirds either do not use the boxes or are unsuccessful in fledging their young, you can assume that cats are a problem. If you used to have turkeys nesting in a particular area but they stopped using it after feral hogs moved in, it is fair to assume that the hogs are probably part of the problem.

NATIVE PREDATOR MANAGEMENT

Systematic and long-term removal is necessary to control invasive exotic feral hogs, fire ants, and domestic cats, as well as Brown-headed Cowbirds and white-tailed deer overpopulation. Dealing with problems caused by native predators, such as raccoons, jays, coyotes, or rat snakes, is different because of their role in the natural predator/prey relationship. Habitat improvement (rather than direct predator control) pre-

Remember: Both predator and prey are necessary for the dynamic equilibrium of native populations.

vents or reduces the effectiveness of native predators. When basic habitat elements such as nest sites, food, water, and/or protective cover are improved, the natural predator/prey dynamic equilibrium improves on its own, and native predators will continue to live in the improved, ecologically sustainable habitat.

Why not just get rid of the natural predators if they are the problem?

Although getting rid of all the predators may seem like the easiest and most effective approach, native predators actually play an important role in the natural fluctuations of prey populations. Removing a native predator population can produce a huge (though usually temporary) rebound in predator numbers. It is possible that when a predator population is removed, its prey will increase to an unusually high level. Then, the predators living nearby move back and feast off a superabundant food supply. With plenty of food, their health and body condition improves dramatically, they reproduce rapidly, and the predator population increases, at least for a time, to a much higher level than was present before removal. For example, in the southern plains during the Dust Bowl, intense coyote control led to a dramatic increase in rabbit populations followed by a larger coyote population before the both predator and prey numbers could return to lower, natural levels. Also, fox populations have been documented to rebound, at least temporarily, far beyond pretrapping levels.

How do we tell if our bird habitat is out of ecological equilibrium?

This is not always easy. When managing for specific birds, be sure to reduce cowbirds, feral hogs, red imported fire ants, deer, and domestic cats. If you do not see population growth or the birds are barely present and everything else indicates they should be more abundant, ask yourself, "Do we have adequate food, water, cover, and nest sites for these species?"

What are the fundamentals of restoring dynamic equilibrium in natural habitats?

- Always manage by habitat type (grassland, riparian, woodland, and so on) and foster healthy native plant communities on your land to promote bird populations. A bird species is found where plant communities provide the habitat it needs and where it can find its basics for survival: food, water, nest sites, shelter from bad weather and predation, and enough space for a sustainable population.

Habitat mosaic

- Shape habitats in a mosaic pattern. It is easy for most predators to search linear habitats such as fencerows. This is an example of an ecological trap. Thus, an irregular patchwork or mosaic of different habitats thwarts native predators and favors bird populations. Habitat in irregular mosaic patterns contributes to a stable equilibrium between predator and prey populations.
- Leave some habitat undisturbed. When mowing, baling, burning, or restoring an old field to grassland, to avoid creating an ecological trap, leave some block-shaped habitat undisturbed for breeding birds to nest and rear their young. In this process you create a habitat mosaic that deters predation.
- Mix meadows and brush. To promote sustainable predator/prey populations in a large pasture, you can manage for a mosaic mixture of meadows and shrubby thickets. This is varied habitat that will support different bird species. Such a mixed habitat creates adjacent and sustainable systems that do not exist to the detriment of each other. And birds are safer here because in a habitat mosaic, predators have a harder time finding their prey.
- Find the happy medium. To prevent an ecological trap near water used by birds, allow cover for the birds but not so dense that their predators can hide and ambush. For example, at a guzzler, provide bird cover such as shrubs and/or open brush piles that supply plenty of low perches. Do not surround your water source

Remember: When managing habitat for sustainable predator/prey populations, *think mosaic.*

with dense cedar or garden plants that might give cats, snakes, skunks, or coyotes a place to hide. Also, remember that no cover at all will make the drinking or bathing spot an easy hunting place for aerial predators.
- Monitor nest boxes for predation. If you have nest boxes, monitor the boxes weekly throughout the nesting season. This is the only way to know the nesting success rate and to identify predators. If eggs are missing and there is no nest disturbance or nest material is compressed, the predator is probably a snake. Install a predator guard. If eggs are missing and the nest is disturbed, the predator is probably a raccoon or cat and a good predator guard can help.

Native Predators of Selected Birds

Little research has been done on bird predation by native species. Some studies show that native predators on high- or midcanopy nesting birds are usually Sharp-shinned Hawks, Cooper's Hawks, or American Kestrels. Their primary impact is on the birds nesting at these levels of the canopy as well as birds concentrated near bird feeders.

SUMMER TANAGER

Summer Tanagers nest in high canopy and have been observed to aggressively defend their nests against raccoons, rat snakes, squirrels, Blue Jays, and Cooper's Hawks. Also, in predation studies, there are very few instances of any kind of predation on Summer Tanager adults, fledglings, nestlings, or eggs. Thus, it seems that this species is well adapted to survival in the high canopy.

GOLDEN-FRONTED WOODPECKER

Golden-fronted Woodpeckers nest in mid- to high canopy. They excavate their own nest cavities up to 30 feet high in snags or tall damaged trees. There is limited information on Golden-fronted Woodpecker predation by native species. A study conducted in northern Mexico, where both Golden-fronted Woodpeckers and the Aplomado Falcon are common, found that the falcon's diet included very few woodpeckers. During this study, a researcher observed either a Sharp-shinned or Cooper's Hawk catching a single adult Golden-fronted Woodpecker in flight and occasionally observed a snake in Golden-fronted Woodpecker cavities preying on nestlings or eggs.

SCISSOR-TAILED FLYCATCHER

Scissor-tailed Flycatchers nest in midcanopy. One study of this species reported two incidents of raptor predation by a Cooper's Hawk and an

American Crow. They also observed some possible nest predation by snakes.

GREATER ROADRUNNER

Greater Roadrunners commonly nest in midcanopy in relatively short trees such as stunted live oaks, shin oaks, cedar, and mesquite. Their nests are usually found within 10 feet of the ground. Studies on roadrunner nests show that coyotes, raccoons, and striped skunks are their most common native predators. Rat snake, coachwhip snake, and bull snake are occasional predators, and less frequent predators are crows and ravens. Also, adult roadrunners are occasionally captured by Red-tailed or Cooper's Hawks.

FIELD SPARROW

In one study that used a trail camera, 72% of Field Sparrow nestling predation was by a snake (mostly rat snakes). This is to be expected because Field Sparrows nest on the ground where nests are most easily found by snakes. The remaining 18% of nestling predation was by a mammal or a raptor.

Prey of Selected Native Predators

Among the native predators of birds in the Hill Country are raccoons, skunks and foxes, coyotes, mountain lions and bobcats, and other birds.

RACCOON

Raccoons are known to prey on ground-nesting birds, including turkey and bobwhite. In rural settings they usually are not numerous enough to

Raccoon (Photo by Jonah Evans)

Skunk (Photo from TPWD)

cause a problem. An exception is around deer feeders or buildings where cat, dog, or bird food is readily available. Raccoons can be a more serious problem in urban or suburban areas, where they commonly prey on bird eggs and nestlings. Here significant predator/prey imbalance is created by cat and dog food, birdseed, and suet left outside at many homes in a small area. Lots of good food means too many raccoons. You can discourage raccoons by bringing the cat, dog, and bird food in at night. If this does not work, you may have to trap and euthanize the raccoons. *Do not move them to another location.*

SKUNK AND FOX
Skunk and fox populations may also contribute to predator/prey imbalance in suburban and urban settings where their populations are higher than normal. Like raccoons, they may increase substantially where they find plenty of good food and protection from predators. Skunks and foxes are omnivores that eat pet and bird food and fruit from domestic trees in urban and suburban neighborhoods. And these places usually are free of their most dangerous predators—Great Horned Owls and coyotes. Skunks and foxes are much less of a problem than raccoons but can be discouraged by bringing pet and bird food in at night.

COYOTE
Coyotes are opportunistic omnivores. They consume just about anything edible. In the Hill Country, coyotes are primarily rural, and their pres-

Gray fox (Photo from TPWD)

Coyote (Photo from TPWD)

Mountain lion (Photo from TPWD)

ence usually contributes to sustainable predator/prey populations. They eat an array of prey animals, including jackrabbits, cottontails, cotton rats, snakes, raccoons, incapacitated deer as well as young fawns and, to a much lesser extent, nesting birds. They also eat fruit, such as persimmon and pricklypear, and insects like grasshoppers. As with other native predators in a natural system with no human intervention, coyotes successfully rear their pups on a variety of prey animals depending on what is available.

MOUNTAIN LION AND BOBCAT

Meat-eating predators such as mountain lions and bobcats have large territories. Mountain lions have exceptionally large territories and are very uncommon in the Hill Country, so they are rarely seen or discovered even though your property may be part if a mountain lion territory. Unless you have a huge supply of "easy pickings," these cats routinely come and go from your property, which is a relatively small part of their territory. Thus, these species are not problem predators for turkey or quail, and they contribute to the natural fluctuation of their prey populations.

AMERICAN KESTREL

American Kestrels live in the Hill Country and throughout South Texas during cold months. In winter, their primary prey animals are birds and mice. Although not common breeders here, kestrels also nest in a portion of the Hill Country from Fredericksburg north. In summer during their nesting season, kestrels feed their young many insects plus bats,

Bobcat (Photo from TPWD)

mice, lizards, frogs, and possibly birds. Kestrels and other avian predators play a natural role in the predator/prey equilibrium.

BALD EAGLE, GOLDEN EAGLE, AND CRESTED CARACARA

All of these large avian predators nest in the Hill Country. They hunt and kill their prey but also readily compete with vultures for carrion. The

Crested Caracara
(Photo by Lora L. Render)

Savannah with forbs, grasses and scattered trees and shrubs

Crested Caracara is becoming more and more common in our area. Because it is most easily seen when eating carrion, caracaras are mistakenly thought of as scavengers, but they regularly hunt a large variety of live animals from insects to fish, snakes, and lizards. Caracara feeding at a roadside kill may actually be more interested in the insects associated with carrion than they are with the dead animal itself.

WESTERN SCRUB-JAY
Western Scrub-Jays occasionally feed small animals such as mice and birds to their young. One study showed that birds made up only 18% of their diet during nesting season. Most bird predation by Western Scrub-

Bobcat tracks (Photo by Jonah Evans)

Jays is limited to eggs, nestlings, and fledglings and contributes to a natural predator/prey dynamic equilibrium.

Seasonal Fluctuations for Predators

Numbers of some predators, both native and exotic, may seem to increase at certain times of the year. This is a reason to get out there and look regularly and to keep dated records of all sightings. Fluctuation due to season or temperature is normal and needs to be factored into your understanding of the predator/prey populations in your area. For example, skunks are active when they come out of hibernation in late winter, Brown-headed Cowbirds gather during spring nesting season, and fire ants appear when ambient temperature rises to above 75°F.

Tools Used to Indicate Presence of Predators

You can use track identification and trail cameras to determine the presence of predators.

TRACK IDENTIFICATION

Since many predators are relatively uncommon and naturally reclusive, it is necessary to recognize signs of their presence. Recognizing tracks of feral hogs, deer, coyotes, raccoons, ringtail cats, skunks, foxes, bobcats, domestic cats, and mountain lions gives valuable information about the predators living on your land. TPWD has printed information and offers classes on wildlife tracks identification. Also, the *iTrack Wildlife* app is an easy-to-use, comprehensive digital field guide for animal tracks and

Trail camera

other signs. It is limited to use on iPhones and iPads and is available online.

TRAIL CAMERAS

More and more land stewards are using outdoor, motion-sensitive trail cameras to learn more about the wildlife on their property. You can buy them online or at an outdoor store.

Bird Summaries: Featured Birds

BROWN-HEADED COWBIRD

OUTLAW BIRD

ALL YEAR

Successful brood parasite on many songbird species in the Texas Hill Country; can be a serious threat to many of these; females can lay up to 70 eggs per summer; cowbird eggs hatch faster than those of other species; most hosts do not react to presence of cowbird eggs.

Identification: About 8 inches head to tail; male is shiny black with brown head and neck; female is brown all over; distinct song has burble notes with a high-pitched, ascending whistle.

Habitats: Old field, plateau prairie, river valley prairie, pocket prairie, mixed wooded slope, post oak savannah, shin oak savannah, live oak savannah, river, creek, canyon, springs, seep, tank, pond, lake, and backyard.

Cover: Mix of low brush and grass.

Breeding Territory: Area between breeding, feeding, and roosting sites can range from 172 to over 2,500 acres.

Nest: Does not build a nest; lays eggs in all kinds of nests and occasionally even in a box or other type of cavity; tends to lay its eggs in nests with eggs smaller than its own.

Food: Mostly seeds plus about 25% insects usually caught around grazing animals.

Control: Use ongoing habitat management and cowbird population control techniques to minimize parasitism.

Brown-headed Cowbird

- Manage large woodland blocks to limit fragments especially susceptible to cowbird parasitism.
- Trap and euthanize and/or shoot cowbirds to reduce parasitism. Trapping and euthanizing requires certification from Texas Parks and Wildlife Department and is limited to March 1–May 31; monitor trap daily to feed and water bait birds (cowbirds kept in the trap to attract others).

COMMON NIGHTHAWK

PRIORITY SPECIES

SPRING, SUMMER, AND FALL

Breeds in open country and has adapted to urban areas with flat, gravel roofs for nesting; well-known for male's acrobatic breeding flight display with steep dive and sudden reverse that makes a boomlike sound in its wing feathers; in Texas declining most significantly since 1980.

Identification: About 9 inches head to tail; long, pointed wings with bright white bands; easy to see and identify on the wing but nondescript and well camouflaged when perching.

Habitats: Old field, plateau prairie, river valley prairie, post oak savannah, shin oak savannah, live oak savannah, tank, pond, lake, and backyard.

Cover: Trees and shrubs for roosting; bare, gravelly areas for nesting; flat gravel roofs.

Breeding Territory: Varies but may average about 26 acres.

Nest: Lays two eggs on bare ground, rock, wood chips, sand, fallen leaves, perennial rosettes; also on tar paper, gravel, cinders on urban rooftops; no material added to nest.

Food: Catches flying insects on the wing in low light of early morning and late evening.

Stewardship: Create and maintain safe, high-quality nest habitat and protect their food supply.
- Avoid broad-spectrum herbicides and insecticides to promote insect diversity.
- Keep cats indoors, and control stray and feral cats to reduce predation.

Common Nighthawk

- Mow only in fall or winter to maintain short grass. Do *not* mow during the nesting season.
- Use moderate grazing after the breeding season, which could be beneficial in maintaining short grass for nesting.
- Use prescribed burns to control cedar and maintain diverse grassland community.

Other Potentially Useful Management:
- Control feral hogs to reduce possible nest predation.
- Limit cultivation (such as occasional light disking) to non-nesting season (October–March) to increase bare ground for nesting.
- Control red imported fire ants with ant-specific insecticides to increase insect diversity.

CRESTED CARACARA

UNDER-THE-RADAR BIRD

ALL YEAR

Big raptor that looks like an eagle or hawk, hangs out with vultures, and is in the falcon family; Texas has largest breeding population in the United States; common in South Texas brush country and moving north into the Hill Country. Sometimes mistaken for a Bald Eagle because of similar coloring and sometimes called the Mexican Eagle.

Identification: About 21 inches head to tail; long legs; red skin on face; black cap, back, and legs; white neck; white band on wings and at end of tail.

Habitats: Old field, plateau prairie, river valley prairie, pocket prairie, post oak savannah, shin oak savannah, and live oak savannah.

Cover: Scattered shrubs and trees interspersed with grasses and forbs.

Breeding Territory: Shows strong territory defense at about 300 feet; active nests may be only 5,000 feet apart.

Nest: Well-woven with sticks, forb stems, and grass; nest cup lined with fine material, including grass, cotton, twigs, and moss; in trees or shrubs, usually in highest vegetation available; about 5–55 feet above the ground and in South Texas average about 18 feet high.

Food: Insects, all types of carrion, vertebrates, and eggs; searches for food flying, perched, and walking on the ground.

Stewardship: Encourage healthy shrub and small tree diversity interspersed with grasses, forbs, and scattered large trees.

- Add brush piles where brushy habitat is lacking to provide important cover for small mammals.

Crested Caracara

- Use prescribed burns in savannahs to increase forbs and maintain diverse woody plant communities.
- Control regrowth cedar using mechanical or chemical methods to maintain shrubs and woody plant diversity.
- Control white-tailed deer and other browsing animals to encourage woody plant recruitment, healthy shrubs, and forbs.
- Use rotational grazing at moderate levels to encourage a balance of grasses and forbs.
- Control red imported fire ants with an ant-specific pesticide to maintain healthy supply of insects.
- Avoid broad-spectrum herbicides and insecticides to increase forbs and promote insect diversity.

DOMESTIC CAT
UNDER-THE-RADAR PREDATOR
ALL YEAR

Even well-fed cats instinctively hunt small creatures, including birds; inexperienced fledglings and exhausted migrant birds most vulnerable.

Identification: Any cat living all or part of its time outdoors.
Habitats: Old field, plateau prairie, river valley prairie, pocket prairie, mixed wooded slope, post oak savannah, shin oak savannah, live oak savannah, river, creek, canyon, seep, springs, pond, tank, lake, and backyard.
Cover: Hunts from excessively dense cover too close to bird feeders and water; found in nearly all rural, suburban, and urban areas.
Home Range: About 78–563 acres with males having larger home ranges than females.
Reproduction: Average young produced per female about 1.75–4.2 per year.
Food: May prefer small mammals and lizards but kills billions of birds in the United States each year.

Domestic Cat

Control: Prevent cat predation on birds and other wildlife.
- Keep cats indoors, and extirpate feral cat populations to stop cats from killing birds.
- Do not support trap, neuter, and return (TNR) programs to control feral cats. TNR programs do not work because they require 75%–95% neutering and no immigration of other cats from the surrounding areas.
- Manage grasslands in large blocks rather than strips to protect ground-nesting birds. Blocks are harder for any predator to hunt.
- Provide safe water sources to minimize cat predation.

FERAL HOG

COMMON PREDATOR

ALL YEAR

Very mobile; travels in corridors that include water sources such as creek bottoms and ponds; may be plentiful but rarely seen; its signs include disturbed ground from rooting, wallows, rubs, trails, droppings, and beds; good to eat, but always use gloves when field-dressing and cook meat to 160°F internal temperature.

Identification: Color varies greatly from splotchy black, white, brown, or red to fairly uniform dark brown; average weight of adults 75–250 pounds; tracks resemble deer tracks but rounder with blunt toes and sometimes with 1 or 2 dew claw marks; depending on what they eat, droppings vary in shape and consistency from loose tube to formless patty or large round pellets.

Habitats: Old field, plateau prairie, river valley prairie, pocket prairie, mixed wooded slope, post oak savannah, shin oak savannah, live oak savannah, river, creek, canyon, seep, springs, pond, tank, lake, and backyard.

Cover: Rests in dense vegetation and lies in wallows near water during hot weather; rapidly expanding into all parts of the Hill Country, including urban, suburban, and rural areas.

Home Range: Averages about 166–2,471 acres; European boar range can be significantly larger; beds often are among thorny vines, shrubs, downed trees near damp wallows, and brush and are best found by following a hog trail.

Reproduction: Average litter size about 4.4–6.3 and is dependent on a number of factors, including age, body condition, and habitat quality; average number of litters per year is 1.22 in southern Texas.

Food: Plant and animal material; varies from season to season and place to place; often roots for plant tubers and roots as well as invertebrates that live in soil; food preference probably best described as opportunistic; eats almost anything it comes across.

Feral Hog

Control: Minimize damage to sensitive habitats and reduce hog predation on ground-nesting birds by consistent use of the most effective hog control practices.
- Identify hog travel corridors to make control measures most effective.
- Use the following control measures:
  - Intense, year-round trap-and-dispatch program
  - Hunters who shoot hogs on sight
  - Exclosure fences around small areas such as a spring, seep, or small planting
  - Deer-friendly exclusion fences around all deer feeders to prevent hogs from getting feed on the ground or turning over the feeders
  - Combination exclusion fencing with conventional and electric fence (most effective but costly)

ORCHARD ORIOLE

PRIORITY BIRD

SPRING, SUMMER, AND FALL

Small oriole with more buoyant flight and darker coloration than others. Prefers habitat near water and can be an important pollinator. Has declined greatly in Texas during recent decades; recent breeding bird surveys suggest that current population less than 10% of 1966 population. Nests heavily parasitized by cowbirds; experts predict this species may no longer breed here in near future.

Identification: About 6.5 inches head to tail; male has distinct black head, throat, back, and tail with chestnut below; female has olive-green back and bright green-yellow below; both have wing bar; immature similar to female; first-year male may have black throat.

Habitats: Post oak savannah, shin oak savannah, live oak savannah, river, creek, canyon, springs, tank, pond, lake, and backyard.

Cover: Deciduous trees and shrubs, mid- and low-level shrubs for perching and foraging for insect food on leaves and twigs.

Breeding Territory: Solitary in marginal habitat; semi-colonial in ideal breeding habitat where nests may be in the same tree.

Nest: Hanging open cup of woven grasses usually 10–20 feet above the ground; suspended from fork of small branches far from main trunk where adult will shade nest from hot sun; in both rural and suburban areas.

Food: Eats and feeds young insects and spiders gleaned from twigs and leaves; adults also consume small ripe fruit such as mulberries or black cherries and nectar from flowers and feeders.

Stewardship: Encourage open, diverse woodlands and savannahs with small to medium-sized deciduous trees interspersed with or adjacent to diverse grasslands.
- Control spread of regrowth cedar with appropriate mechanical or chemical methods to maintain diverse savannah and riparian woodlands.
- Exclude livestock from riparian areas, and in other areas use rotational grazing at light to moderate levels to maintain structure and a diverse plant community.
- Control white-tailed deer and other browsing animals to encourage woody plant recruitment and healthy shrubs.
- Avoid broad-spectrum herbicides and insecticides to increase forbs and promote insect diversity.
- Trap and remove cowbirds to reduce nest parasitism and increase nesting success.

Other Potentially Useful Management:
- Use prescribed burns in savannahs to maintain diverse woody plant communities.
- Use deer exclosures to encourage woody plant recruitment and protection of fruit-producing native trees.
- Control red imported fire ants with an ant-specific insecticide to increase insect diversity.
- Maintain healthy watersheds with deep and sustainable ground cover to protect riparian health.

Orchard Oriole

8 Deer Management

Overabundant deer populations can have tremendous negative impacts on native habitats. Excessive deer browse can lead to plant death and loss of plant diversity and may ultimately affect the survival of wildlife species dependent on those plant communities.

On the other hand, deer populations in balance with their native habitat help sustain healthy plants and habitats that benefit many other wildlife species from painted buntings to Texas horned lizards.—Alan Cain, TPWD White-tailed Deer Program Leader

INTRODUCTION

Migrant, resident, and nesting birds depend on diverse plant communities for their food, water, and cover. Today, throughout the Texas Hill Country, the health and stability of these native plant communities are threatened by an overpopulation of native white-tailed deer (*Odocoileus virginianus*) and introduced browsers, especially axis deer. Although native plants are adapted to being eaten by deer, plant populations that deer prefer cannot maintain themselves when consumed by too many deer. Today, the pressure of deer overpopulation is causing a shift in Hill Country plant communities because there are 3 to 15 times more deer on Hill Country land than it can sustain. Plant species favored by deer are decreasing; plant species that deer do not like are increasing. With high deer density in most of the region, this change in natural plant communities is happening right before our eyes. But the damage goes unnoticed unless the observer is watching and can recognize many different plant species.

LIFE HISTORY

To realize the enormous negative impacts of white-tailed deer overpopulation and effectively reduce their numbers to a sustainable level, it is necessary to know their life history.

Breeding and Reproduction

White-tailed deer have a high reproductive rate. They breed in the fall, and gestation is about 200 days. In the Texas Hill Country, a young doe will probably give birth for the first time when she is two years old. The first birth is usually a single fawn, and after that, if she gets enough food, a doe normally produces twins each year. In the Hill Country, where food is often inadequate, approximately 50% or more of does give birth to a single fawn, and fewer than 40% give birth to twins. In exceptionally advantageous circumstances, a doe may have triplets, but this is rare.

Fawns

In the Hill Country, white-tailed deer fawns normally are born between May 15 and July 1. On the far western edge of the region, fawns may be born as late as the third week of July. A fawn weighs between 4 and 7 pounds when born and remains inactive, isolated, and bedded down for the first few days. Its mother returns periodically to nurse and groom it. The fawn's coloration, behavior, and lack of scent are its protection from predators. For a few weeks, a fawn continues to remain hidden most of the day and does not wander far from the spot where it was born. When three or four months old, the fawn sheds its spotted coat and becomes the color of an adult. It accompanies its mother for greater distances and for longer periods. By fall, it goes everywhere with her, and they remain together until the doe prepares to give birth again. If the fawn is a buck, the doe will drive it away before the next breeding season. Female offspring often remain with the doe, forming a matriarchal group of several of her female offspring. Fawns that survive their first few months rarely live longer than five and one-half years.

Antlers

White-tailed deer antlers are bone (calcium); bucks grow and shed them annually, using them to compete with other males for females. At the end of the breeding season, as the male breeding hormones decline, the bone at the base of the antler erodes, and the bucks shed their antlers. If a buck has been eating well and is in good condition, it may begin growing new antlers almost immediately after shedding the old ones. New antlers are covered with velvet; they grow rapidly and reach full size in 12 to 16 weeks. At this point, the tissue under the velvet solidifies, and the buck rubs the velvet off on bushes and small trees. The size and shape of a buck's antlers depend on age, genetics, and the quality and quantity of food eaten.

Deer Behavior and Survival

With considerable sensory and athletic abilities that contribute to their survival, deer are elusive prey animals that challenge the hunting skills of their traditional predators. They have keen senses of smell, hearing, and vision for moving objects. When a deer identifies danger, it tends to either run or "hold tight." During hunting season, a deer often stands so still that a hunter may walk right by and never see it. Deer can run at a speed of 35 miles per hour for short distances and about 20 miles per hour for up to about 3 miles. They can jump 8-foot-high fences and clear a 25-foot-wide stream. They are excellent swimmers.

Habitat

White-tailed deer prefer savannah habitat (a mixture of grasses, forbs, trees, and shrubs), where they find food, shelter, and abundant places to hide. They also need a reliable water supply in their home range. With current conditions in the Texas Hill Country, no more than one deer per 25 to 30 acres sustains both a healthy plant community and healthy deer. This is an appropriate population density for the typical Hill Country landscape with a mixture of open grassland and treed areas including cedar thickets and low shrubs.

Food

Plant diversity is an important component of good deer habitat. The plants that deer eat are divided into four categories: forbs, browse, mast, and grasses. High-quality deer habitat has a mixture of plants in all four categories. Depending on a number of environmental and physiological conditions, white-tailed deer need 10% to 20% protein in their diet. Consequently, they mostly eat browse and forbs (plus acorns when available), which are relatively high in protein. They consume very little grass, usually only in early spring when grass is green and tender and has higher protein content.

Forbs: Wildflowers and other broadleaf herbaceous flowering plants are forbs, the food white-tailed deer and goats prefer to eat. In addition to delicious garden flowers, some of their favorite native forbs are rain lily, dayflower, spiderwort, primrose, and mistletoe.

Browse: Deer browse is composed of the twigs, leaves, and shoots of woody trees, shrubs, and vines. This type of food is important because abundance of preferred forbs varies seasonally and they are not always available. As forbs decrease during late summer and early autumn or even disappear altogether during a severe drought, deer shift to browse. They eat browse from trees, shrubs, and vines within 3 or 4 feet of the ground.

Mast: The edible nuts, especially acorns, and fruit produced by trees and shrubs is called mast. When present, mast is important deer food in late summer and fall. However, in drought years, the amount of mast is greatly reduced and deer have to depend more heavily on browse.

Grasses: White-tailed deer are not grazers. Unlike cows, they cannot survive by eating grass. Grasses never play a major role in the diet of white-tailed deer and usually are only a small percentage of their annual

---

Remember: Where there are too many deer, hunting is essential so that the survivors can find adequate food, remain healthy, and not destroy their own habitat.

Deer browsing in unhealthy savannah

food source. Although white-tailed deer do eat some fresh, green grass in the early spring, deer density does not have a serious impact on grasses growing in the Hill Country.

OVERPOPULATION

In recent history, the Edwards Plateau region supports a higher density of white-tailed deer than any other dryland area of the United States. The Hill Country portion of the Edwards Plateau covers 12 million acres and is home to more than 1.5 million white-tailed deer. Deer density in the Hill Country averages one deer per 15 acres, but is much higher in many locations. For example, Fair Oaks Ranch, Texas, a suburban town north of San Antonio (in northern Bexar, southern Kendall, and western Comal Counties), has about one deer per 3 acres. A few counties (e.g., Val Verde, Sutton, and Edwards) on the western edge of Edwards Plateau have regular outbreaks of anthrax, which cause dramatic deer population crashes. Thus, density numbers for these counties moderate overall Hill Country statistical averages.

Causes of Deer Overpopulation

Three factors contributing to white-tailed deer overpopulation are control of screwworm fly, habitat change, and decreased predation.

1. The screwworm fly once was a parasite that severely limited fawn survival. With the introduction of effective biological control in the 1960s, this parasite has been virtually eliminated.

2. In the last 150 years, the Hill Country has changed from a grassland-dominated landscape to a savannah or woodland, which greatly increased white-tailed deer habitat.
3. Throughout the Hill Country, the white-tailed deer's natural predators—coyote, wolf, and mountain lion—have been greatly reduced or eliminated. Consequently, they have little impact on today's deer overpopulation. And since increasing the number of predators is not a practical option, hunters play the most important role in regulating deer numbers.

Signs of Deer Overpopulation

In the Hill Country, signs that the deer population is higher than the land can support are everywhere. To determine if you have a deer browse problem in woody habitats, use the deer browse evaluation chart in appendix 1.

- The fact that we nearly always ask if the deer will eat it before buying a new landscape plant is perhaps the most common indication of deer overpopulation.
- A distinct browse line is a sure sign of deer overpopulation. Look for a line about 4 feet from the ground. Above the browse line you will see lots of leaves and normal plant growth. Below the line you see no leaves or new growth because everything is removed by the deer.
- A mass of tiny seedlings under a mature tree but no young trees or saplings is another sign. Without saplings, when the mature trees reach the end of their life span, there will be no others to replace them. The species is unable to maintain itself, and this eventually leads to its loss from the plant community, thus reducing plant diversity and stability.
- Individual plants are stunted, have small clusters of miniature leaves and no slender branches in areas where deer density is especially high. Because they are browsed repeatedly, these plants grow only 1 to 3 feet tall and will never mature and reproduce.

Signs of Good Deer Population Management

1. An abundance of preferred forbs and browse plants grow on well-managed land.
2. Many low branches on trees and shrubs survive where deer density is at a sustainable level.
3. Young saplings of elm, sugar hackberry, Carolina buckthorn, Eve's necklace, Spanish oak, Texas madrone, red mulberry,

Deer browse line on live oaks

Deer browse line on mustang grape vine

Texas mulberry, and creek plum grow where the deer population is controlled. And the deer living in this area will be getting appropriate nutrition. According to research conducted at TPWD's Kerr Wildlife Management Area, Hill Country white-tailed deer prefer these species over all other native trees, shrubs, and vines.

Heavily browsed mass of live oak seedlings under mature tree

4. The sumacs, post oak, Lacey oak, Texas redbud, escarpment black cherry, possumhaw, and netleaf hackberry plus poison ivy, Carolina snailseed, clematis, grape, and Virginia creeper vines indicate healthy habitat for deer, birds, and other wildlife. These plants provide browse moderately preferred by white-tailed deer. They are even more important to the well-being of white-tailed deer than their preferred foods because the somewhat less preferred plant species are usually in greater supply.

Am I part of the problem or the solution?

- Do I feed deer and exotics or hunt them?
- Am I managing my property for the number of deer (natives and exotics) that the land can sustain, or am I allowing or even promoting overpopulation?
- We love the deer, but do we prefer white-tailed deer above all other wildlife?
- Are we willing to witness the slow decline of other native wildlife, including birds and plant species?
- Do we accept the inevitable loss of a piece of our natural heritage?

If the answer to any or all of these questions is no, then we need to figure out how to become better land managers, work together, and stop the crisis of deer overpopulation.

Heavily browsed sapling stunted by deer

DEER AND BIRDS

Deer overpopulation is having a critical and widespread impact on bird habitat. Birds affected include many songbirds, birds of prey, game birds, and endangered species.

Songbirds

Deer overbrowsing on understory shrubs is a major cause of songbird declines throughout woodlands and forests in the United States and

Europe due to destruction of nesting cover and reduction of food supplies. Locally, songbirds such as finches, sparrows, and buntings consume seeds from goldenrod, sunflower, and wood sorrel as well as other forbs preferred by deer. In areas of high deer density, these plants are consumed before going to seed, and an important source of food for birds is being eliminated. Eastern Bluebird, Bewick's Wren, and Painted Bunting must feed insects to their nestlings. Thus, as deer remove plants where insects normally live, these breeding birds are also negatively impacted by deer overpopulation.

Birds of Prey

Obviously, birds of prey do not eat plants, but being high on the food chain does not insulate them from the negative effects of plant loss. An example is the familiar Red-shouldered Hawk. This raptor preys on snakes and lizards that eat insects that in turn feed on plants. Thus, through its food chain, the Red-shouldered Hawk can be diminished by plant loss due to deer overpopulation. Small mammals also require understory shrubs for shelter and food. Deer compete with squirrels and rodents for acorns, and changes in cover may alter species composition of rodent communities, which provide food for raptors.

Endangered Bird Species

In the Hill Country, chronic overbrowsing by deer and goats has contributed to the significant decline of the endangered Black-capped Vireo. These birds require low shrubby cover about 1.5 to 6 feet above the ground for nesting. And they cannot survive in areas where the woody plant understory is consistently eaten by deer. Another endangered species, the Golden-cheeked Warbler, may also be impacted by low recruitment of Spanish oak seedlings in areas of high deer density. This bird requires mature cedar trees for nest material but feeds on insects in deciduous trees, where it also often builds its nest. In many places, too many hungry deer are eliminating the Spanish oak seedlings long before they have a chance to mature into trees large enough to keep many insects alive or for nesting birds. Not enough food or nest sites means fewer golden-cheeks.

> *What do impaired plant succession and woodland structure mean to Wild Turkey populations?*
>
> In short, it will be devastating. According to research by Francis L. Russell and Norma L. Fowler, on the eastern edge of the Edwards Plateau, commonly called the Balcones Canyonlands, deer overpopulation is eliminating Spanish oak and plateau live oak seedlings. Unless deer

overpopulation is reduced, there will be no recruitment of new trees, and these woodlands will die out. The lack of these trees and their acorns, which provide a much-needed high-energy fall food for wildlife, including birds, will have long-term effects on the nutrition of future wildlife populations. For example, Wild Turkeys eat acorns, and as the oak tree seedlings are removed by deer overpopulation, fewer and fewer saplings survive to grow into mature acorn-bearing trees.

DEER MANAGEMENT

The Texas Parks and Wildlife Department is the state agency charged with white-tailed deer management. Because TPWD regulates deer hunting, it might seem obvious that the agency should make the deer hunting season longer and increase the annual bag limit. But these solutions have been tried in other places with no significant decrease in deer overpopulation. In the future, we may look to TPWD for further leadership in addressing this complex wildlife management problem, but today deer overpopulation requires multiple approaches. We all can help in our own way.

Do Not Feed the Deer

Addressing all the problems that result from deer overpopulation starts with one simple rule: *do not feed the deer*. Supplemental food can increase deer reproductive rates and intensify the problems of deer overabundance. Concentrating deer in small areas maximizes damage to landscape plants and increases the risk of disease and parasite transmission between herd members. Feeder-fed deer also become less fearful of people, which increases the risk of vehicle collisions and aggressive behavior toward

Deer feeder used for backyard wildlife viewing

people. Deer may also eat at low-hanging bird feeders. This can be dealt with by raising backyard bird feeders or bringing them in at night.

Become a Conservation Hunter

Conservation hunters harvest deer to improve habitat, increasing forb and woody plant diversity, which leads to a greater diversity of birds and other wildlife. Lower deer numbers also mean healthier deer with better body condition and more and healthier fawns. Conservation hunters also provide a sustainable "free-range" meat for their family's dinner table.

Work with Your Neighbors and Intensify Harvest

Because deer move over a huge area, it is extremely difficult for the owner of small acreage to control a deer population. A doe in the Hill Country typically ranges over 320 acres in a single year, and a buck uses 640 acres or more annually. This means that the deer on your 50 or 100 acres are spending more time on your neighbors' property than on yours. Nonetheless, every deer harvested helps, and just because your neighbor does not harvest enough, the deer you remove adds to the overall balance. It is best to manage deer by working with your neighbors to create and implement common goals for managing the deer in your area. Your local TPWD biologist can provide information on how to organize a wildlife co-op/association, help design surveys to monitor the population, and make appropriate harvest recommendations.

To control the deer population in the Hill Country where there are so many more animals present than the food supply can support, we need to remove two of every three deer. Such a drastic reduction will take many years. Now is the time to start. Landowners who understand the presence and consequences of deer overpopulation must intensify annual deer hunting on their property.

Harvest Does

Trophy buck hunting is embedded in deer hunter psychology and values. Years of trophy buck hunting have produced a white-tailed deer population that has five times more does than bucks. A normal sex ratio would be 1:1. In such an imbalanced population, doe harvest is key to population management. The annual removal of large numbers of does is essential to reducing density and improving deer health and size.

Practice Routine Urban Trap and Removal

In suburban areas such as Fair Oaks Ranch and Hollywood Park, hunting may be unsafe and is prohibited. For deer overpopulation control in urban or suburban settings, a city or community can conduct rou-

> ### A PERSONAL STORY: FAMILY HUNTING
> DUSTY BRUNS
>
> My family moved to the Hill Country from Houston in late 1948, just in time for a record cold winter and the 1950s drought. It was hard times, and we pretty much lived off our garden, free-range chickens, and wild game (mostly rabbit, squirrel, and some deer). The cattle were too valuable to eat. Venison was always a welcome treat, but there weren't many deer back then. It was during this time that my great uncle was a frequent visitor for hunting. Many of the bucks he took during the 1950s dressed well over 100 pounds.
>
> After the late 1960s and 70s, when the screwworm flies were eradicated, we started to have more and more deer. Now, we have so many that their food is getting eaten up and our hardwood trees are not being replaced by young ones. With scarcer and scarcer food, the deer weigh less and less. Now a decent buck might dress out at 75 pounds. Our entire family is involved in deer hunting and processing the venison into sausage, steaks, jerky, and salami. Our son-in-law showed a picture of a deer to the twin two-year-old granddaughters and asked if they knew what it was. They both quickly replied, "Jerky!"

tine trap and removal. To be effective, it must be a long-term, population maintenance program. Currently the two available options are to relocate the deer or slaughter and provide the meat to organizations such as food banks. Relocation may seem more palatable; however, careful consideration of what sustainable habitat is available places extreme limitations on this option. Trapping and slaughtering are conducted to standards similar to those in the commercial meat industry, and meat must be provided to nonprofit organizations like food banks. For information about available deer trap and removal options, contact your local TPWD biologist.

Forget Deer Contraceptives

Towns and suburbs with high deer density are not limited areas with a known number of confined deer. Unlike domestic animals, deer cannot be counted, caught, and treated. Contraceptives and an annual booster would have to be administered by rifle. Because of the obvious difficulties and exceptionally high cost, birth control as a method of controlling deer overpopulation is not a practical solution. Thus, no contraceptives are currently labeled or available for use in deer in Texas.

Be a Conservation Gardener

Not everyone is temperamentally suited to be a deer hunter. If you are a gardener and not a hunter, there is still something that you can do to counteract the harm done by deer overpopulation. Native plant propagation and protection might be for you. Are you landscaping with plants the deer love or hate? Conservation gardening calls for propagating the native plants that deer love. By landscaping with these "priority species," whose populations are declining, your garden increases native plant diversity and, as a result, bird diversity. Your garden also becomes a nursery of native plant seeds that can spread into the surrounding environment and flourish as soon as the deer population is reduced.

Do you know a priority native plant when you see it? To be an effective conservation gardener, you must know the priority species. If you live in an area with excessively high deer density, these are the native plants that are now uncommon—probably because they are white-tailed deer favorites. Learn to love the plants that deer love, find them, plant them, and protect them from the deer. Many of these species are the cornerstone of diverse and productive bird habitat. If these species are not available at local nurseries, ask for them and encourage your favorite nursery to propagate them. Convince your neighbors, friends, and relatives to request and garden with them too. If there is a demand, commercial growers will make them available. These are some of the priority Hill Country natives. For a complete list, see appendix 3.

- American smoke tree
- Bigtooth maple
- Blanco crabapple
- Canyon mock orange
- Escarpment black cherry
- Hawthorn
- Mountain mahogany
- Redroot
- Rusty blackhaw
- Sycamore-leaf snowbell
- Texas and red mulberry

Use Deer Exclosures

Once planted, priority species must be protected. Unless you live behind a high deer fence, this means surrounding these plants with a deer exclosure. In conservation gardening, we must learn to live with deer exclosures. A handsome wire-fence circle around a lovely tree or cluster of trees and shrubs is not so far-fetched. Deer exclosures can be both cre-

HOW TEXAS HILL COUNTRY PRIORITY PLANTS BENEFIT BIRDS

PRIORITY PLANT SPECIES	PARTS EATEN BY BIRDS	BIRD SPECIES BENEFITED	
Blanco crabapple	Small, green apple Ripens October	Cedar Waxwing Sparrows	
Carolina buckthorn	Yellow-red-black fruit Ripens late summer–fall	Carolina Chickadee Gray Catbird Northern Mockingbird Thrashers Thrushes Wrens	
Escarpment black cherry	Black drupe Ripens June–October	American Robin Eastern Bluebird Northern Flicker Northern Cardinal Northern Mockingbird	Thrashers Thrushes Vireos Western Scrub-Jay Woodpeckers
Hackberries	Small red fruit Ripens early fall	American Robin Black-crested Titmouse Blue Grosbeak Brown Thrasher Bullock's Oriole Cedar Waxwing Common Raven Eastern Bluebird Eastern Phoebe	Golden-fronted Woodpecker Gray Catbird Northern Bobwhite Northern Cardinal Northern Flicker Northern Mockingbird Rio Grande Turkey Thrushes Yellow-bellied Sapsucker
Hawthorn	Red-orange rosehips	Cedar Waxwing Northern Mockingbird Rio Grande Turkey	
Passionflower vines	Insects that live on leaves	Bewick's Wren Yellow-billed Cuckoo	
Red mulberry and Texas Mulberry	Red to purple fruit Ripens May–August Red to black fruit Ripens April–May	Black-crested Titmouse Cedar Waxwing Gray Catbird Northern Cardinal	Northern Mockingbird Orioles Summer Tanager Western Scrub-Jay
Rusty blackhaw	Small black fruit Ripens early fall	Cedar Waxwing Hermit Thrush Northern Cardinal Northern Mockingbird Thrashers	
Virginia creeper	Dark blue fruit Ripens late summer	Black-crested Titmouse Carolina Chickadee Eastern Bluebird Northern Flicker	Northern Mockingbird Thrushes Vireos Woodpeckers

Deer exclosure cages protecting Texas redbuds (Photo by Tyra Cox Kane)

ative and conservative. Perhaps in time, it will be common to see attractive exclosures sold at nurseries right next to priority plants.

HABITAT RECOVERY

Habitat recovers slowly after deer density is reduced. Where long-term deer overpopulation has reduced plant cover and diversity, it has disrupted the natural succession of plant communities and nutrient cycling. In this situation, habitat restoration is a very slow process. And the cascading effects of plant loss extend to other groups, including insects, mammals, and birds. Although ecosystem revitalization is possible after deer density is reduced, resurgence of bird and other animal populations can only follow the slow return of understory vegetation.

A PERSONAL STORY: HIGH-FENCE NATIVE PLANT RECOVERY

SUE TRACY

(Sue Tracy owns 90 acres northwest of Medina and built a high fence around her property to exclude deer.) I'd say that how long it takes to see plant recovery depends on where and how closely you're looking, size of the protected area, and condition of the surrounding mature vegetation. Along our creek and under large shade trees, within a year of the first high fencing I noticed little cherries and Texas ash seedlings coming up everywhere. And I realized that each spring before the fence the seedlings must have been sending out leaves that were devoured by deer.

Within three to five years of installing the fence, those little trees were a couple of feet tall, and I also started seeing small Texas redbuds, Texas madrones, bigtooth maples, lindens, and chinkapin oaks under large shade trees, which surprised me. Although there were a few examples of each species on the surrounding 500+ acres, before the fence, there were none growing in the area we fenced. I assume the seeds had been planted by birds some time long before, but the seedlings had just never had a chance to regenerate.

It did take a few years to totally exclude the deer—with the occasional intrusion by hogs tearing holes or floods washing away whole sections of fence. The deer that had retained a "sense of place" would get back in, so ongoing vigilance is essential. We have to shoot these periodic invaders. There is one downside to excluding deer—the exponential proliferation of greenbrier. Even so, that's a trade-off worth taking. And even 15, 20 years later, I find occasional surprises of plants not noticed before—woody and herbaceous perennials as well as trees.

For a few years before we installed the high fence, I had been planting 30 to 50 trees per year. While most of these survived and in many cases thrived, they remain individuals that have not naturalized or reproduced themselves. Nature seems to know the appropriate soil/microclimate better than I did, and just protecting the habitat and allowing natural regeneration is a far more cost-effective way to go.

A PERSONAL STORY: NATIVE PLANT RECOVERY IN SMALL EXCLOSURES

TYRA BANKS

We have lived on 200 acres outside of Comfort, Texas since 1995. From the beginning, we were very anxious to improve our property, and we have learned from our mistakes. At first, we planted and lost bald cypress, chinkapin oak, blue oak, and smoke trees, and the two bigtooth maples we planted grew less than 10 inches in 15 years!

Then, using the book *Trees, Shrubs, and Vines of the Texas Hill Country,* we took the time to identify priority plants that are already growing on our property. We were amazed to find white honeysuckle, Virginia creeper, elbow bush, pink mimosa, possumhaw, roughleaf dogwood, rusty blackhaw, western soapberry, Texas kidneywood, Carolina buckthorn, Eve's necklace, Texas redbud, bracted passionflower, hawthorn, Texas mulberry, escarpment black cherry, and a Blanco crabapple. Unfortunately, we also found that many of our "volunteer" priority plants were being damaged by deer. At the same time, we saw them growing within agarita, cactus, cedar breaks, and brush piles where they were in much better shape. At this point, we decided to look into the idea of caging the plants we wanted to protect.

After years of trial and error, we developed a cage-building system that works for us. We buy 48-inch-wide, 14-gauge, welded wire cut to length for each plant cage. For stakes, we buy 20-foot pieces of quarter-inch rebar that we have cut into 30- to 32-inch lengths. We use two to three stakes per cage depending on its size and find that they adequately protect the plants from deer. When a caged plant grows over 3 feet high, our cows sometimes eat the leaves and break branches. If this happens, we attach a 24-inch strip of wire to the top of the original cage. Because some plants don't survive periodic droughts, to save money, we prefer building a 48-inch-high cage and adding more wire when needed. Estimated cost for cages and stakes is about six dollars, and cages can always be reused.

Cages do work! We have caged and protected 300 to 400 plants, including 111 immature Texas red bud trees in a 21-acre pasture. And caging priority plants means no shopping, no purchasing, no hauling, no digging, no watering, and no guilt if a plant dies. So, when do you remove the cages? Maybe never. Again, we learned from our mistakes. About a week after I ceremoniously removed the cage of a "saved plant," it was run over by our tractor! After we removed the cages from 26 little Texas redbud trees on our entry lane, white-tailed deer used them as rubs. Luckily, though scarred, they survived. For now, most of our cages remain in place as we search for other options such as plastic tree guards that could work in some situations.

Living in the Hill Country has expanded our knowledge, and on our walks, we still have the thrill of identifying a new priority plant worthy of being—you guessed it—caged!

9 Cedar Management

There is always a well-known solution to every human problem—
neat, plausible, and wrong.—H. L. Mencken

INTRODUCTION

Although there are six species of the genus *Juniperus* native to Texas, only *J. ashei* and *J. pinchotii* grow naturally in the greater Texas Hill Country counties covered in this book. Common names for *J. ashei* are cedar, mountain cedar, or Ashe juniper. Common names for *J. pinchotii* are redberry juniper and Pinchot juniper. Staying with the most common usage, in this book we will call *J. ashei* cedar or Ashe juniper and *J. pinchotii* redberry juniper.

The largest stands of cedar are in the Texas Hill Country, but cedar also grows in disjunctive populations in the Ozark Mountains along the Missouri/Arkansas border, in the Arbuckle Mountains of south-central Oklahoma, and in Coahuila, Mexico. Redberry juniper is limited to the western edge of the Hill Country, and it extends west, also in disjunctive populations, into southwestern New Mexico, southern Arizona, and parts of Mexico. While cedar is by far the most common of the two junipers in the Hill Country, it is extremely important from a stewardship perspective to know which one you have on your property. On the far western edge of the Hill Country, redberry juniper is the dominant species, not cedar.

Fossilized cedar or Ashe juniper pollen from the last ice age (14,000–20,000 years ago) was found in the Friesenhahn Cave in northern Bexar County. Accounts written by explorers and settlers in the 1700s–1900s describe this plant growing in the Hill Country. In 1930, John Buchholz named it as a separate species for William Willard Ashe, a forester with the US Forest Service. Unless otherwise indicated, throughout this chapter what we say about cedar applies to redberry juniper as well. Information on history and management that applies specifically to redberry juniper is at the end of this chapter.

HISTORY

After European settlement of this region in the mid- to late 1800s, cedar was logged extensively for fuel, fence posts, railroad ties, construction, telegraph poles, and oil. Cedar chopping was big business. Early settlers and businessmen were cutting trees 20 to 25 feet tall and with a 1.0- to

1.5-foot trunk diameter. These big trees were plentiful, and the wood was especially valued for its durability. In an 1874 Austin newspaper, a representative of the Central Railroad reported that 200,000 cedar ties had been shipped from Austin in the previous two years.

> *Why do cedar and redberry juniper create special problems for land managers in the Hill Country?*
> Both are invaders of grassland communities and considered noxious pests by ranchers because both reduce biodiversity and limit the land's carry capacity for livestock and wildlife.

MANAGEMENT

Even if you were raised to believe that the only good cedar is a dead cedar, it is not too late to change a little and begin to manage your cedar in a new way that benefits Hill Country birds. Cedar is an important component of Hill Country woodland habitat. When managing cedar, you must consider its benefits to birds and other wildlife. You can learn about the birds that need cedar to survive and decide if you want to work with the cedar on your land to benefit these species.

Cedar and Birds

Cedar provides bird food, nest sites, warmth, and cover. In fall and winter, female cedar trees produce the blue cedar berries eaten by Northern Cardinal, House Finch, American Goldfinch, American Robin, Eastern Bluebird, Cedar Waxwing, and Western Scrub-Jay. In winter, when insects are scarce, these and other wild birds depend on this abundant food source. In spring and summer, cedar trees provide nest sites for many birds, including Golden-cheeked Warbler, Greater Roadrunner, Western Scrub-Jay, and American Robin.

- The Golden-Cheeked Warbler is an endangered species that nests exclusively in the Hill Country. It breeds in the sheltered canyons and east- or north-facing slopes of the Edwards Plateau and builds its nest using slender strips of bark picked off older cedar trees. This rare species eats insects from the leaves and branches of cedar and hardwood trees within its habitat.
- Greater Roadrunners live in habitat with a combination of open space and compact cover of sturdy, low shrubs. In the Hill Country, roadrunners commonly hide their large, flat, stick nests in dense, second-growth cedar trees. And each pair needs a huge area (60–250 acres) of good habitat to feed and rear their young.

Mixed wooded slope, Golden-cheeked Warbler habitat

- The Western Scrub-Jays in the Hill Country live in thickets of cedar and oak trees; their nests are well hidden, and the adult birds are very secretive around their nests.
- Some American Robins also nest in the Hill Country. They prefer low shrubs, including cedar, growing on land near water. The water provides moist soil for earthworm and insect food, and the thick, second-growth cedar bushes make a good hiding place for their nests.

"Ball cedar," young cedar growing in full sun

In winter, flocks of finches, sparrows, robins, titmice, and chickadees gather in Hill Country mixed woodland and savannah habitats. Round cedar bushes, sometimes called ball cedar, provide excellent shelter for these winter birds. Ball cedar, with its heavy branches and leaves growing all the way to the ground, creates an enclosed microhabitat that is much better shelter than tall, deciduous trees or small, slender shrubs with thin branches at the top. Birds disappear into ball cedar where they can roost out of the cold north wind.

Woodland birds also hide from predators in cedar's dense, evergreen foliage, especially in winter when the deciduous trees and shrubs have dropped their leaves. At this time of year Cooper's Hawk and Sharp-shinned Hawk, both agile, woodland bird hunting specialists, are in the Hill Country looking for food.

Cedar and Other Animals

White-tailed deer, ring-tailed cat, gray fox, raccoon, white-ankled mouse, and even rock squirrel eat cedar berries. White-tailed deer hide among cedar trees with low branches. Oak and cedar trees growing on bluffs and rocky slopes are preferred habitat for the Texas spiny lizard. Rough green snakes and lizards hunt insects in cedar trees, especially those covered with grape vines. And cedar is host plant to the bright, olive-green juniper hairstreak butterflies.

Cedar and Water

Like any tree, cedar requires water to survive. Water must be taken in by the tree roots and lifted to the leaves, where photosynthesis converts sunlight into food. Some water is lost through transpiration in a tree's leaves, and some trees such as cedar are better than others at preventing excess transpiration.

Does cedar waste water?

The short answer is no. Cedar, like other trees, needs a lot of water to live, but cedar does not waste water. If cedars did use an inordinate amount of water, they could not survive and spread so widely across the Hill Country's relatively dry landscape. Cedar is, in fact, a xeric plant that retains and uses water very efficiently. This is possible because a cedar tree has small leaves covered with a waxy coating that reduces transpiration or water loss. A tree such as sycamore, which has big, thin leaves with a large surface area, is a water guzzler because of the high rate of transpiration or water loss through its leaves. Thus, sycamores grow in riparian zones where there is plenty of soil moisture.

The cedar's reputation as a greedy water hog has been refuted by research done by Jim Heilman in the Department of Soil and Crop Sciences at Texas A&M University. He conducted tests on trees growing in a live oak/cedar woodland near San Marcos, and his results showed that cedar uses less water than live oaks. Cedar roots are shallower than oak roots, and the internal plumbing of cedar greatly restricts the amount of water they can take up.

Why does cedar have a greedy water hog reputation?

Dense cedar thickets create a closed canopy where the overhead cover catches or intercepts precipitation. In a light rain, it can prevent all or nearly all the water from reaching the ground. In moderate to heavy rain, some will reach the ground, but some does not. Much of the rainwater captured by cedar trees in this way evaporates back into the atmosphere, so it cannot be absorbed by the soil. Because steep wooded slopes are especially vulnerable to erosion, a cedar thicket on these sites can help reduce soil loss by preventing large amounts of rainwater getting to the ground all at once and rushing down the hillside. Thus, the soil on a steep cedar-covered slope is more likely to absorb the rainwater as it slowly makes its way to the ground than the soil on a cleared slope.

What is the evidence for clearing cedar to conserve water?

Based on his research on the Freeman Ranch near San Marcos, Jim Heilman has said, "The idea of brush removal to save water is a case of where 'policy gets ahead of science.'" In addition, in a 2009 paper he coauthored, Heilman and his colleagues reported, "A brush control project was begun in the North Concho watershed 10 years ago, and even though this period included one of the wettest years on record (2007), there has been no evidence of increased flow. In other words, brush control has been tried and has not worked in terms of increasing stream flows to the North Concho." They concluded, "There is no compelling evidence at this time that it is a viable strategy for increasing water supply. In the case of the North Concho watershed, there is strong evidence that it does not increase water supply."

Some land management agencies support cedar clearing on a landscape scale to enhance stream flow and aquifer recharge. Although cedar is essentially a watertight species, dense cedar thickets catch rainwater and prevent it from reaching the ground. Beyond just being a large plant that needs water and dissolved nutrients to grow, this is the main reason dense cedar thickets are cleared on a wide scale—to allow more rain to reach the soil where it can benefit more "desirable" plants.

CEDAR-CLEARING GUIDELINES

Always keep bird habitat conservation in mind, and take your time. Cedar clearing can be an endless project, but if you stay with a plan, it will get easier as time goes on.

- Focus on regrowth cedar. Benefits will be greatest, and cost will be least.
- Do not cut old-growth cedar, and do no more than thin cedar growing on steep slopes, where cedar provides especially important wildlife habitat.
- Clear in phases. Cedar management can be overwhelming and seem impossible. Divide the job into small parcels. Put your cedar-clearing efforts on a rotation cycle. When making a cedar-clearing schedule, consider your physical and financial constraints. Cutting modest-sized plots instead of clearing everything at once is less stressful on you and the environment.
- Remove young cedar seedlings when they are very small. Dense shade under cedar inhibits normal growth of its seedlings. When clearing opens up an area, shade is replaced by sunlight that stimulates rapid growth of existing cedar seedlings (which may be present in huge numbers). These must be removed to prevent a thicker cedar forest than existed before clearing. If prescribed burning is an option, it is an excellent method to control this early regrowth cedar. Once you get ahead of the cedar, your annual work will diminish.
- Plan for long-term and regular cedar removal. If you do not, it will come back to haunt you. To improve habitat for birds and other

A PERSONAL STORY: TRAIL BUILDING WITH CEDAR

BILL KENNON

Trail Building with Cedar:
The Prune, Chip, Mulch Method

1. Plan the course of your trail. It will depend on the terrain and ability to get a tractor or truck pulling a chipper safely through the woods. I prune a 100-foot-wide strip first and then mark the best route later.

   Tip: A serpentine trail is more aesthetically pleasing than a straight one. A trail perpendicular to a slope will control erosion.

2. Prune lower limbs as high as you can reach with the chainsaw (about 8 feet). Save limbs larger than 3 inches in diameter to line the trail.

   Tip: Cutting off the smaller branches on these limbs while they are still on the tree is easier than when they are lying on the ground. The Stihl 14-inch chainsaw is my favorite for pruning. Pruning is the most time-consuming part but also the most rewarding as you transform an impenetrable cedar break into a walkable woodland. Since I do this part time, slash is often left on the ground for a year or more. This helps prevent erosion since more rain can get to the ground. Downside is that chipping old dried cedar is very dusty.

3. Hire a crew to drag and stack the slash along both sides of the trail before renting a chipper. This will dramatically increase productivity during chipping.

   Tip: A four-man crew produces maximum productivity (two draggers and two stuffers).

4. Save large branches for lining the trail. Work alongside the crew so you are in control of the whole process. Provide crew with eye/ear protection and good-quality dust masks.

Trail building with cleared cedar chips and rails (Photo by Bill Kennon)

5. Rent a chipper. I have always rented a Vermeer 1000 chipper, which produces a fine mulch but can handle large branches. A smaller chipper may suffice as long as you are saving the large branches to line the trail. Weekend rates are cheaper. Keep a chainsaw handy for last-minute cutting. Aim the chute down to blow the mulch into distinct piles in a cleared spot accessible to a tractor bucket.

6. Rake the trail clear of rocks, and cut stumps level to the ground.

7. Place mulch piles near trail with a tractor. Allow only 3 to 4 feet between piles to make spreading easier. Rake mulch piles flat with a four-pronged claw rake. Mulch will compact to about half its depth after a year, so 4 to 6 inches deep is ideal.

8. Line trail with the large branches that were saved.

   Tip: Buying native seed is not necessary. To speed up regrowth of native grasses, rake cedar needle mats down to soil level to release stored grass seeds and allow water to reach the soil.

9. Have fun walking your new woodland trail.

wildlife, write a plan that makes a difference but is something that you can actually do year after year.

- Following cedar removal, protect seedlings of other native plants from the deer. Young woody plants or their seeds are on the ground beneath the cedar. They are ready to sprout and grow when cedar is removed. They may not be able to reestablish themselves unless protected by piles of slash or deer exclosures. You also can use slash to create a large exclosure to protect areas of an acre or more.
- When cutting cedar, make the most of a plentiful natural resource. There are many uses for cut cedar: mulch for gardens and trails, posts for fencing, slash for erosion control, brush piles, and exclosure corrals.

CEDAR MANAGEMENT PLAN

A cedar management plan depends on your land stewardship goals, age of the cedar, soil condition, aspect and slope of the land, size of your pocketbook, and species of juniper.

Planning Steps

Many people approach cedar with emotion and prejudice. To overcome these obstacles, be methodical. Plan and work in steps. Cedar clearing does not have a single "no thinking required" strategy.

1. Ask, "What do we have now?"

- Map your property and plot the plant communities present. Google Earth (www.earth.google.com) is a good resource. Another is the NRCS Web Soil Survey (http://websoilsurvey.nrcs.usda.gov/app/HomePage.htm).
- Label streams and their riparian zones, ponds or tanks and lakes, springs and seeps, as well as grasslands, mixed wooded slopes, live oak savannahs, shin oak savannahs, and post oak savannahs.
- Assess each area named on the map to determine where cedar is present. Make notes on the age and abundance of cedar in each area.

2. Ask, "What do we want?"

- Record the desired use for each habitat area. Some common uses in addition to bird habitat are deer hunting, livestock grazing, and family recreation. You may want some bird habitats to support multiple uses. In most but not all cases, this should be possible.
- Write short-term goals that may be accomplished in one year.

- Write long-term goals that address what you want your land to look like in 5, 10, or even 20 years.

3. Select cedar management that will benefit bird habitat.

- Think mosaic. When clearing cedar for bird and wildlife habitat, leave block-shaped areas with some cedar standing. Patches of cedar thicket where branches grow to the ground are important protection for birds and other wildlife. Dense cedar thickets also exclude or limit deer access and may protect some forbs and woody plants from being eaten.
- Decide where extensive clearing will benefit bird habitat and where selective removal is best for birds and other wildlife.
- Healthy trees do not necessarily have a neat, traditional umbrella-like form. But if you want to trim lower branches on some cedar trees to give them a classic tree shape, be sure to leave plenty of woody understory and ground cover nearby. Birds and other wildlife need low-level hiding places to avoid predation and escape inclement weather.
- Be sure that your land stewardship practices can lead you to your goals. You may have a pet project in mind that you just want to do. Be careful. Never do cedar clearing just because it is easiest, what has always been done, or what your neighbor suggests. It could be a waste of time and money. Only when your goals are clear can you take the steps necessary to reach them.

Why not get rid of it all?

Cedar in moderation is a natural component of most Hill Country habitats. Clearing every stick of cedar is rarely the best choice for environmental quality or stability or for wildlife. For nearly all birds and other wildlife, acres of parklike landscape with an expanse of short grass under a monoculture of live oaks is a poverty-stricken wasteland. Birds eat cedar berries, nest in cedar branches, and find shelter from weather and predation in thick cedar trees.

If your land presents a case of indiscriminate or reckless cedar clearing, convert to cedar management using preferred clearing methods done by a well-informed and cooperative operator. For areas already extensively cleared, look at where young cedar trees are regenerating and decide which ones to keep and which might be culled.

Cedar-Clearing Instructions in Hill Country Habitats

Cedar clearing differs a great deal from one Hill Country habitat to another. Keep in mind that many birds and other wildlife eat cedar berries.

> **A PERSONAL STORY: CEDAR-CLEARING MISTAKE**
> DICK PARK
>
> Much of our land was choked with cedar. The hardwood trees struggled for space, and no sunlight or rain penetrated the cedar canopy to support undergrowth. We wanted our land to become more bird friendly and imagined free-standing oaks, elm, cherry, grasses, and shrubs. We hired a crew to clear cedar but leave the largest old cedar and those on steep canyon sides. Unfortunately, we could not be present to make sure that our orders were followed, and almost every stick of cedar was removed. Our land was left bare, and we lost a lot of soil.

Take the following into consideration to create a sustainable stewardship plan.

CEDAR CLEARING ON MIXED WOODED SLOPES

Mixed wooded slopes are also called oak/juniper woodlands. This name recognizes that cedar is a major component of the plant community. Wet mixed wooded slopes will typically have a mix of cedar and hardwoods such as escarpment black cherry, bigtooth maple, Carolina buckthorn, and Texas ash. And dry mixed wooded slopes might have Spanish oak, cedar, mountain laurel, gum bumelia, and Mexican persimmon.

On mixed wooded slopes, do only selective cedar clearing. Here, thinning cedar to improve plant diversity may be desirable, but be careful to maintain at least 50% canopy cover—70%–80% is better. Place slash in contact with the ground to hold the soil until the grass returns naturally. In addition, new grasses need time to produce well-developed root systems and ground cover that can catch rainwater and prevent erosion.

CEDAR CLEARING ON SAVANNAHS AND GRASSLANDS

If you wish to enhance the habitat value of live oak savannah or maintain or promote its natural mosaic character, keep some cedar in patches or clumps large enough to provide escape cover for birds and large animals. The patches can be in wide strips. Do not keep narrow strips of cedar because this does not give the protection needed by bird or other animals. The best cedar patches for birds are a meandering shape with irregular edges and varying widths. Avoid a neat geometric pattern.

Post oak savannah, when healthy, is composed of tall post oaks over mostly medium high grasses and forbs. Many birds depend on these big trees and the open habitat beneath them. In some post oak savannah

Western Scrub-Jay in cedar growing on mixed wooded slope (Photo by Lora L. Render)

sites, cedar has moved in and become so thick that there is little grass, and movement through the thicket is almost impossible. In this situation, removing most or all of the cedar will restore the habitat to a healthy condition. Post oaks are slow to regenerate. Leaving cedar slash on the ground gives post oak seedlings protection from deer browse and assures their slow but sure return.

Shin oak savannah is found on dry, rocky terrain with shallow soil and is covered with short grasses, forbs, and scattered mottes of low-growing shin oaks and shrubs where roadrunners and jackrabbits are common. To maintain its normal structure and inherent value for bird and wildlife habitat, remove all or almost all the cedar.

CEDAR CLEARING ALONG RIVERS AND CREEKS

Near Hill Country rivers and creeks, cedar clearing should never be all or nothing. Along healthy streams, cedar is a natural but minor component of riparian habitat, and big cedar trees, essential in Golden-cheeked Warbler habitat, often grow adjacent to dry streambeds. Riparian soils are easily eroded, and the root systems of large trees often hold the soil and prevent erosion. If you have a solid stand of regrowth cedar near a stream, clear alternating strips of 25%–30% cleared and 75%–80% cleared along the length of the stream.

By focusing on removing regrowth cedar and leaving other species, you will protect plant diversity and improve wildlife habitat. Use the

Cedar-choked post oak savannah

Healthy post oak savannah

most selective methods possible. In this way, you minimize the potential for erosion and maintain the existing variety of trees, shrubs, vines, and herbaceous plants. Hand clearing is the best technique, and hydraulic tree shears on a Bobcat with soft tires may be useful on level places. And as elsewhere, take your time. Make a plan and follow it. When one cleared section begins to reestablish a healthy plant community, clear the next section.

CEDAR CLEARING IN CANYONS

Historically, typical Hill Country canyons included cedar as part of their natural and diverse woody plant community. Fire that limited cedar in other habitats was probably not as common in canyons because of their steep slopes, bare rock, high moisture, and lower air temperatures. Consequently, these canyons were and still are excellent Golden-cheeked Warbler habitat, an avian specialist that breeds in this habitat. However, today, cedar thickets dominate many canyons because long-term overbrowsing by deer has eliminated natural competition by the deer's favorite food plants. Follow these guidelines when clearing cedar in steep canyons:

- *Use extreme caution when cutting cedar in a canyon or near a spring or seep.*
- Preserve dense overhead canopy.
- Remove cedar only from September through February to avoid bird breeding season.
- Clear only with hand tools to protect fragile soil in steep terrain.
- Leave large cedar trees.
- Cut only the smaller, regrowth cedar.

When clearing cedar at the mouth of a canyon or in shallow canyons, where the land is wide and flat with limited canopy, follow these guidelines:

- Clear only from September through February.
- Leave large mature cedar trees.
- Remove most or all of the smaller, regrowth cedar.
- Use appropriate equipment, including Bobcat with tree shears and hand tools.

CEDAR CLEARING AT SPRINGS OR SEEPS

Springs are unique microhabitats that often contain a variety of aquatic and terrestrial plants and animals that make these sites rare and highly productive habitat, providing shelter, food, and water to a host of wildlife. Seeps are moist microhabitats with a limited water supply and are especially important to many birds of the Hill Country. Unfortunately, today dense stands of cedar can reduce the diversity of a spring or seep and impact its productivity. Thus, careful removal of cedar can encourage the growth of native plants intrinsic to these special places.

- Take your time.
- Assess the area to determine the best approach.
- Survey and identify birds using the area.

- Use only hand tools such as chainsaw, lopper, or metal blade trimmer to protect the desirable plants and delicate soils around springs.
- Near springs or seeps on flat or rolling land or on south- or west-facing hillsides with few trees other than cedar:
    - focus on dense stands of cedar, especially regrowth cedar
    - leave some large cedar trees in the vicinity to provide important shelter and berries for birds and other wildlife
- At a spring or seep found in a canyon or on a wooded slope that may be Golden-cheecked Warbler habitat:
    - focus on small regrowth cedar
    - cut only from September through February
    - leave larger cedar trees needed by species that breed in this habitat
    - maintain a closed canopy
- Many plants that normally grow at an undisturbed seep or spring are not readily available from seed companies, but with good land management practices most native plants will return on their own.
- Be patient, and take time to enjoy the recovery process.

CEDAR CLEARING NEAR CONSTRUCTED TANKS, PONDS, AND LAKES

Remove all cedar growing on a dam. Tree roots can create structural problems for dams and should be removed. Depending on the slope of the dam, cut cedar by hand or with hydraulic tree shears on a Bobcat. Be selective when cutting cedar around the perimeter of a tank or lake. Leave some cedar to provide protective cover for wildlife coming to water. When working around an existing tank or lake or building a new one, remember that cedar left on the bottom or at the shoreline becomes valuable fish or bird habitat.

CEDAR CLEARING IN BACKYARDS

Hand tools are best for removing cedar in a yard. Before removing cedar in your landscape, consider its value as winter cover for birds and as visual screen, privacy for you and your family. A common mistake is removal of lower branches or limbing up. This dramatically reduces a tree's benefit to wildlife because it removes low-level shelter for many winter birds such as Northern Cardinal, finches, Pine Siskins, and sparrows that live near the ground. Make space between the cedar trees that you keep, so their lower branches get enough sun to thrive. This prevents the unsightly die-off of lower limbs commonly seen in cedar thickets.

Cutting small cedar with long-handled tree trimmer

Cedar Removal Methods

Cedar can be removed by mechanical tools, prescribed burns, chemical means, or even goats.

MECHANICAL TOOLS

Chainsaws, nippers, and loppers are the hand tools recommended for small jobs or where adequate human labor is available. Although labor intensive, hand clearing allows the most careful selection of trees to be removed and those to be saved.

Hydraulic tree shears are preferred for removing cedar 2 to 14 inches in diameter. Shears are usually mounted on the front of a rubber-tired Bobcat and will nip off a tree at or near ground level. Using a Bobcat with shears is favored because it is less ground disturbing and more selective than other methods, such as a bulldozer. However, Bobcats are not safe on steep slopes where cedar should not be cleared anyway because it damages habitat and yields little range benefit.

A hydro-axe is a machine that chews a tree from its top down to the ground. This machine is less selective than hydraulic shears because it often consumes other tree species growing close to the cedar. A hydro-axe creates a deep mulch pile that must be spread to about 2 inches deep so grass and forbs can grow there. It has its place in Ashe juniper clearing, especially when you are working on an old field with solid regrowth cedar and you want to restore it to grassland. A hydro-axe is not effective on redberry juniper, which resprouts after treatment.

An excavator is a specialized piece of equipment that usually moves

Cutting cedar at ground level with hydraulic tree shears

around on tracks and has a large arm with a bucket or scoop. Excavators are used to control mature redberry juniper on generally flat terrain with relatively deep soil. With the arm and bucket, an excavator can uproot mature redberry juniper, including the bud zone.

A bulldozer might be used on a large ranch or when making a wide path through a cedar thicket, but it will flip big rocks onto the surface and chew up the ground. The bulldozer's size and power make it almost impossible to save any desirable plants in the path of the machine. If you must use a bulldozer, hire an operator who will follow your instructions. It is safest to be present during the work. Tag trees to be saved so the tags are clearly visible to the operator well in advance of approaching the trees.

Chaining is a land-clearing method in which a massive chain is dragged across the land between two huge Caterpillars. If chaining is used to turn a cedar-covered pasture into a grassland for cattle, all tree species are destroyed indiscriminately, and no desirable shade trees remain. Chaining also profoundly disturbs the soil so that grass regrowth is inhibited. Thus, chaining is never recommended.

FIRE/PRESCRIBED BURNS

On rural property with second-growth Ashe juniper under 5 feet tall, fire is a very good way to remove cedar. *Do not attempt to burn old-growth cedar.* The crown fires too easily get out of control. If you wish to burn

Removing mesquite with excavator bucket (Photo by Dale Schmidt)

a parcel that is not within a subdivision or city limits, contact your local prescribed burn association, TPWD, the NRCS, or your local Texas AgriLife Extension agent.

CHEMICALS

Chemicals that control cedar through foliar application are useful only for cedar less than 3 feet tall. Some experts in integrated brush control discourage chemical treatment for cedar management in the Hill Country and recommend that small cedar be grubbed out or cut. If you must use a chemical to kill small cedars, two herbicides, Velpar and Pronone Power Pellet (both with the active ingredient hexazinone), are the least

Goats eating low branches of mesquite tree

expensive. Hexazinone has above-average soil mobility and is nonselective (meaning it kills all tree species). Thus, to protect nontarget trees, the treatment area must be three to five drip lines away from species such as oak, elm, sycamore, cottonwood, and cherry. On a slope, where most cedar grows, hexazinone will move through the soil even farther downhill from the treatment site. This adds to the danger for nontarget trees. Hexazinone degrades more slowly in poorly drained soil. Thus, never use chemical treatment to remove cedar growing in clay or near marshy places such as a seep, spring, or fen. *Follow the label directions exactly.*

GOATS

In some very limited circumstances, goats might be used to control cedar. A cedar plant will die when all its green growth is removed. Thus, to kill a cedar, the goats must eat all of it. When working on a pasture with early second-growth cedar at a height accessible to the goats, pasture a large herd of goats during December–February. The disadvantage in this cedar-control method is that goats will eat most other woody plants before they consume the plant you want most to control. Since cedar is low on the goats' preference list, small nontarget evergreen species such as live oak and the tender shoots and branches of small deciduous shrubs and hardwoods will be eaten first.

REDBERRY JUNIPER

Before European settlement and widespread ranching, frequent wildfires prevented the spread of redberry juniper. Thus, until the mid-1800s, redberry juniper was restricted to rocky cliffs and north slopes. In the late 1800s and early 1900s, it moved to adjacent grasslands and now grows in a variety of soils other than rocky hillsides. Overgrazing also may have played a role in its spread by creating more bare ground for seedlings and reducing competition from perennial grasses.

Because of its unique biology, even some prescribed burns used to control cedar are not as effective with redberry juniper. Unlike Ashe juniper, redberry juniper is a "basal resprouter," meaning that it will resprout prolifically when cut below the lowest-growing branch but above the bud zone, a region on the plant where resprouting occurs after the body of the plant is removed.

Redberry juniper growth is remarkably slow—approximately 2 to 3 inches a year for the first 10 years. In that first decade of growth, the bud zone of most trees is just above the soil, where it is vulnerable to fire, which will kill about 70% of them. But after a redberry juniper tree is about 10 years old (or 2 to 3 feet tall), the soil beneath it has built up enough to cover and protect its bud zone from fire. Thus, burning older redberry juniper trees, which are usually multitrunked, does not kill them. Redberry juniper is drought tolerant, but some trees, especially on hillsides where there is very little soil, may die in historically dry years, such as 2011.

Redberry Juniper and Hill Country Birds

Redberry juniper provides food for wintering American Robins and Cedar Waxwings. During some winters, thousands of American Robins feed in stands of redberry juniper until all available food is consumed before they move on. Where redberry juniper grows on shallow, rocky soils at the heads of draws, it tends to stay shrublike. Thus, it adds to the mix of low-growing shrubs providing habitat for Black-capped Vireos. Consequently, although not known to be a nest site for these threatened birds, redberry juniper may provide important protective cover or insect-feeding places. Resident birds, including the Western Scrub-Jay, Canyon Towhee, Rufous-crowned Sparrow, and Northern Mockingbird, use redberry juniper for shelter, and wintering Sage Thrashers are found almost exclusively in stands of redberry juniper. Where it is mixed with other woody species like plateau live oak, it also provides wintering habitat for Bushtits and Spotted Towhees.

Redberry Juniper Removal Methods

The cost of controlling redberry juniper can be high because it is often necessary to retreat many times. This is especially true for mature trees, which usually require the added expense of applying both mechanical and chemical treatments.

MECHANICAL CONTROL

In order to kill young plants whose bud zones are still above the soil surface, hand-cut seedlings or saplings at ground level. Hand-grub plants up to 28 inches high when they are small enough to be uprooted. When using mechanical control in shallow rocky soils, use an eight-year treatment cycle to remove regrowth before it gets too big. Plants grow faster in deeper soils, so more frequent treatment every six or fewer years is necessary.

CHEMICAL AND MECHANICAL CONTROL COMBINED

For mature redberry juniper trees where the bud zone is below ground, combine chemical and mechanical methods for effective control. Cut trees and chemically treat stumps using Tordon 22K. Foliar herbicides are effective on trees under 3 feet tall, as they are for Ashe juniper. For complete information on the use of chemicals to control redberry juniper, see the Texas A&M AgriLife Extension publication "Brush Busters: How to Master Cedar," available at https://oaktrust.library.tamu.edu/bitstream/handle/1969.1/87766/pdf_722.pdf?sequence=1&isAllowed=y.

PRESCRIBED BURNS

Prescribed burning is effective only on redberry juniper seedlings and saplings when their bud zones are above the ground. Even then fire often kills no more than 70% of trees. A mature redberry juniper tree can be top-killed by fire but will regrow from the base. After burning regrowth redberry juniper, burning must be repeated before the trees become 5 feet tall. This will knock the trees back again but not kill them.

Appendix 1. Deer Browse Evaluation

This deer browse chart is designed to evaluate deer browse on wooded slopes, savannahs, and riparian habitat in the Texas Hill Country. To evaluate habitat, find three to five plants in each category and answer "yes" or "no" to the two questions for each.

Moderate browse problem: First- and second-choice plants present but hedged and not reproducing
Severe browse damage: Second-choice plants with heavy or moderate browse and third-choice plants with any browse
Woody plant community impaired and not sustainable: Either a moderate browse problem or severe browse damage

COMMON NAME	SCIENTIFIC NAME	BOTH SEEDLINGS AND SAPLINGS PRESENT?	PLANTS BROWSED OR HEDGED?
	First-choice plants for deer		
Carolina buckthorn	*Rhamnus caroliniana* var. *caroliniana*		
Cedar elm	*Ulmus crassifolia*		
Eve's necklace	*Sophora affinis*		
Goldenball leadtree	*Leucaena retusa*		
Guayacan, soapbush	*Guajacum angustifolium*		
Hackberries	*Celtis* spp.		
Hawthorn	*Crataegus* sp.		
Huisache	*Acacia minuata*		
Kidneywood	*Eysenhardtia texana*		
Mountain mahogany	*Cercocarpus montanus*		
Old man's beard	*Clematis drummondii*		
Rose pavonia	*Pavonia lasiopetala*		
Rusty blackhaw	*Viburnum rufidulum*		
Slippery elm	*Ulmus rubra*		
Southwest bernardia	*Bernardia myricifolia*		
Spanish oak	*Quercus buckleyi*		
Texas madrone	*Arbutus texana*		
Texas mulberry	*Morus microphylla*		
White honeysuckle	*Lonicera albiflora* var. *albiflora*		
	Second-choice plants for deer		
Aromatic sumac	*Rhus aromatica*		
Blackjack oak	*Quercus marilandica*		
Carolina snailseed	*Cocculus carolinus*		
Catclaw, Roemer's acacia	*Acacia roemeriana*		
Cenizo	*Leucophyllum frutescens*		
Chinkapin oak	*Quercus muehlenbergii*		
Cowitch vine	*Cissus incisa*		

COMMON NAME	SCIENTIFIC NAME	BOTH SEEDLINGS AND SAPLINGS PRESENT?	PLANTS BROWSED OR HEDGED?
Desert sumac	*Rhus microphylla*		
Escarpment black cherry	*Prunus serotina* var. *oblongifolia*		
Evergreen sumac	*Rhus virens*		
Flame-leaf sumac	*Rhus lanceolata*		
Grapes	*Vitis* spp.		
Greenbrier	*Smilax bona-nox*		
Gum bumelia	*Bumelia lanuginosa* var. *oblongifolia*		
Hog plum	*Colubrina texensis*		
Lacey oak	*Quercus laceyi*		
Live oak	*Quercus virginiana* var. *virginiana*		
Netleaf hackberry	*Celtis reticulata*		
Poison ivy	*Rhus toxiocendron* var. *vulgaris*		
Possumhaw	*Ilex decidua*		
Post oak	*Quercus stellata* var. *stellata*		
Shin oak	*Quercus durandii* var. *breviloba*		
Texas redbud	*Cercis canadensis* var. *texensis*		
Virginia creeper	*Parthenocissus quinquefolia*		
	Third-choice plants for deer		
Agarita	*Berberis trifoliolata*		
Cedar	*Juniperus ashei*		
Elbow bush	*Forestiera* sp.		
Lindheimer's silktassel	*Garrya ovata* var. *lindheimeri*		
Honey mesquite	*Prosopis glandulosa* var. *glandulosa*		
Lotebush	*Condalia obtusifolia*		
Mexican buckeye	*Ungnadia speciosa*		
Pink mimosa	*Mimosa borealis*		
Poverty bush	*Baccharis neglecta*		
Pricklypear	*Opuntia lindheimeri*		
Texas persimmon	*Diospyros texana*		
Texas walnut	*Juglans microcarpa*		
Tickle-tongue	*Zanthoxylum hirsutum*		
Western soapberry	*Sapindus drummondii*		
Whitebrush	*Alyosia lyicoides*		
Yucca	*Yucca* sp.		

Source: Adapted from draft of a deer browse list for the Kerr Wildlife Management Area in Central Texas.

Appendix 2. Master Bird Chart

CHAPTER[a]	COMMON NAME	HABITATS
1, 3, **6**, 9	American Goldfinch	Old field, river valley prairie, pocket prairie, post oak savannah, shin oak savannah, live oak savannah, river, creek, seep, springs, and backyard
1, 7	American Kestrel	Old field, plateau prairie, river valley prairie, pocket prairie, post oak savannah, shin oak savannah, live oak savannah, river, creek, canyon, springs, seep, tank, pond, lake, and backyard
2	American Pipit	Old field, plateau prairie, river valley prairie, river, creek, tank, pond, and lake
1, 3, 4, 6	American Redstart	Mixed wooded slope, post oak savannah, shin oak savannah, live oak savannah, river, creek, and canyon
2, 4, **6**, 8, 9	American Robin	Old field, pocket prairie, mixed wooded slope, post oak savannah, shin oak savannah, live oak savannah, river, creek, canyon, seep, springs, and backyard
3, 5	American Wigeon	River, creek, tank, pond, and lake
1	Ash-throated Flycatcher	Mixed wooded slope, post oak savannah, shin oak savannah, live oak savannah, river, creek, and canyon
5	Barn Swallow	Old field, plateau prairie, river valley prairie, post oak savannah, shin oak savannah, live oak savannah, river, creek, canyon, springs, tank, pond, lake, and backyard
1, 2, 3, **4**, 7	Bell's Vireo	Old field, post oak savannah, shin oak savannah, live oak savannah, river, creek, canyon, seep, and springs
3, 5	Belted Kingfisher	River, creek, springs, tank, pond, and lake
1, 2, 3, 4, **6**, 8	Bewick's Wren	Pocket prairie, mixed wooded slope, post oak savannah, shin oak savannah, live oak savannah, river, creek, canyon, seep, springs, and backyard
3	Black Phoebe	River, creek, canyon, seep, springs, tank, pond, and lake
3, **4**, 6	Black-and-white Warbler	Mixed wooded slope, river, creek, canyon, and backyard
3, 5	Black-bellied Whistling-Duck	Post oak savannah, live oak savannah, river, creek, canyon, seep, springs, tank, pond, lake, and backyard
1, 7, 8	Black-capped Vireo	Shin oak savannah and live oak savannah
1, 3, 4, 6	Black-chinned Hummingbird	Old fields, pocket prairie, mixed wooded slope, post oak savannah, live oak savannah, river, creek, canyon, seep, springs, and backyard
1, 3, 4, **6**, 8	Black-crested Titmouse	Mixed wooded slope, post oak savannah, shin oak savannah, live oak savannah, river, creek, canyon, seep, springs, and backyard
1, 3	Black-throated Green Warbler	Mixed wooded slope, post oak savannah, shin oak savannah, live oak savannah, river, creek, and canyon

SEASONS PRESENT	CONSERVATION STATUS	PREDATION BY COWBIRDS?[b]	HARMED BY DEER OVERPOPULATION?	HARMED BY FERAL HOGS?	HARMED BY RIFA?	KILLED BY CATS?[c]
Winter						Yes
All year			Yes		Yes	
Winter			Yes		Yes	Yes
Spring			Yes		Yes	
Winter						Yes
Fall, winter, early spring						
Spring, summer			Yes		Yes	Yes
Spring, summer		*				Yes
Spring, summer	Priority	High				Yes
All year						Yes
All year	Priority	*				Yes
All year			Yes	Yes		Yes
Spring, summer, fall		Low				Yes
All year						Yes
Spring, summer	Priority	High	Yes		Yes	Yes
Spring, summer	Priority					Yes
All year	Priority					Yes
Spring					Yes	

MASTER BIRD CHART

CHAPTER[a]	COMMON NAME	HABITATS
1, 2	Black-throated Sparrow	Plateau prairie, river valley prairie, shin oak savannah, and live oak savannah
6, 7, 8	Blue Grosbeak	Old field, river valley prairie, pocket prairie, post oak savannah, shin oak savannah, live oak savannah, river, creek, seep, springs, and backyard
6, 7	Blue Jay	Mixed wooded slope, post oak savannah, shin oak savannah, live oak savannah, river, creek, seep, springs, and backyard
1, 7	Blue-gray Gnatcatcher	Mixed wooded slope, post oak savannah, shin oak savannah, live oak savannah, river, creek, canyon, springs, and seep
1, 3	Blue-headed Vireo	Mixed wooded slope, post oak savannah, shin oak savannah, live oak savannah, river, creek, and canyon
5	Blue-winged Teal	River, creek, tank, pond, and lake
1, 2, 3, 4, 5, 6, 7	Brown-headed Cowbird	Old field, plateau prairie, river valley prairie, pocket prairie, mixed wooded slope, post oak savannah, shin oak savannah, live oak savannah, river, creek, canyon, seep, springs, pond, tank, lake, and backyard
3	Bufflehead	River, creek, tank, pond, and lake
1, 3	Bullock's Oriole	Old field, pocket prairie, post oak savannah, shin oak savannah, live oak savannah, river, creek, seep, springs, and backyard
1	Cactus Wren	Shin oak savannah, live oak savannah, canyon, and backyard
1, 3, 9	Canyon Towhee	Mixed wooded slope, shin oak savannah, live oak savannah, river, creek, and canyon
1, 3, 4	Canyon Wren	Mixed wooded slope, river, creek, canyon, seep, and springs
1, 3, 6, 8	Carolina Chickadee	Mixed wooded slope, post oak savannah, shin oak savannah, live oak savannah, river, creek, canyon, seep, springs, and backyard
1, 2, 3, 6, 7	Carolina Wren	Pocket prairie, mixed wooded slope, post oak savannah, live oak savannah, river, creek, canyon, seep, springs, and backyard
1, 2	Cassin's Sparrow	Old field, plateau prairie, river valley prairie, pocket prairie, post oak savannah, shin oak savannah, and live oak savannah
1, 3, 4, 7, 8, 9	Cedar Waxwing	Old field, river valley prairie, pocket prairie, mixed wooded slope, post oak savannah, shin oak savannah, live oak savannah, river, creek, canyon, seep, springs, tank, pond, lake, and backyard
5, 6	Chimney Swift	River valley prairie, mixed wooded slope, post oak savannah, shin oak savannah, live oak savannah, river, creek, tank, pond, lake, and backyard
1, 3, 4, 5, 6	Chipping Sparrow	Mixed wooded slope, post oak savannah, shin oak savannah, live oak savannah, river, creek, canyon, seep, springs, pond, tank, lake, and backyard
1, 2, 3, 6	Clay-colored Sparrow	Old field, plateau prairie, river valley prairie, pocket prairie, mixed wooded slope, post oak savannah, shin oak savannah, live oak savannah, river, creek, canyon, seep, springs, and backyard

SEASONS PRESENT	CONSERVATION STATUS	PREDATION BY COWBIRDS?[b]	HARMED BY DEER OVERPOPULATION?	HARMED BY FERAL HOGS?	HARMED BY RIFA?	KILLED BY CATS?[c]
All year	Priority	Moderate	Yes		Yes	Yes
Spring, summer		Moderate				Yes
All year		†				Yes
Spring, summer		Moderate	Yes		Yes	Yes
Spring, fall						
Spring, summer, fall						Yes
All year						Yes
Fall, winter						
Spring, summer, fall		Low	Yes		Yes	Yes
All year	Priority				Yes	Yes
All year	Priority		Yes			Yes
All year	Priority					Yes
All year	Priority	*	Yes		Yes	Yes
All year		Low	Yes		Yes	Yes
All year	Priority		Yes		Yes	Yes
Fall, winter, spring		*†				Yes
Spring, summer, fall	Priority					
All year	Priority	Moderate				Yes
Spring, fall			Yes		Yes	Yes

CHAPTER[a]	COMMON NAME	HABITATS
1, 2, 5, 6, **7**	Common Nighthawk	Old field, plateau prairie, river valley prairie, post oak savannah, shin oak savannah, live oak savannah, tank, pond, lake, and backyard
1	Common Poorwill	Old field, plateau prairie, river valley prairie, pocket prairie, shin oak savannah, and live oak savannah
3, 7, 8	Common Raven	Old field, plateau prairie, river valley prairie, pocket prairie, mixed wooded slope, post oak savannah, shin oak savannah, live oak savannah, river, creek, canyon, seep, springs, tank, pond, lake, and backyard
4, 5, 6	Common Yellowthroat	River valley prairie, pocket prairie, live oak savannah, river, creek, canyon, seep, springs, tank, pond, lake, and backyard
7	Crested Caracara	Old field, plateau prairie, river valley prairie, pocket prairie, post oak savannah, shin oak savannah, and live oak savannah
5	Dark-eyed Junco	Old field, river valley prairie, pocket prairie, river, creek, canyon, seep, springs, tank, pond, lake, and backyard
2, 7	Dickcissel	Old field, river valley prairie, pocket prairie, and creek
1, **2**, 7, 8, 9	Eastern Bluebird	Old field, plateau prairie, river valley prairie, pocket prairie, post oak savannah, shin oak savannah, live oak savannah, river, creek, and backyard
1, 2	Eastern Meadowlark	Old field, plateau prairie, river valley prairie, pocket prairie, post oak savannah, shin oak savannah, and live oak savannah
1, 3, **4**, 5, 6, 8	Eastern Phoebe	Pocket prairie, mixed wooded slope, post oak savannah, shin oak savannah, live oak savannah, river, creek, canyon, seep, springs, tank, pond, lake, and backyard
1, 6	Eastern Screech-Owl	Mixed wooded slope, post oak savannah, shin oak savannah, live oak savannah, river, creek, canyon, and backyard
1	Eastern Wood-Pewee	Mixed wooded slope, live oak savannah, river, creek, canyon, and backyard
5	Egyptian Goose	River, creek, springs, tank, pond, and lake
6	European Starling	Old field, post oak savannah, shin oak savannah, live oak savannah, river, creek, seep, springs, and backyard
1, **2**, 7	Field Sparrow	Old field, plateau prairie, river valley prairie, pocket prairie, post oak savannah, shin oak savannah, and live oak savannah
3, **5**	Gadwall	River, creek, tank, pond, and lake
1, 3, 4, 7, 8, 9	Golden-cheeked Warbler	Mixed wooded slope, post oak savannah, shin oak savannah, live oak savannah, river, creek, and canyon
1, **4**	Golden-crowned Kinglet	Mixed wooded slope, post oak savannah, shin oak savannah, live oak savannah, river, creek, canyon, springs, seep, tank, pond, lake, and backyard
1, 3, 4, **6**, 7, 8	Golden-fronted Woodpecker	Mixed wooded slope, post oak savannah, shin oak savannah, live oak savannah, river, creek, canyon, and backyard

SEASONS PRESENT	CONSERVATION STATUS	PREDATION BY COWBIRDS?[b]	HARMED BY DEER OVERPOPULATION?	HARMED BY FERAL HOGS?	HARMED BY RIFA?	KILLED BY CATS?[c]
Spring, summer, fall	Priority					Yes
Spring, summer, fall				Yes		Yes
All year						
All year	Priority	Moderate				Yes
All year						
Winter		Low–moderate				Yes
Spring, summer	Priority	High*		Yes	Yes	Yes
All year		Low	Yes		Yes	Yes
All year	Priority	Moderate	Yes		Yes	Yes
All year		Low				Yes
All year						Yes
Spring, summer		Low*	Yes			
All year						
All year		*				Yes
All year	Priority	Moderate	Yes		Yes	Yes
Winter						
Spring, summer	Priority	Moderate	Yes	Yes	Yes	Yes
Winter						Yes
All year	Priority					Yes

CHAPTER[a]	COMMON NAME	HABITATS
1, **2**, 7	Grasshopper Sparrow	Old field, plateau prairie, river valley prairie, and shin oak savannah
6, 8	Gray Catbird	Mixed wooded slope, post oak savannah, shin oak savannah, live oak savannah, river, creek, canyon, seep, springs, pond, tank, lake, and backyard
3, 5	Great Blue Heron	River, creek, canyon, seep, springs, tank, pond, lake, and backyard
3, 4, 5	Great Egret	River, creek, canyon, seep, springs, tank, pond, and lake
1, 7, 9	Greater Roadrunner	Old field, plateau prairie, pocket prairie, post oak savannah, shin oak savannah, live oak savannah, river, creek, and canyon
6	Great-tailed Grackle	Old field, plateau prairie, river valley prairie, pocket prairie, mixed wooded slope, post oak savannah, shin oak savannah, live oak savannah, river, creek, canyon, seep, springs, tank, pond, lake, and backyard
3, 5	Green Heron	River, creek, springs, tank, pond, and lake
3, 5	Green Kingfisher	River, creek, tank, pond, and lake
3, **5**	Green-winged Teal	River, creek, tank, pond, and lake
2	Harris's Sparrow	Old field, river valley prairie, pocket prairie, river, creek, canyon, seep, and springs
3, **4**, 8	Hermit Thrush	Mixed wooded slope, post oak savannah, shin oak savannah, live oak savannah, river, creek, canyon, seep, and springs
1, **6**, 7, 9	House Finch	Old field, river valley prairie, pocket prairie, post oak savannah, shin oak savannah, live oak savannah, river, creek, canyon, seep, springs, and backyard
1, 6	House Sparrow	Backyard
1	Hutton's Vireo	Mixed wooded slope, live oak savannah, river, creek, and canyon
1, **4**	Indigo Bunting	Old field, pocket prairie, post oak savannah, shin oak savannah, live oak savannah, river, creek, canyon, seep, and springs
1, 2, 3, **5**	Killdeer	Old field, plateau prairie, shin oak savannah, live oak savannah, river, creek, tank, pond, and lake
1, 3, 6	Ladder-backed Woodpecker	Mixed wooded slope, post oak savannah, shin oak savannah, live oak savannah, river, creek and backyard
2	Lark Bunting	Old field, plateau prairie, river valley prairie, pond, and tank
1, **2**, 7	Lark Sparrow	Old field, plateau prairie, river valley prairie, pocket prairie, post oak savannah, shin oak savannah, and live oak savannah
2	Le Conte's Sparrow	Old field, river valley prairie, pocket prairie, and post oak savannah
5	Least Sandpiper	River, creek, seep, tank, pond, and lake
1, **2**, 3, 4, 6	Lesser Goldfinch	Old field, river valley prairie, pocket prairie, post oak savannah, shin oak savannah, live oak savannah, river, creek, seep, springs, and backyard

SEASONS PRESENT	CONSERVATION STATUS	PREDATION BY COWBIRDS?[b]	HARMED BY DEER OVERPOPULATION?	HARMED BY FERAL HOGS?	HARMED BY RIFA?	KILLED BY CATS?[c]
All year	Priority	Moderate	Yes		Yes	Yes
Spring			Yes			Yes
All year			Yes			
Spring, summer, fall						
All year			Yes			Yes
All year						Yes
Spring, summer, fall			Yes	Yes		yes
All year	Priority		Yes	Yes	Yes	Yes
Winter						
Winter	Priority		Yes		Yes	
Winter						Yes
All year		Moderate				Yes
All year		*				Yes
All year		*	Yes		Yes	Yes
Early spring, summer, fall		High				Yes
All year	Priority					Yes
All year	Priority		Yes		Yes	Yes
Winter						Yes
All year	Priority	High	Yes		Yes	Yes
Winter	Priority		Yes			Yes
Spring, fall				Yes		
All year	Priority	*	Yes			Yes

CHAPTER[a]	COMMON NAME	HABITATS
3, **5**	Lesser Scaup	River, creek, tank, pond, and lake
1, **6**	Lincoln's Sparrow	Old field, river valley prairie, pocket prairie, post oak savannah, shin oak savannah, live oak savannah, river, creek, canyon, seep, springs, and backyard
1, 2, 3	Loggerhead Shrike	Old fields, plateau prairie, river valley prairie, post oak savannah, shin oak savannah, live oak savannah, river, and creek
3, 4, **7**	Louisiana Waterthrush	River, creek, and springs
1, 3, 4	MacGillivray's Warbler	Mixed wooded slope, post oak savannah, shin oak savannah, live oak savannah, river, creek, canyon, and backyard
2	Mississippi Kite	Old field, plateau prairie, river valley prairie, pocket prairie, mixed wooded slope, post oak savannah, shin oak savannah, and live oak savannah
1, **2**	Mourning Dove	Old field, plateau prairie, river valley prairie, pocket prairie, post oak savannah, shin oak savannah, live oak savannah, seep, springs, tank, pond, lake, and backyard
1, 3, 4	Mourning Warbler	Mixed wooded slope, post oak savannah, live oak savannah, river, creek, and canyon
5	Muscovy Duck	River, creek, springs, tank, pond, lake, and backyard
1, 3, 4	Nashville Warbler	Pocket prairie, mixed wooded slope, post oak savannah, shin oak savannah, live oak savannah, river, creek, canyon, springs, seep, tank, pond, lake, and backyard
1, 2, 3, 7, 8	Northern Bobwhite	Old field, plateau prairie, river valley prairie, pocket prairie, shin oak savannah, live oak savannah, river, and creek
1, **6,** 7, 8	Northern Cardinal	Old field, river valley prairie, pocket prairie, mixed wooded slope, post oak savannah, shin oak savannah, live oak savannah, river, creek, canyon, seep, springs, and backyard
1, 8	Northern Flicker	Old field, pocket prairie, mixed wooded slope, post oak savannah, shin oak savannah, live oak savannah, and backyard
2	Northern Harrier	Old field, plateau prairie, and river valley prairie
6, 8, 9	Northern Mockingbird	Old field, river valley prairie, pocket prairie, post oak savannah, shin oak savannah, live oak savannah, river, creek, canyon, seep, springs, and backyard
3	Northern Parula	River, creek, and canyon
5	Northern Pintail	Tank, pond, and lake
1, **2,** 3, 4, 6	Orange-crowned Warbler	Old field, mixed wooded slope, post oak savannah, shin oak savannah, live oak savannah, river, creek, canyon, seep, springs, and backyard
1, 3, 4, 5, 6, **7**	Orchard Oriole	Post oak savannah, shin oak savannah, live oak savannah, river, creek, canyon, springs, tank, pond, lake, and backyard

SEASONS PRESENT	CONSERVATION STATUS	PREDATION BY COWBIRDS?[b]	HARMED BY DEER OVERPOPULATION?	HARMED BY FERAL HOGS?	HARMED BY RIFA?	KILLED BY CATS?[c]
Winter	Priority					
Winter						Yes
Spring, summer	Priority		Yes		Yes	Yes
Spring, Summer	Priority	Moderate	Yes	Yes	Yes	Yes
Spring, fall			Yes		Yes	
Spring, fall	Priority					
All year	Priority		Yes			Yes
Spring, fall			Yes		Yes	
All year						
Spring, fall		*	Yes		Yes	Yes
All year	Priority		Yes	Yes	Yes	Yes
All year		Moderate				Yes
Fall, winter					Yes	Yes
Winter	Priority					
All year		*				Yes
Spring, summer		*	Yes		Yes	Yes
Winter	Priority			Yes		
Fall, winter	Priority	*				Yes
Spring, summer, fall	Priority	Moderate				Yes

CHAPTER[a]	COMMON NAME	HABITATS
1, 2, 3, 4, 5, 6, 7	Painted Bunting	Old field, river valley prairie, pocket prairie, post oak savannah, shin oak savannah, live oak savannah, river, creek, springs, seep, and backyard
5	Pectoral Sandpiper	Tank, pond, and lake
5	Pied-billed Grebe	River, creek, springs, tank, pond, and lake
1, 6, 9	Pine Siskin	Old field, plateau prairie, river valley prairie, pocket prairie, mixed wooded slope, post oak savannah, shin oak savannah, live oak savannah, and backyard
5, **6**	Purple Martin	Old field, river valley prairie, post oak savannah, shin oak savannah, live oak savannah, river, creek, seep, springs, tank, pond, lake, and backyard
1	Pyrrhuloxia	Old field, plateau prairie, pocket prairie, shin oak savannah, live oak savannah, river, creek, seep, springs, pond, tank, lake, and backyard
5	Red-breasted Merganser	Tank, pond, and lake
1, 7	Red-eyed Vireo	Mixed wooded slope, post oak savannah, river, creek, canyon, and backyard
5	Redhead	Tank, pond, and lake
2, **3**, 5, 8	Red-shouldered Hawk	River valley prairie, mixed wooded slope, post oak savannah, shin oak savannah, live oak savannah, river, creek, canyon, tank, pond, and lake
1	Red-tailed Hawk	Old field, plateau prairie, river valley prairie, pocket prairie, mixed wooded slope, post oak savannah, shin oak savannah, live oak savannah, river, creek, canyon, springs, seep, tank, pond, lake, and backyard
3, 4, **5**, 7	Red-winged Blackbird	Old field, river valley prairie, river, creek, canyon, seep, springs, tank, pond, and lake
3, **5**	Ring-necked Duck	River, creek, springs, tank, pond, and lake
1, **2**, 3, 4, 7, 8	Rio Grande Turkey	Old field, plateau prairie, river valley prairie, post oak savannah, shin oak savannah, live oak savannah, river, creek, and canyon
1, 6	Ruby-crowned Kinglet	Old field, pocket prairie, mixed wooded slope, post oak savannah, shin oak savannah, live oak savannah, river, creek, canyon, seep, springs, pond, tank, lake, and backyard
1, 3, 4, 6	Ruby-throated Hummingbird	Old field, pocket prairie, mixed wooded slope, post oak savannah, live oak savannah, river, creek, canyon, seep, springs, and backyard
6	Rufous Hummingbird	Backyard
1, 2, 6, 9	Rufous-crowned Sparrow	Plateau prairie, shin oak savannah, live oak savannah, canyon, springs, seep, and backyard
5	Sandhill Crane	Old field, tank, pond, and lake
2	Savannah Sparrow	Old field, plateau prairie, river valley prairie, pocket prairie, post oak savannah, shin oak savannah, and live oak savannah

SEASONS PRESENT	CONSERVATION STATUS	PREDATION BY COWBIRDS?[b]	HARMED BY DEER OVERPOPULATION?	HARMED BY FERAL HOGS?	HARMED BY RIFA?	KILLED BY CATS?[c]
Spring, summer	Priority	Moderate	Yes		Yes	Yes
Spring, fall				Yes		
Winter, spring						
Winter, spring						Yes
Spring, summer						Yes
All year		Low*	Yes			Yes
Fall, winter						
Spring, summer		Moderate	Yes		Yes	
Winter	Priority					
All year	Priority		Yes			
All year			Yes			
All year	Priority	Moderate				Yes
Winter						
All year	Priority			Yes		Yes
Winter, spring, fall			Yes		Yes	
Fall			Yes			
Winter						
All year	Priority		Yes	Yes	Yes	Yes
Spring, fall						
Winter		Low–moderate	Yes		Yes	Yes

MASTER BIRD CHART

443

CHAPTER[a]	COMMON NAME	HABITATS
4	Say's Phoebe	Old field, plateau prairie, river valley prairie, shin oak savannah, live oak savannah, river, creek, and canyon
1, **2**, 3, 7	Scissor-tailed Flycatcher	Old field, plateau prairie, river valley prairie, post oak savannah, shin oak savannah, live oak savannah, river and creek
1	Scott's Oriole	Post oak savannah, shin oak savannah, live oak savannah, river, creek, canyon, springs, seep, tank, and pond
3, **4**, 5	Sedge Wren	River, creek, seep, springs, tank, pond, and lake
2, 4	Song Sparrow	Old field, river valley prairie, pocket prairie, post oak savannah, shin oak savannah, live oak savannah, river, creek, canyon, seep, and springs
3, 4, 5	Sora	River, creek, canyon, seep, springs, tank, pond, and lake
5	Spotted Sandpiper	River, creek, tank, pond, and lake
1, 4, 6	Spotted Towhee	Mixed wooded slope, post oak savannah, shin oak savannah, live oak savannah, river, creek, canyon, seep, springs, and backyard
1, 3, 4, 7, 8	Summer Tanager	Mixed wooded slope, post oak savannah, river, creek, and canyon
2	Swainson's Hawk	Old field, plateau prairie, and river valley prairie
1, 3, 4	Swainson's Thrush	Mixed wooded slope, post oak savannah, shin oak savannah, live oak savannah, river, creek, and canyon
1, 4	Tennessee Warbler	Mixed wooded slope, post oak savannah, shin oak savannah, live oak savannah, river, creek, canyon, seep, springs, and backyard
2	Upland Sandpiper	Old field, plateau prairie, river valley prairie, post oak savannah, shin oak savannah, and live oak savannah
1, 2, 3, 4, **5**, 6	Vermilion Flycatcher	Old field, post oak savannah, shin oak savannah, live oak savannah, river, creek, canyon, seep, springs, tank, pond, lake, and backyard
1, 2	Vesper Sparrow	Old field, plateau prairie, river valley prairie, pocket prairie, and shin oak savannah
1	Western Kingbird	Old field, plateau prairie, river valley prairie, post oak savannah, shin oak savannah, live oak savannah, and backyard
1, 7, 8, 9	Western Scrub-Jay	Mixed wooded slope, post oak savannah, shin oak savannah, live oak savannah, river, creek, canyon, springs, seep, and backyard
1, 2, 4, 6	White-crowned Sparrow	Old field, mixed wooded slope, post oak savannah, shin oak savannah, live oak savannah, canyon, springs, seep, and backyard
1, **4**, 6, 7	White-eyed Vireo	Mixed wooded slope, post oak savannah, shin oak savannah, live oak savannah, river, creek, canyon, seep, springs, and backyard
1	White-throated Sparrow	Old field, pocket prairie, post oak savannah, shin oak savannah, live oak savannah, river, creek, canyon, seep, springs, pond, tank, lake, and backyard

SEASONS PRESENT	CONSERVATION STATUS	PREDATION BY COWBIRDS?[b]	HARMED BY DEER OVERPOPULATION?	HARMED BY FERAL HOGS?	HARMED BY RIFA?	KILLED BY CATS?[c]
Winter, spring		Low			Yes	
Spring, summer, fall	Priority	*	Yes		Yes	Yes
Spring, summer		Low				Yes
Winter	Priority					Yes
Winter			Yes			Yes
Fall, winter, spring	Priority					Yes
All year	Priority					Yes
Winter		*	Yes		Yes	Yes
Spring, summer	Priority	High	Yes		Yes	Yes
Spring, fall	Priority				Yes	
Spring			Yes		Yes	
Spring						
Spring, fall					Yes	Yes
All year	Priority	*				Yes
Winter			Yes		Yes	Yes
Spring, summer		Low!			Yes	
All year			Yes		Yes	Yes
Winter		Low–moderate	Yes			Yes
Spring, summer		Moderate				Yes
Winter			Yes			Yes

CHAPTER[a]	COMMON NAME	HABITATS
6	White-winged Dove	Old field, post oak savannah, shin oak savannah, live oak savannah, river, creek, seep, springs, and backyard
5	Wilson's Phalarope	Tank, pond, and lake
3, **4**, 5	Wilson's Snipe	River, creek, canyon, seep, springs, tank, pond, and lake
1, **3**, 4	Wilson's Warbler	Mixed wooded slope, post oak savannah, shin oak savannah, live oak savannah, river, creek, canyon, seep, springs, and backyard
3	Wood Duck	River, creek, springs, tank, pond, and lake
1, **3**, 4	Yellow Warbler	Mix wooded slope, post oak savannah, live oak savannah, river, creek, canyon, seep, and springs
1, **3**, 8	Yellow-billed Cuckoo	Post oak savannah, shin oak savannah, live oak savannah, river, creek, canyon, and springs
1	Yellow-breasted Chat	Pocket prairie, mixed wooded slope, post oak savannah, shin oak savannah, live oak savannah, river, creek, canyon, seep, and springs
3, 5	Yellow-crowned Night-Heron	River, creek, springs, tank, pond, lake, and backyard
2	Yellow-headed Blackbird	Old field, river valley prairie, pocket prairie, post oak savannah, shin oak savannah, live oak savannah, seep, springs, pond, tank, and lake
1, **3**, 4, 6	Yellow-rumped Warbler	Old field, pocket prairie, mixed wooded slope, post oak savannah, shin oak savannah, live oak savannah, river, creek, canyon, seep, springs, tank, pond, and backyard
4	Yellow-throated Vireo	Post oak savannah, live oak savannah, river, creek, canyon, and backyard

[a]Chapter numbers indicate where the bird is mentioned, and a boldface number indicates where the bird summary can be found. Not all birds mentioned have a bird summary.

[b]Although cowbird parasitism has been documented in many research projects, it has by no means been studied thoroughly, and not enough is known about all its phases in individual species. The results shown here are from different projects that documented various aspects of the parasitism process.

*Known to have reared cowbirds

†Known to have ejected cowbird eggs

High: 60%–100% of found nests parasitized

Moderate: 30%–59% of found nests parasitized

Low: 1%–29% of found nests parasitized

Low–moderate: From 1% to 59% incidence of parasitism

[c]Killed by cats: Yes = cat predation documented, but many other bird species susceptible

SEASONS PRESENT	CONSERVATION STATUS	PREDATION BY COWBIRDS?[b]	HARMED BY DEER OVERPOPULATION?	HARMED BY FERAL HOGS?	HARMED BY RIFA?	KILLED BY CATS?[c]
Spring, summer						Yes
Spring, fall				Yes		
Winter						Yes
Spring, fall			Yes		Yes	Yes
All year						
Spring, fall	Priority		Yes	Yes	Yes	Yes
Spring, summer	Priority		Yes			
Spring, summer		Moderate*†	Yes		Yes	Yes
Spring, summer, fall	Priority					
Spring, fall			Yes			
Winter		*	Yes			Yes
Spring, summer	Priority	Moderate				Yes

Appendix 3. Priority Woody Plants of the Texas Hill Country

These valuable native plants contribute to habitat diversity. Protect them and use them in your landscape.

VINES

Bracted passionflower*	*Passiflora affinis*
Carolina snailseed	*Cocculus carolinus*
Coral honeysuckle	*Lonicera sempervirens*
Lindheimer's morning glory*	*Ipomoea lindheimeri*
Trumpet creeper	*Campsis radicans*
Virginia creeper	*Parthenocissus quinquefolia*
White honeysuckle	*Lonicera albiflora*

SHRUBS

Black dalea*	*Dalea frutescens*
Canyon mock orange*	*Philadelphus ernestii*
Creek plum	*Prunus rivularis*
Dwarf palmetto or bush palmetto*	*Sabal minor*
Elbow bush	*Forestiera pubescens*
False indigo*	*Amorpha fruticosa*
Hawthorn*	*Crataegus* sp.
Mexican plum	*Prunus mexicana*
Pink mimosa	*Mimosa borealis*
Possumhaw	*Ilex decidua*
Redroot*	*Ceanothus herbaceus*
Roughleaf dogwood	*Cornus drummondii*
Rusty blackhaw*	*Viburnum rufidulum*
Spicebush*	*Lindera benzoin*
Sycamore-leaf snowbell*	*Styrax platanifolius*
Texas kidneywood	*Eysenhardtia texana*
Texas mulberry*	*Morus microphylla*
Witch hazel*	*Hamamelis virginiana*

TREES

American elm*	*Ulmus americana*
American smoke tree*	*Cotinus obovatus*
Bigtooth maple*	*Acer grandidentatum* var. *sinuosum*
Blanco crabapple*	*Malus ioensis* var. *texana*
Carolina buckthorn	*Frangula caroliniana*
Cedar elm	*Ulmus crassifolia*
Escarpment black cherry*	*Prunus serotina* var. *eximia*
Eve's necklace	*Styphnolobium affine*

Hackberries	*Celtis* sp.
Red mulberry	*Morus rubra*
Shin oak	*Quercus sinuata* var. *breviloba*
Slippery elm*	*Ulmus rubra*
Soapberry	*Sapindus saponaria* var. *drummondii*
Spanish oak	*Quercus buckleyi*
Texas madrone*	*Arbutus xalapensis*
Texas redbud	*Cercis canadensis* var. *texensis*

*May be considered Top Priority because they are uncommon.

Appendix 4. Master Plant List

COMMON NAME	SCIENTIFIC NAME
Trees	
American elm	*Ulmus americana*
American smoke tree	*Cotinus obovatus*
American sycamore	*Platanus occidentalis*
Bald cypress	*Taxodium distichum*
Bigtooth maple	*Acer grandidentatum* var. *sinuosum*
Black willow	*Salix nigra*
Blackjack oak	*Quercus marilandica*
Blanco crabapple	*Malus ioensis* var. *texana*
Carolina buckthorn	*Frangula caroliniana*
Cedar, Ashe juniper, mountain cedar	*Juniperus ashei*
Cedar elm	*Ulmus crassifolia*
Chinaberry*	*Melia azedarach*
Chinese tallow*	*Triadica sebifera*
Chinkapin oak	*Quercus muehlenbergii*
Eastern cottonwood	*Populus deltoides*
Elm	*Ulmus* sp.
Escarpment black cherry	*Prunus serotina* var. *eximia*
Eve's necklace	*Styphnolobium affine*
Evergreen sumac	*Rhus virens*
Flameleaf sumac, prairie flameleaf sumac	*Rhus lanceolata*
Gum bumelia	*Sideroxylon lanuginosum*
Hackberries	*Celtis* spp.
Lacey oak	*Quercus laceyi*
Linden, Carolina basswood	*Tilia americana* var. *caroliniana*
Live oak, plateau live oak	*Quercus fusiformis*
Mesquite, honey mesquite	*Prosopis glandulosa* var. *glandulosa*
Mexican pinyon	*Pinus cembroides*
Netleaf hackberry	*Celtis laevigata* var. *reticulata*
Oaks	*Quercus* spp.
Pecan	*Carya illinoinensis*
Post oak	*Quercus stellata*
Red mulberry	*Morus rubra*
Redberry juniper, Pinchot juniper	*Juniperus coahuilensis*
Shin oak	*Quercus sinuata* var. *breviloba*
Slippery elm	*Ulmus rubra*
Soapberry	*Sapindus saponaria* var. *drummondii*
Spanish oak	*Quercus buckleyi*

COMMON NAME	SCIENTIFIC NAME
Sugar hackberry	*Celtis laevigata* var. *laevigata*
Texas ash	*Fraxinus albicans*
Texas madrone	*Arbutus xalapensis*
Texas mulberry	*Morus microphylla*
Texas redbud	*Cercis canadensis* var. *texensis*
Texas walnut	*Juglans microcarpa*
Walnuts	*Juglans* spp.

Shrubs and cactus

Agarita	*Berberis trifoliolata* (*Mahonia trifoliolata*)
American beautyberry	*Callicarpa americana*
Aromatic sumac	*Rhus aromatica*
Black dalea	*Dalea frutescens*
Brickell-bush	*Brickellia* sp.
Canyon mock orange	*Philadelphus ernestii*
Catclaw acacia	*Acacia greggii*
Cenizo	*Leucophyllum frutescens*
Common buttonbush	*Cephalanthus occidentalis*
Coyotillo	*Karwinskia humboldtiana*
Creek plum	*Prunus rivularis*
Desert sumac	*Rhus microphylla*
Dwarf palmetto, bush palmetto	*Sabal minor*
Elbow bush, spring herald	*Forestiera pubescens*
Evergreen sumac	*Rhus virens*
False indigo	*Amorpha fruticosa*
Fig*	*Ficus* sp.
Flameleaf sumac	*Rhus lanceolata*
Goldenball leadtree	*Leucaena retusa*
Green condalia	*Condalia viridis*
Guayacan, soapbush	*Guaiacum angustifolium*
Hawthorn	*Crataegus* sp.
Hog plum	*Colubrina texensis*
Huisache	*Acacia farnesiana*
Ligustrums, privets*	*Ligustrum* spp.
Lindheimer's silktassel	*Garrya ovata* subsp. *lindheimeri*
Lotebush	*Ziziphus obtusifolia*
Mexican buckeye	*Ungnadia speciosa*
Mexican plum	*Prunus mexicana*
Mountain mahogany	*Cercocarpus montanus*
Nolina	*Nolina* sp.
Pale buckeye	*Aesculus pavia* var. *flavescens*

COMMON NAME	SCIENTIFIC NAME
Pink mimosa	*Mimosa borealis*
Possumhaw	*Ilex decidua*
Poverty bush	*Baccharis neglecta*
Pricklypear	*Opuntia* sp.
Pyracantha*	*Pyracantha* sp.
Redroot	*Ceanothus herbaceus*
Roemer's acacia	*Acacia roemeriana*
Rose pavonia	*Pavonia lasiopetala*
Roughleaf dogwood	*Cornus drummondii*
Rusty blackhaw	*Viburnum rufidulum*
Salt cedar*	*Tamarix* sp.
Skunkbush sumac	*Rhus trilobata* var. *trilobata* (*Rhus aromatica* var. *flabelliformis*)
Southwest bernardia	*Bernardia myricifolia*
Spicebush	*Lindera benzoin*
Sumacs	*Rhus* spp.
Sycamore-leaf snowbell	*Styrax platanifolius*
Tasajillo	*Opuntia leptocaulis*
Texas almond	*Prunus minutiflora*
Texas kidneywood	*Eysenhardtia texana*
Texas lantana	*Lantana urticoides*
Texas mountain laurel	*Sophora secundiflora*
Texas mulberry	*Morus microphylla*
Texas persimmon	*Diospyros texana*
Texas pricklypear	*Opuntia engelmannii* var. *lindheimeri*
Texas sotol	*Dasylirion texanum*
Tickle-tongue	*Zanthoxylum hirsutum*
Twist-leaf yucca	*Yucca rupicola*
Western soapberry	*Sapindus saponaria* var. *drummondii*
Whitebrush, common beebush	*Aloysia gratissima*
Witch hazel	*Hamamelis virginiana*
Yucca	*Yucca* sp.

Vines

Balsam gourd, Lindheimer's globeberry	*Ibervillea lindheimeri*
Bracted passionflower	*Passiflora affinis*
Carolina snailseed	*Cocculus carolinus*
Clematis	*Clematis* sp.
Coral honeysuckle	*Lonicera sempervirens*
Cowitch vine	*Cissus trifoliata*
Grapes	*Vitis* spp.
Greenbrier	*Smilax bona-nox*

COMMON NAME	SCIENTIFIC NAME
Japanese honeysuckle*	*Lonicera japonica*
Leatherflower	*Clematis* sp.
Lindheimer's morning glory	*Ipomoea lindheimeri*
Milkweed vines	*Matelea* spp.
Mustang grape	*Vitis mustangensis*
Old man's beard	*Clematis drummondii*
Passionflower vines	*Passiflora* spp.
Pearl milkweed	*Matelea reticulata*
Plateau milkvine	*Matelea edwardsensis*
Poison ivy	*Toxicodendron radicans*
Trumpet creeper	*Campsis radicans*
Vinca*	*Vinca* sp.
Virginia creeper	*Parthenocissus quinquefolia*
White honeysuckle	*Lonicera albiflora*

Forbs

Annual sunflower	*Helianthus annuus*
Antelope horns	*Asclepias asperula*
Arrowhead	*Sagittaria* sp.
Bastard cabbage*	*Rapistrum rugosum*
Beebalm	*Monarda* sp.
Beggarticks	*Bidens* spp.
Big red sage	*Salvia penstemonoides*
Black samson echinacea, purple cone-flower	*Echinacea angustifolia*
Blue mistflower	*Conoclinium coelestinum*
Bluebonnet	*Lupinus* sp.
Brazilian vervain*	*Verbena brasiliensis*
Brown-eyed Susan	*Rudbeckia hirta*
Bundleflower	*Desmanthus* sp.
Bur-clover*	*Medicago* sp.
Bush sunflower	*Simsia calvia*
Cedar sage	*Salvia roemeriana*
Chile pequin	*Capsicum annuum*
Columbine	*Aquilegia* sp.
Common elderberry	*Sambucus nigra* subsp. *canadensis*
Common four o'clock*	*Mirabilis jalapa*
Croton	*Croton* sp.
Crow poison	*Nothoscordum bivalve*
Damianita	*Chrysactinia mexicana*
Dayflower	*Commelina* sp.
Dock	*Rumex* sp.

COMMON NAME	SCIENTIFIC NAME
Drummond's wild petunia	*Ruellia drummondiana*
Elephant ear*	*Colocasia esculenta*
Engelmann daisy, cutleaf daisy	*Engelmannia peristenia*
Engelmann's evening-primrose	*Oenothera engelmannia*
Firewheel	*Gaillardia* sp.
Frostweed	*Verbesina virginica*
Gayfeather	*Liatris* sp.
Goldenrod	*Solidago* sp.
Goldenwave	*Coreopsis tinctoria*
Greenthread	*Thelesperma* sp.
Hairyfruit chervil	*Chaerophyllum tainturieri* var. *tainturieri*
Illinois bundleflower	*Desmanthus illinoensis*
Malta star-thistle*	*Centaurea melitensis*
Maximilian sunflower	*Helianthus maximiliani*
Mealy blue sage	*Salvia farinacea*
Mexican hat	*Ratibida columnifera*
Milkpea	*Galactia* sp.
Mint*	*Mentha* sp.
Mistletoe	*Phoradendron* sp.
Musk thistle*	*Carduus nutans*
Penstemons	*Penstemon* spp.
Pigeonberry	*Rivina humilis*
Pigweeds	*Amaranthus* spp.
Pink evening-primrose	*Oenothera speciosa*
Pink smartweed	*Polygonum pensylvanicum*
Plateau goldeneye	*Viguiera dentata*
Prairie paintbrush	*Castilleja purpurea*
Primroses	*Oenothera* spp.
Purple cone-flower	*Echinacea angustifolia*
Purple prairie clover	*Dalea purpurea*
Ragweed	*Ambrosia* sp.
Rain lily	*Cooperia* sp.
Rock daisy	*Perityle lindheimeri*
Sensitive briar	*Mimosa* sp.
Snoutbean	*Rhynchosia* sp.
Spiderwort	*Tradescantia* sp.
Straggler daisy	*Calyptocarpus vialis*
Sunflowers	*Helianthus* spp.
Tall goldenrod	*Solidago altissima*
Texas bush-clover	*Lespedeza texana*
Texas milkweed	*Asclepias texana*

COMMON NAME	SCIENTIFIC NAME
Texas star	*Lindheimera texana*
Texas thistle	*Cirsium texanum*
Thin-leaf brookweed	*Samolus valerandi* subsp. *parviflorus*
Tropical sage	*Salvia coccinea*
Turk's cap	*Malvaviscus arboreus* var. *drummondii*
Water pennywort	*Hydrocotyle* sp.
Western ironweed	*Vernonia baldwinii*
Western ragweed	*Ambrosia psilostachya*
White boneset	*Eupatorium serotinum*
White heath aster	*Symphyotrichum ericoides*
Wild onion	*Allium* sp.
Wild petunias	*Ruellia* spp.
Wood-sorrel	*Oxalis* sp.
Yarrow	*Achillea millefolium*
Zexmenia, wedelia	*Wedelia texana*

Grasses, sedges, and rushes

Aparejo muhly	*Muhlenbergia utilis*
Bearded sprangletop	*Leptochloa fusca* subsp. *fascicularis*
Bermudagrass*	*Cynodon dactylon*
Big bluestem	*Andropogon gerardii*
Bristlegrasses	*Setaria* spp.
Broadleaf woodoats	*Chasmanthium latifolium*
Buffalograss	*Buchloe dactyloides*
Bulrushes	*Schoenoplectus* spp.
Bushy bluestem	*Andropogon glomeratus*
Canada wildrye, prairie wildrye	*Elymus canadensis*
Cane bluestem, silver bluestem	*Bothriochloa* sp.
Cedar sedge	*Carex planostachys*
Curly mesquite	*Hilaria belangeri*
Dropseeds	*Sporobolus* spp.
Eastern gamagrass	*Tripsacum dactyloides*
Fall witchgrass	*Digitaria cognata*
Flatsedges	*Cyperus* spp.
Giant reed*	*Arundo donax*
Gramas	*Bouteloua* spp.
Green sprangletop	*Leptochloa dubia*
Hairy grama	*Bouteloua hirsuta*
Hall's panicum	*Panicum hallii* var. *hallii*
Jamaican saw-grass	*Cladium mariscus* subsp. *jamaicense*
Johnsongrass*	*Sorghum halepense*

COMMON NAME	SCIENTIFIC NAME
King Ranch bluestem, KR bluestem*	*Bothriochloa ischaemum* var. *songarica*
Lindheimer rosettegrass	*Dichanthelium acuminatum* var. *lindheimeri*
Lindheimer's muhly	*Muhlenbergia lindheimeri*
Little bluestem	*Schyzachyrium scoparium*
Milo, grain sorghum†	*Sorghum bicolor*
Oats†	*Avena sativa*
Panicgrasses	*Dichanthelium* spp.
Paspalum	*Paspalum pubiflorum*
Plains bristlegrass	*Setaria leucopila*
Plains lovegrass	*Eragrostis intermedia*
Purpletop	*Tridens flavus* var. *flavus*
Red grama	*Bouteloua trifida*
Rice cutgrass	*Leersia oryzoides*
Rushes	*Juncus* spp.
Sedges	*Carex* spp.
Seep muhly	*Muhlenbergia reverchonii*
Sideoats grama	*Bouteloua curtipendula* var. *curtipendula*
Slim tridens	*Tridens muticus*
Southwestern bristlegrass	*Setaria scheelei*
Spikerushes, spike sedges	*Eleocharis* spp.
Switchgrass	*Panicum virgatum*
Tall dropseed	*Sporobolus compositus* var. *compositus*
Tall grama	*Bouteloua pectinata*
Texas cupgrass	*Eriochloa sericea*
Texas wintergrass	*Nassella leucotricha*
Threeawns	*Aristida* spp.
Vine mesquite	*Panicum obtusum*
Wild millet, barnyard grass, coast cockspur grass	*Echinochloa walteri*
Wildrye	*Elymus* sp.
Yellow indiangrass	*Sorgastrum nutans*

Ferns

Lindheimer's maiden fern	*Thelypteris ovata* var. *lindheimeri*
Southern maidenhair fern	*Adiantum capillus-veneris*

True aquatics

Coon's tail	*Ceratophyllum demersum*
Duckweed	*Lemna valdiviana*
Filamentous algae	
Hydrilla*	*Hydrilla verticillata*
Pondweeds	*Potamogeton* spp.
Southern water nymph, southern naiad	*Najas guadalupensis*

COMMON NAME	SCIENTIFIC NAME
Stoneworts	*Nitella* spp.
Water lettuce*	*Pistia stratiotes*
Water milfoil	*Myriophyllum* sp.
Water primrose	*Ludwigia* sp.
Watercress*	*Nasturtium officinale*

*Invasive exotic
†Agricultural exotic

Appendix 5. Purple Martin Monitoring Data Sheet

Place dates across the top row. Key provides sample entries for the chambers.
Place notes and observations on the back of the data sheet (e.g., 5/5/17, gourd, 3, has one broken egg).

Year_____

Date / Chamber number														
1														
2														
3														
4														
5														
6														
7														
8														
9														
10														
11														
12														
13														
14														
15														
16														
17														
18														
19														
20														
21														
22														
23														
24														
25														

KEY

X = no activity	Y# = young	IH = interference by human
C = claim straw	F# = fledges	ER = egg removed
P = partial nest	IS = interference by snake	NR = nest removed
N = complete nest	IM = interference by mammal	
E# = eggs	IB = interference by bird	

Glossary

aflatoxin. Poison produced by mold found in deer corn and is toxic to birds at low levels.

allelopathy. Release of substances by a plant that influence germination, growth, and survival of other plants

alluvial. Relating to soil material carried and then deposited by streams

arthropod. Animal with a segmented body, jointed legs, and an exoskeleton, including insects, spiders, scorpions, and centipedes

Balcones Canyonlands. A portion of the Texas Hill Country, on the south and east edge of the Edwards Plateau, with deep limestone canyons carved by flowing water

bale (v.). To use agricultural equipment to tightly bind hay with twine or wire

bankfull. Type of high-water event or flood in which water reaches the point just before spilling onto the lower floodplain; often occurs every year or two

basal resprouter. Plant that can regrow when cut below the lowest-growing branch but above the *bud zone*

bench. Flat area found in parallel bands or stair steps along Hill Country hillsides

bird habitat. Area that provides all the basic components for survival and successful reproduction of a bird, e.g., place to nest, the right kind of food, accessible water, and protection from bad weather and predators

breeding bird. A bird that can reproduce successfully in a specific area; reproduction includes mating behavior that leads up to copulation, nest building, egg laying, brooding, feeding offspring, fledging, and protecting young

browse. Woody plants such as trees, shrubs, or vines that are eaten by animals

browse line. Height below which woody plants obviously have been eaten by animals such as deer or goats

bud zone. Region on a plant where resprouting or regrowth occurs

cedar. Evergreen tree (*Juniperus ashei*) native to the Hill Country; also called Ashe juniper or mountain cedar

community. Interacting populations in the same general area

conservation hunter. Hunter who harvests animals with a focus on improving habitat conditions through population control

dabbling duck. Duck that feeds in shallow water by tipping its head and part of its body beneath the water surface to feed on submerged invertebrates and plant vegetation

decompose. To break down by natural processes

deposition. Geologic process in which rocks, gravel, or soil is moved by forces such as water or wind and accumulates in an area over time, e.g., soil or gravel dropped by water as it slows after running quickly during a flood or high-water event

depredation. Damage or loss caused by an animal

detritus. Dead pieces of organic matter

diving duck. Duck that swims underwater when feeding

dry lot. Fenced enclosure bare of food plants; relatively small confined area lacking natural habitat

duff. Small bits of organic material in various states of decomposition, on the ground and under trees

ecosystem. Complex of living organisms, their physical environment, and all their interrelationships within a specific area

edge. Area where two habitat types meet and blend, often rich in species and structural diversity

endangered species. Designation for a species with such small numbers that it is at risk of extinction throughout or in a large portion of its range and thus is listed and given special protection by the US Fish and Wildlife Service

EPBBS. Edwards Plateau Breeding Bird Survey

ephemeral stream. A stream that flows inconsistently, usually dry but with short periods of flow after heavy rain

erosion. Process of soil being picked up and carried by fast-moving water or wind

extirpated. Eradicated or wiped out from a specific geographical region but still existing in other areas

fen. Type of wetland with neutral or alkaline pH that is mineral-rich with a diverse plant community usually dominated by grasses and sedges

flotsam. Both whole plant material and pieces left behind after a flood

forb. Broadleaf flowering herbaceous plant that may be considered a weed or wildflower

fragmentation. Process of being broken into smaller pieces such as a large block of habitat cut up into smaller, usually discontinuous pieces

gallinaceous bird. Native bird that is a member of the order Galliformes, including turkey and quail

girdling. Restricting the growth of or killing a tree by cutting a ring around a large branch or its trunk that is wide and deep enough to prevent the cambium layer from regrowing

grassland. Plant community dominated by a mixture of grasses and forbs

habitat. Place where an organism lives and finds its food, water, shelter, and space

harvest. To gather a resource for human use

hedged. Shaped or sculpted by browsing animals such as deer or goats, occurring on woody plants heavily used by browsing animals

herbaceous. Referring to a plant that has leaves and stems that die at end of the growing season

Hill Country. An area on the central and eastern portion of the Edwards Plateau that has been carved into canyons and hills by running water

holistic land management. A system of sustainable land stewardship; interdisciplinary approach to land stewardship that addresses all aspects of the land, including conservation of wildlife and wildlife habitat, forestry management, water conservation, ranching, and farming practices

hunt. Pursuit of an animal for recreation and/or food

hydrology. Study of occurrence, distribution, movement, and properties of both surface and groundwater

hydrophilic. Referring to a water-loving plant

impervious ground cover. Anything that does not allow water to penetrate the soil

invasive species. Plant or animal that moves beyond its historic range and causes harm to ecosystems; the invasion can threaten native ecosystems as well as commercial, agricultural, or recreational activities dependent on these ecosystems

jetsam. Human-made materials like plastic and metal left behind by a flood

karst. Sedimentary limestone structure with porous characteristics that consists of fissures and holes made by mechanical and chemical dissolution

limiting factor. Component such as food, water, or predation that inhibits population growth

litter. Scattering of plant pieces on the ground and in process of decay

Llano Uplift. A region of the Edwards Plateau dominated by igneous rock; also referred to as the central mineral region

mast. Edible nuts and fruit of trees or shrubs consumed by wildlife

metamorphosis. Change; in insects the physical transformation from egg to adult

microhabitat. Small area supporting diverse vegetation that is noticeably different from the surrounding habitat; often facilitated by factors such as water, slope, or soils

migrant or migratory bird. A bird that moves long distances, often hundreds of miles, between breeding and wintering areas annually

morphogenesis. Biological process that causes change in shape

mosaic. Free-form pattern composed of many components, e.g., patchwork of different vegetative types in well-managed savannah or prairie

motte. Cluster of trees or shrubs growing in savannah, grassland, or prairie, often dominated by a single species such live oak or shin oak

mulch. Plant material (usually in small pieces) spread over the ground naturally or placed there by humans

NRCS. Natural Resources Conservation Service, a US federal agency within the US Department of Agriculture whose mission is to work with farmers and ranchers to improve the health of the nation's private lands while sustaining and enhancing the productivity of American agriculture and natural resources

nymph. Immature phase of insect with incomplete metamorphosis

parasitic bird. Bird that lays its eggs in the nest of another bird species or host that incubates and rears its young

passerine. Bird in the order Passeriformes, which

includes 60% of all bird species; also called perching bird and songbird

perennial stream. A stream that usually runs in all seasons

pervious ground cover. Anything covering the soil such as plants or mulch that allows water to percolate into the soil

PIF. Partners in Flight, a collaboration between state, federal, nonprofit organizations, and private individuals dedicated to the conservation of landbird (ones that use terrestrial habitat) populations

plain. Flat or gently rolling land

population. All organisms of a single species living and interbreeding in the same geographical area

poult. Young turkey able to walk and feed itself but remains under the protection of an adult bird

poverty bush. One of many common names for *Baccharis neglecta,* a loose-limbed, deep-rooted shrub of the Hill Country, that plays an important ecological role as a colonizer or pioneer plant and is common in disturbed areas

prairie. Plant community dominated by grasses and forbs, with few trees

predator. Animal that kills and consumes other animals

priority bird. A species of special conservation concern because its population is in significant decline and/or has a highly limited range

rail. Shy, marsh bird in Rallidae family; has round body, short tail, and wings

raptor. Bird of prey such as hawk, eagle, or owl

reach. Short section of a stream with natural variations such as pools, runs, and riffles

recruitment. In botany, replacing old trees with young saplings

resident bird. A bird that lives in the same region throughout the year

riparian. Referring to a zone or plant community along a stream that is influenced by the stream's moisture and is a transition between the stream itself and the drier uplands

savannah. Land with a mixture of grasses, forbs, trees, and shrubs where the trees, which may be shrublike, do not form a closed canopy

seasonal stream. A creek that usually runs annually during wet seasons

seep. Area with water emerging from the ground slowly or saturating the soil surface but has less consistent water than a spring and contains plants tolerant of wet soils

sheeting. Process of water moving across the ground in a thin layer that can carry small soil particles and detritus

slash. Large plant debris, usually branches from fallen trees; also trimmings left behind after brush control

snag. A standing, partly or completely dead tree, often missing the top or small branches

songbird. A bird in the order Passeriformes, which includes 60% of all bird species; also called perching bird

succession. Natural process of change in the species composition of a plant community over time

sustainable. Able to continue into the future; when applied to land stewardship, involves conserving ecological balance (e.g., predator/prey numbers) and natural resources (e.g., protecting a spring by keeping its natural hydrology and vegetation)

Texas Hill Country. The greater Texas Hill Country that includes three Edwards Plateau *subregions,* Balcones Canyonland, Edwards Plateau Woodland, and Llano Uplift, and covers all or part of 28 counties in Central Texas

thatch. Layer of leaves and other plant material on the ground

threatened species. Designation for a species that is likely to become endangered throughout or in a large portion of its range and thus is listed and given special protection by US Fish and Wildlife Service

TPWD. Texas Parks and Wildlife Department; state agency whose mission is to manage and conserve the natural and cultural resources of Texas and provide hunting, fishing, and outdoor recreation opportunities

understory. Plants that grow below larger plants; often woody or herbaceous plants growing in the shade of larger trees

under-the-radar bird. A bird species that is usually common but relatively unknown

weed. Forb or broadleaf plant growing where it is unwanted

wildflower. Forb or broadleaf plant with pretty flowers and usually uncultivated

xeric. Adapted to survive on small amount of water; drought tolerant

References

Actkinson, M. A., W. R. Kuvlesky Jr., C. W. Boal, L. A. Brennan, and F. Hernandez. 2007. "Nesting Habitat Relationships of Sympatric Crested Caracaras, Red-tailed Hawks, and White-tailed Hawks in South Texas." *Wilson Journal of Ornithology* 119 (4): 570–578.

Alden, Harry A. 1983. *Hardwoods of North America.* USDA Forest Service General Technical Report FPL-GTR-83. Madison, WI: Forest Products Laboratory. http://www.fpl.fs.fed.us/documnts/fplgtr/fplgtr83.pdf.

Allombert, Sylvain, Steve Stockton, and Jean-Louis Martin. 2005. "A Natural Experiment on the Impact of Overabundant Deer on Forest Invertebrates." *Conservation Biology* 19 (6): 1917–1929.

American Bird Conservancy. "Domestic Cat Predation on Birds and Other Wildlife." Cats Indoors Campaign. http://abcbirds.org/wp-content/uploads/2015/05/CatPredation2011.pdf.

Ammon, Elisabeth M. 1995. "Lincoln's Sparrow (*Melospiza lincolnii*)." In *The Birds of North America Online,* ed. A. Poole. Ithaca, NY: Cornell Lab of Ornithology. http://bna.birds.cornell.edu/bna/species/191; doi:10.2173/bna.191.

Arcese, Peter, Mark K. Sogge, Amy B. Marr, and Michael A. Patten. 2002. "Song Sparrow (*Melospiza melodia*)." In *The Birds of North America Online,* ed. A. Poole. Ithaca, NY: Cornell Lab of Ornithology. http://bna.birds.cornell.edu/bna/species/704; doi:10.2173/bna.704.

Armstrong, W. E., and E. L. Young. 2000. *White-tailed Deer Management in the Texas Hill Country.* Austin: Texas Parks and Wildlife.

Askins, Robert A., Felipe Chávez-Ramírez, Brenda C. Dale, Carola A. Haas, James R. Herkert, Fritz L. Knopf, and Peter D. Vickery. 2007. "Conservation of Grassland Birds in North America: Understanding Ecological Processes in Different Regions." Report of the AOU Committee on Conservation. *Ornithological Monographs* 64:1–46.

Audubon Magazine Action Alert. 2010. "Feral Cat Predation on Birds Costs Billions of Dollars a Year." Audubonmagazine.org.

Augustine, David J., and Samuel J. McNaughton. 1998. "Ungulate Effects on the Functional Species Composition of Plant Communities: Herbivore Selectivity and Plant Tolerance." *Journal of Wildlife Management* 62 (4): 1165–1183.

Austin, Jane E., Christine M. Custer, and Alan D. Afton. 1998. "Lesser Scaup (*Aythya affinis*)." In *The Birds of North America Online,* ed. A. Poole. Ithaca, NY: Cornell Lab of Ornithology. http://bna.birds.cornell.edu/bna/species/338; doi:10.2173/bna.338.

Balcones Canyonlands Preserve Partners. "Plants of the Balcones Canyonlands Preserve." Accessed December 15, 2015. https://www.traviscountytx.gov/images/tnr/Docs/Plants.pdf.

Baltosser, William H., and Stephen M. Russell. 2000. "Black-chinned Hummingbird (*Archilochus alexandri*)." In *The Birds of North America Online,* ed. A. Poole. Ithaca, NY: Cornell Lab of Ornithology. http://bna.birds.cornell.edu/bna/species/495; doi:10.2173/bna.495.

Barber, David R., and Thomas E. Martin. 1997. "Influence of Alternate Host Densities on Brown-headed Cowbird Parasitism Rates in Black-capped Vireos." *The Condor* 99 (3): 595–604.

Beadle, David, and James D. Rising. 2002. *Sparrows of the United States and Canada: The Photographic Guide.* Princeton, NJ: Princeton University Press.

Brigham, R. M., Janet Ng, R. G. Poulin, and S. D. Grindal. 2011. "Common Nighthawk (*Chordeiles minor*)." In *The Birds of North America Online,* ed. A. Poole. Ithaca, NY: Cornell Lab of Ornithology. http://bna.birds.cornell.edu/bna/species/213; doi:10.2173/bna.213.

Brooks, R. T., and W. M. Healy. 1988. "Response to Small Mammal Communities to Silvicultural Treatments in Eastern Hardwood Forests of West Virginia and Massachusetts." In *Management of Amphibians, Reptiles, and Small Mammals in North America,*

General Technical Report RM-166, ed. R. C. Szaro, K. E. Severson, and D. R. Patton, 313–318. Fort Collins, CO: USDA Forest Service Rocky Mountain Forest Experiment Station.

Brown, Charles R., and Mary Bomberger Brown. 1999. "Barn Swallow (*Hirundo rustica*)." In *The Birds of North America Online,* ed. A. Poole. Ithaca, NY: Cornell Lab of Ornithology. http://bna.birds.cornell.edu/bna/species/452; doi:10.2173/bna.452.

Brune, Gunnar M. 2002. *Springs of Texas.* College Station: Texas A&M University Press.

Burns, Robert. 2011. "Busting Feral Hog Myths." *AgriLife Today.* College Station, TX: Texas A&M AgriLife. http://today.agrilife.org/2011/03/24/busting-feral-hog-myths/.

Butler, Robert W. 1992. "Great Blue Heron (*Ardea herodias*)." In *The Birds of North America Online,* ed. A. Poole. Ithaca, NY: Cornell Lab of Ornithology. http://bna.birds.cornell.edu/bna/species/025; doi:10.2173/bna.25.

Cabe, Paul R. 1993. "European Starling (*Sturnus vulgaris*)." In *The Birds of North America Online,* ed. A. Poole. Ithaca, NY: Cornell Lab of Ornithology. http://bna.birds.cornell.edu/bna/species/048; doi:10.2173/bna.48.

Campbell, Linda. 2003. *Endangered and Threatened Animals of Texas: Their Life History and Management.* PWD BK W7000-013. Austin: Texas Parks and Wildlife Department, Wildlife Division.

Cardiff, Steven W., and Donna L. Dittmann. 2002. "Ash-throated Flycatcher (*Myiarchus cinerascens*)." In *The Birds of North America Online,* ed. A. Poole. Ithaca, NY: Cornell Lab of Ornithology. http://bna.birds.cornell.edu/bna/species/664; doi:10.2173/bna.664.

Carey, Michael, D. E. Burhans, and D. A. Nelson. 2008. "Field Sparrow (*Spizella pusilla*)." In *The Birds of North America Online,* ed. A. Poole. Ithaca, NY: Cornell Lab of Ornithology. http://bna.birds.cornell.edu/bna/species/103; doi:10.2173/bna.103.

Carr, Bill. 2008. "Texas Endemics of the Edwards Plateau and Llano Uplift." Draft manuscript. The Nature Conservancy of Texas.

Casey, D., and D. Hein. 1983. "Effects of Heavy Browsing on a Bird Community in Deciduous Forest." *Journal of Wildlife Management* 47 (3): 829–836.

Cathey, James C., Kyle Melton, Justin Dreibelbis, Bob Cavney, Shawn L. Locke, Stephen J. DeMaso, T. Wayne Schwertner, and Bret Collier. 2007. *Rio Grande Wild Turkey in Texas: Biology and Management.* Agricultural Communications B-6196. College Station: Texas A&M University System. http://wildlife.tamu.edu/files/2012/05/77906723-Rio-Grande-Wild-Turkey-in-Texas-Biology-and-Management.pdf.

Chasko, Gregory G., and J. Edward Gates. 1982. "Avian Habitat Suitability along a Transmission-Line Corridor in an Oak-Hickory Forest Region." *Wildlife Monographs* 82:3–41.

Chilton, G., M. C. Baker, C. D. Barrentine, and M. A. Cunningham. 1995. "White-crowned Sparrow (*Zonotrichia leucophrys*)." In *The Birds of North America Online,* ed. A. Poole. Ithaca, NY: Cornell Lab of Ornithology. http://bna.birds.cornell.edu/bna/species/183; doi:10.2173/bna.183.

Cimprich, D. A., and K. Comolli. 2010. *Monitoring of the Black-capped Vireo during 2010 on Fort Hood, Texas.* Endangered Species Monitoring and Management at Fort Hood, Texas: 2010 Annual Report. Fort Hood, TX: The Nature Conservancy, Fort Hood Project.

Cink, Calvin L., and Charles T. Collins. 2002. "Chimney Swift (*Chaetura pelagica*)." In *The Birds of North America Online,* ed. A. Poole. Ithaca, NY: Cornell Lab of Ornithology. http://bna.birds.cornell.edu/bna/species/646; doi:10.2173/bna.646.

Clements, J. F., T. S. Schulenberg, M. J. Iliff, D. Roberson, B. L. Sullivan, and C. L. Wood. 2012. "The eBird/Clements Checklist of Birds of the World: v6.7." http://www.birds.cornell.edu/clementschecklist/download/.

Cooper, Susan, ed. 2008 (February 4). "Abundance of White-tailed Deer in the Texas Hill Country: A White Paper." Collaboration for Sustainable Texas Hill Country Deer Population Working Group. Uvalde, TX: TAMU AgriLife Research and Extension Center. Unpublished draft. Available from Rufus.Stephens@tpwd.texas.gov.

Côté, Steve D., Thomas P. Rooney, Jean-Pierre Tremblay, Christian Dussault, and Donald M. Waller. 2004. "Ecological Impacts of Deer Overabundance." *Annual Review of Ecology, Evolution, and Systematics* 35. doi:10.1146/annurev.ecolsys.35.021103.105725.

Cox, Paul W., and Patty Leslie. 1988. *Texas Trees:*

A Friendly Guide. College Station: Texas A&M University Press.

Curry, Robert L., A. Townsend Peterson, and Tom A. Langen. 2002. "Western Scrub-Jay (*Aphelocoma californica*)." In *The Birds of North America Online,* ed. A. Poole. Ithaca, NY: Cornell Lab of Ornithology. http://bna.birds.cornell.edu/bna/species/712; doi:10.2173/bna.712.

Cushman, J. Hall, Trisha A. Tierney, and Jean M. Hinds. 2004. "Variable Effects of Feral Pig Disturbances on Native and Exotic Plants in a California Grassland." *Ecological Applications* 14 (6): 1746–1756.

Davis, Jeff N. 1995. "Hutton's Vireo (*Vireo huttoni*)." In *The Birds of North America Online,* ed. A. Poole. Ithaca, NY: Cornell Lab of Ornithology. http://bna.birds.cornell.edu/bna/species/189; doi:10.2173/bna.189.

Davis, W. E., Jr., and J. A. Kushlan. 1994. "Green Heron (*Butorides virescens*)." In *The Birds of North America Online,* ed. A. Poole. Ithaca, NY: Cornell Lab of Ornithology. http://bna.birds.cornell.edu/bna/species/129; doi:10.2173/bna.129.

deCalesta, David. 1994a. "Deer and Ecosystem Management." http://www.deerandforests.info/resources/Deer%20and%20Ecosystem%20Mgmt.pdf.

———. 1994b. "Effect of White-tailed Deer on Songbirds within Managed Forests in Pennsylvania." *Journal of Wildlife Management* 58 (4): 711–718.

Dellinger, Rachel, Petra Bohall Wood, Peter W. Jones, and Therese M. Donovan. 2012. "Hermit Thrush (*Catharus guttatus*)." In *The Birds of North America Online,* ed. A. Poole. Ithaca, NY: Cornell Lab of Ornithology. http://bna.birds.cornell.edu/bna/species/261; doi:10.2173/bna.261.

DeNicola, Anthony J., Kurt C. VerCauteren, Paul D. Curtis, and Scott E. Hygnstrom. 2000. *Managing White-tailed Deer in Suburban Environments: A Technical Guide.* Ithaca, NY: Cornell Cooperative Extension, the Wildlife Society–Wildlife Damage Management Working Group, and the Northeast Wildlife Damage Research and Outreach Cooperative.

Derrickson, K. C., and R. Breitwisch. 1992. "Northern Mockingbird (*Mimus polyglottos*)." In *The Birds of North America Online,* ed. A. Poole. Ithaca, NY: Cornell Lab of Ornithology. http://bna.birds.cornell.edu/bna/species/007; doi:10.2173/bna.7.

Dickerson, James G., ed. 1992. *The Wild Turkey.* Mechanicsburg, PA: Stackpole Books.

Dickson, James G., Richard N. Conner, and J. Howard Williamson. 1983. "Snag Retention Increases Bird Use of a Clear-Cut." *Journal of Wildlife Management* 47 (3): 799–804.

Dickson, James G., J. Howard Williamson, Richard N. Conner, and Brent Ortego. 1995. "Streamside Zones and Breeding Birds in Eastern Texas." *Wildlife Society Bulletin* 23 (4): 750–755.

Diggs, George M., Jr., Barney L. Lipscomb, and Robert J. O'Kennon. 1999. *Shinner & Mahler's Illustrated Flora of North Central Texas.* SIDA, Botanical Miscellany 16. Fort Worth: Botanical Research Institute of Texas (BRIT).

"Digital Flora of Texas Vascular Plant Image Library." 2011. http://botany.csdl.tamu.edu/FLORA/gallery.htm.

Doughty, Robin. 1998. *The Mockingbird.* Austin: University of Texas Press.

Dunning, John B., Jr., Richard K. Bowers Jr., Sherman J. Suter, and Carl E. Bock. 1999. "Cassin's Sparrow (*Peucaea cassinii*)." In *The Birds of North America Online,* ed. A. Poole. Ithaca, NY: Cornell Lab of Ornithology. http://bna.birds.cornell.edu/bna/species/471; doi:10.2173/bna.471.

Dykstra, Cheryl R., Jeffrey L. Hays, and Scott T. Crocoll. 2008. "Red-shouldered Hawk (*Buteo lineatus*)." In *The Birds of North America Online,* ed. A. Poole. Ithaca, NY: Cornell Lab of Ornithology. http://bna.birds.cornell.edu/bna/species/107; doi:10.2173/bna.107.

Eckerle, Kevin P., and Charles F. Thompson. 2001. "Yellow-breasted Chat (*Icteria virens*)." In *The Birds of North America Online,* ed. A. Poole. Ithaca, NY: Cornell Lab of Ornithology. http://bna.birds.cornell.edu/bna/species/575; doi:10.2173/bna.575.

Ellison, Kevin, Blair O. Wolf, and Stephanie L. Jones. 2009. "Vermilion Flycatcher (*Pyrocephalus rubinus*)." In *The Birds of North America Online,* ed. A. Poole. Ithaca, NY: Cornell Lab of Ornithology. http://bna.birds.cornell.edu/bna/species/484; doi:10.2173/bna.484.

Elphick, Chris, J. B. Dunning Jr., and D. A. Sibley. 2001. *The Sibley Guide to Bird Life & Behavior.* New York: Alfred A. Knopf.

Engelman, C. A. 2010. *Winter Grassland Bird Banding*

2008–2010 on Fort Hood, TX. Endangered Species Monitoring and Management at Fort Hood, Texas: 2010 Annual Report. Fort Hood, TX: The Nature Conservancy, Fort Hood Project.

Enquist, Marshall. 1987. *Wildflowers of the Texas Hill Country*. Austin, TX: Lone Star Botanical.

Farnsworth, George, Gustavo Adolfo Londono, Judit Ungvari Martin, K. C. Derrickson, and R. Breitwisch. 2011. "Northern Mockingbird (*Mimus polyglottos*)." In *The Birds of North America Online,* ed. A. Poole. Ithaca, NY: Cornell Lab of Ornithology. http://bna.birds.cornell.edu/bna/species/007; doi:10.2173/bna.7.

Finity, Leah. 2011. "The Role of Habitat and Dietary Factors in Chimney Swift (*Chaetura pelagica*) Population Declines." Master's thesis, Trent University, Peterborough, Ontario. http://people.trentu.ca/~joenocera/Finity_thesis.pdf.

Flanders, Aron A., William P. Kuvlesky Jr., Donald C. Ruthven III, Robert E. Zaiglin, Ralph L. Bingham, Timothy E. Fulbright, Ridel Hernández, and Leonard A. Brennan. 2006. "Effects of Invasive Exotic Grasses on South Texas Rangeland Breeding Birds." *The Auk* 123 (1): 171–182.

Fleenor, Scott B., and Stephen Welton Taber. 2009. *Plants of Central Texas Wetlands*. Lubbock: Texas Tech University Press.

Flood, Nancy J. 2002. "Scott's Oriole (*Icterus parisorum*)." In *The Birds of North America Online,* ed. A. Poole. Ithaca, NY: Cornell Lab of Ornithology. http://bna.birds.cornell.edu/bna/species/608; doi:10.2173/bna.608.

Freilich, J. E., J. M. Emlen, J. J. Duda, D. C. Freeman, and P. J. Cafaro. 2003. "Ecological Effects of Ranching: A Six-Point Critique." *Bioscience* 53 (8): 759–765.

Friedmann, Herbert. 1928. "Social Parasitism in Birds." *Quarterly Review of Biology* 3 (4): 554–569.

Friend, Milton, and J. Christian Franson, eds. 1999. "Duck Plague." In *Field Manual of Wildlife Diseases: General Field Procedures and Diseases of Birds,* ed. Milton Friend and J. Christian Franson. US Department of the Interior Biological Resources Division and US Geological Survey. http://www.nwhc.usgs.gov/publications/field_manual/chapter_16.pdf.

Fulbright, Timothy Edward, and J. Alfonso Ortega-S. 2006. *White-tailed Deer Habitat: Ecology and Management on Rangelands*. College Station: Texas A&M University Press.

Fuller, R. J., D. G. Noble, K. W. Smith, and D. Vanhinsbergh. 2005. "Recent Declines in Populations of Woodland Birds in Britain: A Review of Possible Causes." *British Birds* 98:116–143.

Gaskill, Melissa. 2011. "War." Texas Co-op Power. http://www.texascooppower.com/texas-stories/life-arts/war.

Gates, J. Edward, and Leslie W. Gysel. 1978. "Avian Nest Dispersion and Fledging Success in Field-Forest Ecotones." *Ecology* 59 (5): 871–883.

Gauthier, Gilles. 1993. "Bufflehead (*Bucephala albeola*)." In *The Birds of North America Online,* ed. A. Poole. Ithaca, NY: Cornell Lab of Ornithology. http://bna.birds.cornell.edu/bna/species/067; doi:10.2173/bna.67.

Gehlbach, Frederick R. 1995. "Eastern Screech-Owl (*Megascops asio*)." In *The Birds of North America Online,* ed. A. Poole. Ithaca, NY: Cornell Lab of Ornithology. http://bna.birds.cornell.edu/bna/species/165; doi:10.2173/bna.165.

Gilbert, W. M., M. K. Sogge, and C. Van Riper III. 2010. "Orange-crowned Warbler (*Vermivora celata*)." In *The Birds of North America Online,* ed. A. Poole. Ithaca, NY: Cornell Lab of Ornithology. http://bna.birds.cornell.edu/bna/species/101; doi:10.2173/bna.101.

Gill, Robin M., and Robert J. Fuller. 2007. "The Effects of Deer Browsing on Woodland Structure and Songbirds in Lowland Britain." *Ibis* 149 (Suppl. 2): 119–127.

Giuliano, William M., Craig R. Allen, R. Scott Lutz, and Stephen Demarais. 1996. "Effects of Red Imported Fire Ants on Northern Bobwhite Chicks." *Journal of Wildlife Management* 60 (2): 309–313.

Goguen, Christopher B., and Nancy E. Mathews. 2001. "Brown-headed Cowbird Behavior and Movements in Relation to Livestock Grazing." *Ecological Applications* 11 (5): 1533–1544.

Golden-Cheeked Warbler Habitat Enhancement Fact Sheet. 2011. Practice Code 645. Environmental Quality Incentives Program (EQIP) and Wildlife Habitat Incentive Program (WHIP), Texas.

Gowaty, Patricia Adair, and Jonathan H. Plissner. 1998. "Eastern Bluebird (*Sialia sialis*)." In *The Birds of North*

America Online, ed. A. Poole. Ithaca, NY: Cornell Lab of Ornithology. http://bna.birds.cornell.edu/bna/species/381; doi:10.2173/bna.381.

Gray, Frank. 1999. "Reducing Cat Predation on Wildlife." *Outdoor California.* https://nrm.dfg.ca.gov/FileHandler.ashx?DocumentID=63784&inline.

Gray, Lawrence J. 1993. "Response of Insectivorous Birds to Emerging Aquatic Insects in Riparian Habitats of a Tallgrass Prairie Stream." *American Midland Naturalist* 129 (2): 288–300.

Greeney, Harold F., and Susan M. Wethington. 2009. "Proximity to Active *Accipiter* Nests Reduces Nest Predation of Black-chinned Hummingbirds." *Wilson Journal of Ornithology* 121 (4): 809–812.

Greenlaw, Jon S. 1996. "Spotted Towhee (*Pipilo maculatus*)." In *The Birds of North America Online,* ed. A. Poole. Ithaca, NY: Cornell Lab of Ornithology. http://bna.birds.cornell.edu/bna/species/263; doi:10.2173/bna.263.

Grzybowski, Joseph A. 1995. "Black-capped Vireo (*Vireo atricapilla*)." In *The Birds of North America Online,* ed. A. Poole. Ithaca, NY: Cornell Lab of Ornithology. http://bna.birds.cornell.edu/bna/species/181; doi:10.2173/bna.181.

Guthery, Fred S. 2000. *On Bobwhites.* College Station: Texas A&M University Press.

Guzy, Michael J., and Gary Ritchison. 1999. "Common Yellowthroat (*Geothlypis trichas*)." In *The Birds of North America Online,* ed. A. Poole. Ithaca, NY: Cornell Lab of Ornithology. http://bna.birds.cornell.edu/bna/species/448; doi:10.2173/bna.448.

Haggerty, Thomas M., and Eugene S. Morton. 1995. "Carolina Wren (*Thryothorus ludovicianus*)." In *The Birds of North America Online,* ed. A. Poole. Ithaca, NY: Cornell Lab of Ornithology. http://bna.birds.cornell.edu/bna/species/188; doi:10.2173/bna.188.

Halkin, Sylvia L., and Susan U. Linville. 1999. "Northern Cardinal (*Cardinalis cardinalis*)." In *The Birds of North America Online,* ed. A. Poole. Ithaca, NY: Cornell Lab of Ornithology. http://bna.birds.cornell.edu/bna/species/440; doi:10.2173/bna.440.

Halls, Lowell K., ed. 1984. *White-tailed Deer—Ecology and Management.* Harrisburg, PA: Stackpole Books.

Hamilton, Robert A., Glenn A. Proudfoot, Dawn A. Sherry, and Steve Johnson. 2011. "Cactus Wren (*Campylorhynchus brunneicapillus*)." In *The Birds of North America Online,* ed. A. Poole. Ithaca, NY: Cornell Lab of Ornithology. http://bna.birds.cornell.edu/bna/species/558; doi:10.2173/bna.558.

Harris, Richard, and Bill Laudenslayer. 1998. "Tips for Creating Snags." *Forestland Steward Newsletter* (Winter). California Forest Stewardship Coordinating Committee. http://calfire.ca.gov/foreststeward/pdf/newslettr5.pdf.

Hatch, Stephan L., Kancheepuram N. Gandhi, and Larry E. Brown. 1990. *Checklist of the Vascular Plants of Texas.* College Station: Texas Agricultural Experiment Station.

Hepp, Gary R., and Frank C. Bellrose. 1995. "Wood Duck (*Aix sponsa*)." In *The Birds of North America Online,* ed. A. Poole. Ithaca, NY: Cornell Lab of Ornithology. http://bna.birds.cornell.edu/bna/species/169; doi:10.2173/bna.169.

Herbel, Carlton, Fred Ares, and Joe Bridges. 1958. "Hand-Grubbing Mesquite in the Semidesert Grassland." *Journal of Range Management* 11 (6): 267–270.

Herkert, James R., Donald E. Kroodsma, and James P. Gibbs. 2001. "Sedge Wren (*Cistothorus platensis*)." In *The Birds of North America Online,* ed. A. Poole. Ithaca, NY: Cornell Lab of Ornithology. http://bna.birds.cornell.edu/bna/species/582; doi:10.2173/bna.582.

Hickman, Karen R., Greg H. Farley, Rob Channell, and Jan E. Steier. 2006. "Effects of Old World Bluestem (*Bothriochloa ischaemum*) on Food Availability and Avian Community Composition with the Mixed-Grass Prairie." *Southwestern Naturalist* 51 (4): 524–530.

Hill, Geoffrey E. 1993. "House Finch (*Carpodacus mexicanus*)." In *The Birds of North America Online,* ed. A. Poole. Ithaca, NY: Cornell Lab of Ornithology. http://bna.birds.cornell.edu/bna/species/046; doi:10.2173/bna.46.

Hohman, William L., and Robert T. Eberhardt. 1998. "Ring-necked Duck (*Aythya collaris*)." In *The Birds of North America Online,* ed. A. Poole. Ithaca, NY: Cornell Lab of Ornithology. http://bna.birds.cornell.edu/bna/species/329; doi:10.2173/bna.329.

Hopp, Steven L., Alice Kirby, and Carol A. Boone. 1995. "White-eyed Vireo (*Vireo griseus*)." In *The Birds of North America Online,* ed. A. Poole. Ithaca, NY:

Cornell Lab of Ornithology. http://bna.birds.cornell.edu/bna/species/168; doi:10.2173/bna.168.

Horton, Jennifer. 2013. "15 Products That Prevent Window Strikes." *BirdWatching Daily.* http://www.birdwatchingdaily.com/featured-stories/15-products-that-prevent-windows-strikes.

Houston, C. Stuart, Cameron R. Jackson, and Daniel E. Bowen Jr. 2011. "Upland Sandpiper (*Bartramia longicauda*)." In *The Birds of North America Online,* ed. A. Poole. Ithaca, NY: Cornell Lab of Ornithology. http://bna.birds.cornell.edu/bna/species/580; doi:10.2173/bna.580.

Hughes, Janice M. 1996. "Greater Roadrunner (*Geococcyx californianus*)." In *The Birds of North America Online,* ed. A. Poole. Ithaca, NY: Cornell Lab of Ornithology. http://bna.birds.cornell.edu/bna/species/244; doi:10.2173/bna.244.

———. 1999. "Yellow-billed Cuckoo (*Coccyzus americanus*)." In *The Birds of North America Online,* ed. A. Poole. Ithaca, NY: Cornell Lab of Ornithology. http://bna.birds.cornell.edu/bna/species/418; doi:10.2173/bna.418.

Hunt, P. D., and David J. Flaspohler. 1998. "Yellow-rumped Warbler (*Dendroica coronata*)." In *The Birds of North America Online,* ed. A. Poole. Ithaca, NY: Cornell Lab of Ornithology. http://bna.birds.cornell.edu/bna/species/376; doi:10.2173/bna.376.

Husak, Michael S. 2000. "Seasonal Variation in Territorial Behavior of Golden-fronted Woodpeckers in West-Central Texas." *Southwestern Naturalist* 45 (1): 30–38.

Husak, Michael S., and Terry C. Maxwell. 1998. "Golden-fronted Woodpecker (*Melanerpes aurifrons*)." In *The Birds of North America Online,* ed. A. Poole. Ithaca, NY: Cornell Lab of Ornithology. http://bna.birds.cornell.edu/bna/species/373; doi:10.2173/bna.373.

Jackson, Bette J., and Jerome A. Jackson. 2000. "Killdeer (*Charadrius vociferus*)." In *The Birds of North America Online,* ed. A. Poole. Ithaca, NY: Cornell Lab of Ornithology. http://bna.birds.cornell.edu/bna/species/517; doi:10.2173/bna.517.

Jackson, Jerome A., and Henri R. Ouellet. 2002. "Downy Woodpecker (*Picoides pubescens*)." In *The Birds of North America Online,* ed. A. Poole. Ithaca, NY: Cornell Lab of Ornithology. http://bna.birds.cornell.edu/bna/species/613; doi:10.2173/bna.613.

James, J. Dale, and Jonathan E. Thompson. 2001. "Black-bellied Whistling-Duck (*Dendrocygna autumnalis*)." In *The Birds of North America Online,* ed. A. Poole. Ithaca, NY: Cornell Lab of Ornithology. http://bna.birds.cornell.edu/bna/species/578; doi:10.2173/bna.578.

Jaworowski, Chris. 2008. "Feral Hogs—Wildlife Enemy Number One." *Outdoor Alabama.* http://www.outdooralabama.com/feral-hogs-wildlife-enemy-number-one.

Johnson, E. H. "Edwards Plateau." *Handbook of Texas Online.* Texas State Historical Association. Accessed December 17, 2012. http://www.tshaonline.org/handbook/online/articles/rxe01.

Johnson, Kevin. 1995. "Green-winged Teal (*Anas crecca*)." In *The Birds of North America Online,* ed. A. Poole. Ithaca, NY: Cornell Lab of Ornithology. http://bna.birds.cornell.edu/bna/species/193; doi:10.2173/bna.193.

Johnson, Kristine, and Brian D. Peer. 2001. "Great-tailed Grackle (*Quiscalus mexicanus*)." In *The Birds of North America Online,* ed. A. Poole. Ithaca, NY: Cornell Lab of Ornithology. http://bna.birds.cornell.edu/bna/species/576; doi:10.2173/bna.576.

Johnson, M. J., C. Van Riper III, and K. M. Pearson. 2002. "Black-throated Sparrow (*Amphispiza bilineata*)." In *The Birds of North America Online,* ed. A. Poole. Ithaca, NY: Cornell Lab of Ornithology. http://bna.birds.cornell.edu/bna/species/637; doi:10.2173/bna.637.

Johnson, R. Roy, and Lois T. Haight. 1996. "Canyon Towhee (*Melozone fuscus*)." In *The Birds of North America Online,* ed. A. Poole. Ithaca, NY: Cornell Lab of Ornithology. http://bna.birds.cornell.edu/bna/species/264; doi:10.2173/bna.264.

Jolley, D. Buck, Stephen S. Ditchkoff, Bill D. Sparklin, Laura B. Hanson, Michael S. Mitchell, and James B. Grand. 2010. "Estimate of Herpetofauna Depredation by a Population of Wild Pigs." *Journal of Mammalogy* 91 (2). http://www.umt.edu/mcwru/personnel/MikeMitchell/PDF%20Mitchell/Mitche112010%20-%20JMamm%2091_2.pdf); doi:10.1644/09-MAMM-A-129.1.

Jones, Peter W., and Therese M. Donovan. 1996. "Hermit Thrush (*Catharus guttatus*)." In *The Birds of North

America Online, ed. A. Poole. Ithaca, NY: Cornell Lab of Ornithology. http://bna.birds.cornell.edu/bna/species/261; doi:10.2173/bna.261.

Jones, Stephanie L., and John E. Cornely. 2002. "Vesper Sparrow (*Pooecetes gramineus*)." In *The Birds of North America Online,* ed. A. Poole. Ithaca, NY: Cornell Lab of Ornithology. http://bna.birds.cornell.edu/bna/species/624; doi:10.2173/bna.624.

Jones-Lewey, Sky, ed. 2010. *Your Remarkable Riparian: A Field Guide to Riparian Plants within the Nueces River Basin of Texas.* Uvalde, TX: Nueces River Authority.

Karr, James R., and Roland R. Roth. 1971. "Vegetation Structure and Avian Diversity in Several New World Areas." *American Naturalist* 105 (945): 423–435.

Kelly, Jeffrey F., Eli S. Bridge, and Michael J. Hamas. 2009. "Belted Kingfisher (*Megaceryle alcyon*)." In *The Birds of North America Online,* ed. A. Poole. Ithaca, NY: Cornell Lab of Ornithology. http://bna.birds.cornell.edu/bna/species/084; doi:10.2173/bna.84.

Kennedy, E. Dale, and Douglas W. White. 1997. "Bewick's Wren (*Thryomanes bewickii*)." In *The Birds of North America Online,* ed. A. Poole. Ithaca, NY: Cornell Lab of Ornithology. http://bna.birds.cornell.edu/bna/species/315; doi:10.2173/bna.315.

Kershner, Eric L., and Walter G. Ellison. 2012. "Blue-gray Gnatcatcher (*Polioptila caerulea*)." In *The Birds of North America Online,* ed. A. Poole. Ithaca, NY: Cornell Lab of Ornithology. http://bna.birds.cornell.edu/bna/species/023; doi:10.2173/bna.23.

Kilpatrick, H. J., and S. M. Spohr. 2000. "Spatial and Temporal Use of a Suburban Landscape by Female White-tailed Deer." *Wildlife Society Bulletin* 28:1023–1029.

Klem, Daniel, Jr. 2010. "Glass: A Deadly Conservation Issue for Birds." *Bird Observer* 34 (2): 73–81.

Knapp, Keith. *Deer Crash: Reducing Deer-Vehicle Collisions through Enhanced Road Safety Practices.* Iowa Local Technical Assistance Program (LTAP), Iowa State University. http://www.deercrash.org/.

Knopf, F. L. 1996. "Prairie Legacies—Birds." In *Prairie Conservation: Preserving North America's Most Endangered Ecosystem,* ed. F. B. Samson and F. L. Knopf, 135–148. Washington, DC: Island Press.

Knopf, Fritz L., and Fred B. Samson, eds. 1997. *Ecology and Conservation of Great Plains Vertebrates.* New York: Springer-Verlag.

Kopachena, Jeffrey G., and Christopher J. Crist. 2000. "Macro-habitat Features Associated with Painted and Indigo Buntings in Northeast Texas." *Wilson Bulletin* 112 (1): 108–114.

Kostecke, R. M. 2010. *The U.S. Army, Fort Hood Garrison and the Nature Conservancy Cooperative Agreement No: DPW-ENV 07-A-001: 2010 Annual Report.* Fort Hood, TX: The Nature Conservancy, Fort Hood Project.

Kostecke, Richard M., Scott G. Summers, Gilbert H. Eckrich, and David A. Cimprich. 2005. "Effects of Brown-headed Cowbird (*Molothrus ater*) Removal on Black-capped Vireo (*Vireo atricapilla*) Nest Success and Population Growth at Fort Hood, Texas." *Ornithological Monographs* 57:28–37.

Kostka, Ken. 2000. "Cricket Tossing: A New Emergency Feeding Technique for Purple Martins." *Purple Martin Update* 9 (4). Reprint, Purple Martin Conservation Association. purplemartin.org.

Kricher, John C. 1995. "Black-and-white Warbler (*Mniotilta varia*)." In *The Birds of North America Online,* ed. A. Poole. Ithaca, NY: Cornell Lab of Ornithology. http://bna.birds.cornell.edu/bna/species/158; doi:10.2173/bna.158.

Kus, Barbara, Steven L. Hopp, R. Roy Johnson, and Bryan T. Brown. 2010. "Bell's Vireo (*Vireo bellii*)." In *The Birds of North America Online,* ed. A. Poole. Ithaca, NY: Cornell Lab of Ornithology. http://bna.birds.cornell.edu/bna/species/035; doi:10.2173/bna.35.

Ladd, Clifton, and Leila Gass. 1999. "Golden-cheeked Warbler (*Setaphaga chrysoparia*)." In *The Birds of North America Online,* ed. A. Poole. Ithaca, NY: Cornell Lab of Ornithology. http://bna.birds.cornell.edu/bna/species/420; doi:10.2173/bna.420.

Ladybird Johnson Wildflower Center. Native Plants Database. https://www.wildflower.org/plants/.

LaFayette, Russell A., John R. Pruitt, and William Zeedyk. 1993. *Riparian Area Enhancement through Road Design and Maintenance.* USDA-Forest Service, Southwestern Region. http://www.fs.fed.us/rm/boise/AWAE/labs/awae_flagstaff/Hot_Topics/ripthreatbib/lafayette_etal_ripareaenhance.pdf.

Lanyon, Wesley E. 1995. "Eastern Meadowlark (*Sturnella magna*)." In *The Birds of North America Online,* ed. A. Poole. Ithaca, NY: Cornell Lab of Ornithology. http://bna.birds.cornell.edu/bna/species/160; doi:10.2173/bna.160.

Lepczyk, Christopher A., Angela G. Mertig, and Jianguo Liu. 2003. *Landowners and Cat Predation across Rural-to-Urban Landscapes.* Elsevier Biological Conservation 115. http://archive.csis.msu.edu/Publications/Landowners_and_cat_predation.pdf.

Leschack, C. R., S. K. McKnight, and G. R. Hepp. 1997. "Gadwall (*Anas strepera*)." In *The Birds of North America Online,* ed. A. Poole. Ithaca, NY: Cornell Lab of Ornithology. http://bna.birds.cornell.edu/bna/species/283; doi:10.2173/bna.283.

Lewis, Clancey, Matt Berg, James C. Cathey, Jim Gallagher, Nikki Dictson, and Mark McFarland. 2009a. *Box Traps for Capturing Feral Hogs.* Texas A&M AgriLife Communications L-5528. College Station: Texas A&M System. http://feralhogs.tamu.edu/files/2010/05/BoxTraps.pdf.

———. 2009b. *Corral Traps for Capturing Feral Hogs.* Texas A&M AgriLife Communications L-5528. College Station: Texas A&M System. http://feralhogs.tamu.edu/files/2010/05/CorralTraps.pdf.

———. 2009c. *Recognizing Feral Hog Sign.* Texas A&M AgriLife Communications L-5528. College Station: Texas A&M System. http://feralhogs.tamu.edu/files/2010/05/RecognizingFeralHogSign.pdf.

———. 2011. *Snaring Feral Hogs.* Texas A&M AgriLife Communications L-5528. College Station: Texas A&M System. http://plumcreek.tamu.edu/media/6641/1-5528-snaring-feral-hogs.pdf.

Lockwood, M. W. 2008. *Birds of the Edwards Plateau: A Field Checklist.* PWD BK P4000-667 (4/08). Austin: Texas Parks and Wildlife Department.

Lockwood, M. W., and B. Freeman. 2004. *The TOS Handbook of Texas Birds.* College Station: Texas A&M University Press.

Loflin, Brian, and Shirley Loflin. 2006. *Grasses of the Texas Hill Country: A Field Guide.* College Station: Texas A&M University Press.

Longcore, Travis, Catherine Rich, and Lauren M. Sullivan. 2009. "Critical Assessment of Claims regarding Management of Feral Cats by Trap-Neuter-Return." *Conservation Biology* 23 (4): 887–894.

Lowther, Peter E. 1993. "Brown-headed Cowbird (*Molothrus ater*)." In *The Birds of North America Online,* ed. A. Poole. Ithaca, NY: Cornell Lab of Ornithology. http://bna.birds.cornell.edu/bna/species/047; doi:10.2173/bna.47.

———. 2001. "Ladder-backed Woodpecker (*Picoides scalaris*)." In *The Birds of North America Online,* ed. A. Poole. Ithaca, NY: Cornell Lab of Ornithology. http://bna.birds.cornell.edu/bna/species/565; doi:10.2173/bna.565.

———. 2005. "Le Conte's Sparrow (*Ammodramus leconteii*)." In *The Birds of North America Online,* ed. A. Poole. Ithaca, NY: Cornell Lab of Ornithology. http://bna.birds.cornell.edu/bna/species/224; doi:10.2173/bna.224.

Lowther, P. E., C. Celada, N. K. Klein, C. C. Rimmer, and D. A. Spector. 1999. "Yellow Warbler (*Steophaga petechia*)." In *The Birds of North America Online,* ed. A. Poole. Ithaca, NY: Cornell Lab of Ornithology. http://bna.birds.cornell.edu/bna/species/454; doi:10.2173/bna.454.

Lowther, Peter E., and Calvin L. Cink. 2006. "House Sparrow (*Passer domesticus*)." In *The Birds of North America Online,* ed. A. Poole. Ithaca, NY: Cornell Lab of Ornithology. http://bna.birds.cornell.edu/bna/species/012; doi:10.2173/bna.12.

Lowther, Peter E., and James L. Ingold. 2011. "Blue Grosbeak (*Passerina caerulea*)." In *The Birds of North America Online,* ed. A. Poole. Ithaca, NY: Cornell Lab of Ornithology. http://bna.birds.cornell.edu/bna/species/079; doi:10.2173/bna.79.

Lowther, Peter E., Scott M. Lanyon, and Christopher W. Thompson. 1999. "Painted Bunting (*Passerina ciris*)." In *The Birds of North America Online,* ed. A. Poole. Ithaca, NY: Cornell Lab of Ornithology. http://bna.birds.cornell.edu/bna/species/398; doi:10.2173/bna.398.

"Macaulay Library." 2012. Ithaca, NY: Cornell Lab of Ornithology. http://macaulaylibrary.org/.

Martin, Alexander C., H. S. Zim, and A. L. Nelson. 1951. *American Wildlife and Plants: A Guide to Wildlife Food Habits.* Reprint, New York: McGraw-Hill, 1961.

Martin, John W., and Jimmie R. Parrish. 2000. "Lark

Sparrow (*Chondestes grammacus*)." In *The Birds of North America Online,* ed. A. Poole. Ithaca, NY: Cornell Lab of Ornithology. http://bna.birds.cornell.edu/bna/species/488; doi:10.2173/bna.488.

Martin, Thomas E., and Deborah M. Finch. 1995. *Ecology and Management of Neotropical Migratory Birds: A Synthesis and Review of Critical Issues.* New York: Oxford University Press.

Massachusetts Audubon. "Bird Window Collisions." Accessed December 27, 2015. http://www.massaudubon.org/learn/nature-wildlife/birds/bird-window-collisions.

Mattsson, Brady J., Terry L. Master, Robert S. Mulvihill, and W. Douglas Robinson. 2009. "Louisiana Waterthrush (*Parkesia motacilla*)." In *The Birds of North America Online,* ed. A. Poole. Ithaca, NY: Cornell Lab of Ornithology. http://bna.birds.cornell.edu/bna/species/151; doi:10.2173/bna.151.

Mayer, John J., and L. Lehr Brisbin Jr. 2008. *Wild Pigs in the United States: Their History, Comparative Morphology, and Current Status.* Athens: University of Georgia Press.

McCarty, John P. 1996. "Eastern Wood-Pewee (*Contopus virens*)." In *The Birds of North America Online,* ed. A. Poole. Ithaca, NY: Cornell Lab of Ornithology. http://bna.birds.cornell.edu/bna/species/245; doi:10.2173/bna.245.

McCoy, Timothy D., Eric W. Kurzejeski, Loren W. Burger Jr., and Mark R. Ryan. 2001. "Effects of Conservation Practice, Mowing, and Temporal Changes in Vegetation Structure on CRP Fields in Northern Missouri." *Wildlife Society Bulletin* 29 (3): 979–987.

McGinty, Allan, and Darrell N. Ueckert. 2001. "The Brush Busters Success Story." *Rangelands* 23 (6): 3–8.

———. 2005. *Brush Busters: How to Beat Mesquite.* Texas Cooperative Extension L-5144. College Station: Texas A&M University System. http://ector.agrilife.org/files/2011/07/bbmesquite_2.pdf.

McGraw, Kevin J., and Alex L. Middleton. 2009. "American Goldfinch (*Carduelis tristis*)." In *The Birds of North America Online,* ed. A. Poole. Ithaca, NY: Cornell Lab of Ornithology. http://bna.birds.cornell.edu/bna/species/080; doi:10.2173/bna.80.

McNeil, Don. "Build an Easy Box for Golden-fronted, Red-headed, and Hairy Woodpeckers." In *The Original Birdhouse Book. Bird Watcher's Digest.* Accessed December 27, 2015. http://www.birdwatchersdigest.com/bwdsite/solve/howto/woodpeckerbox.php.

McShea, William J., and John H. Rappole. 2000. "Managing the Abundance and Diversity of Breeding Bird Populations through Manipulation of Deer Populations." *Conservation Biology* 14 (4): 1161–1170.

McShea, W. J., H. B. Underwood, and J. H. Rappole, eds. 1997. *The Science of Overabundance: Deer Ecology and Population Management.* Washington, DC: Smithsonian Institution Press.

Melvin, Scott M., and James P. Gibbs. 1996. "Sora (*Porzana carolina*)." In *The Birds of North America Online,* ed. A. Poole. Ithaca, NY: Cornell Lab of Ornithology. http://bna.birds.cornell.edu/bna/species/250; doi:10.2173/bna.250.

Middleton, Alex L. 1998. "Chipping Sparrow (*Spizella passerina*)." In *The Birds of North America Online,* ed. A. Poole. Ithaca, NY: Cornell Lab of Ornithology. http://bna.birds.cornell.edu/bna/species/334; doi:10.2173/bna.334.

"The Mockingbird (*Mimus polyglottos*): Beneficial Garden Critter/Florida State Bird." 2011. *Florida Gardener.* http://www.floridagardener.com/critters/Mockingbird.htm.

Moldenhauer, Ralph R., and Daniel J. Regelski. 1996. "Northern Parula (*Setophaga americana*)." In *The Birds of North America Online,* ed. A. Poole. Ithaca, NY: Cornell Lab of Ornithology. http://bna.birds.cornell.edu/bna/species/215; doi:10.2173/bna.215.

Morrison, Joan L., and James F. Dwyer. 2012. "Crested Caracara (*Caracara cheriway*)." In *The Birds of North America Online,* ed. A. Poole. Ithaca, NY: Cornell Lab of Ornithology. http://bna.birds.cornell.edu/bna/species/249; doi:10.2173/bna.249.

Morrison, Michael L., and Martin G. Raphael. 1993. "Modeling the Dynamics of Snags." *Ecological Applications* 3 (2): 322–330.

Moskoff, William. 2002. "Green Kingfisher (*Chloroceryle americana*)." In *The Birds of North America Online,* ed. A. Poole. Ithaca, NY: Cornell Lab of Ornithology. http://bna.birds.cornell.edu/bna/species/621; doi:10.2173/bna.621.

Mostrom, Alison M., Robert L. Curry, and Bernard Lohr. 2002. "Carolina Chickadee (*Poecile carolinensis*)."

In *The Birds of North America Online,* ed. A. Poole. Ithaca, NY: Cornell Lab of Ornithology. http://bna.birds.cornell.edu/bna/species/636; doi:10.2173/bna.636.

Mostyn, Chris M. 2003. "White-tailed Deer Overabundance: A Threat to Regeneration of Golden-cheeked Warbler Habitat." Master's thesis, Texas State University.

Mowbray, Thomas. 1999. "American Wigeon (*Anas americana*)." In *The Birds of North America Online,* ed. A. Poole. Ithaca, NY: Cornell Lab of Ornithology. http://bna.birds.cornell.edu/bna/species/401; doi:10.2173/bna.401.

Mueller, Helmut. 1999. "Wilson's Snipe (*Gallinago delicata*)." In *The Birds of North America Online,* ed. A. Poole. Ithaca, NY: Cornell Lab of Ornithology. http://bna.birds.cornell.edu/bna/species/417; doi:10.2173/bna.417.

Mueller, James M., C. Brad Dabbert, Stephen Demarais, and Andrew R. Forbes. 1999. "Northern Bobwhite Chick Mortality Caused by Red Imported Fire Ants." *Journal of Wildlife Management* 63 (4): 1291–1298.

Nelle, Steve. 2012. *Riparian Notes Number 28: Creek and River Myths.* San Angelo, TX: US Department of the Interior, Bureau of Land Management National Riparian Service Team.

———. 2015. "Holistic Perspective on Juniper." Texas Natural Resources Server. http://texnat.tamu.edu/library/symposia/juniper-ecology-and-management/holistic-perspective-on-juniper/. Last updated 2015.

New Jersey Audubon Society. 2005. "Forest Health and Ecological Integrity—Stressors and Solutions." Policy White Paper. http://pbadupws.nrc.gov/docs/ML0719/ML071980035.pdf.

Nolan, V., Jr., E. D. Ketterson, D. A. Cristol, C. M. Rogers, E. D. Clotfelter, R. C. Titus, S. J. Schoech, and E. Snajdr. 2002. "Dark-eyed Junco (*Junco hyemalis*)." In *The Birds of North America Online,* ed. A. Poole. Ithaca, NY: Cornell Lab of Ornithology. http://bna.birds.cornell.edu/bna/species/716; doi:10.2173/bna.716.

Oberholser, Harry C. 1974a. *The Bird Life of Texas: Volume One.* Austin: University of Texas Press.

———. 1974b. *The Bird Life of Texas: Volume Two.* Austin: University of Texas Press.

Oklahoma Cooperative Extension Service and Oklahoma Conservation Commission. 1998. *Riparian Area Management Handbook.* Fact Sheet E-952. Stillwater: Oklahoma State University. http://pods.dasnr.okstate.edu/docushare/dsweb/Get/Document-2251/e-952.pdf.

Oklahoma Forestry Service. 2009. "Introduction to Road Stream Crossings." Forestry Note. Oklahoma City, OK: Forestry Services Division.

Oring, Lewis W., Elizabeth M. Gray, and J. Michael Reed. 1997. "Spotted Sandpiper (*Actitis macularius*)." In *The Birds of North America Online,* ed. A. Poole. Ithaca, NY: Cornell Lab of Ornithology. http://bna.birds.cornell.edu/bna/species/289; doi:10.2173/bna.289.

Ortega, Catherine. 1998. *Cowbirds and Other Brood Parasites.* Tucson: University of Arizona Press.

Ostfeld, Richard S., Clive G. Jones, and Jerry O. Wolff. 1996. "Of Mice and Mast: Ecological Connections in Eastern Deciduous Forests." *BioScience* 46 (5): 323–330.

Otis, David L., John H. Schulz, David Miller, R. E. Mirarchi, and T. S. Baskett. 2008. "Mourning Dove (*Zenaida macroura*)." In *The Birds of North America Online,* ed. A. Poole. Ithaca, NY: Cornell Lab of Ornithology. http://bna.birds.cornell.edu/bna/species/117; doi:10.2173/bna.117.

Patten, Michael A., and Brenda D. Smith-Patten. 2008. "Black-crested Titmouse (*Baeolophus atricristatus*)." In *The Birds of North America Online,* ed. A. Poole. Ithaca, NY: Cornell Lab of Ornithology. http://bna.birds.cornell.edu/bna/species/717; doi:10.2173/.

Payne, Robert B. 2006. "Indigo Bunting (*Passerina cyanea*)." In *The Birds of North America Online,* ed. A. Poole. Ithaca, NY: Cornell Lab of Ornithology. http://bna.birds.cornell.edu/bna/species/004; doi:10.2173/bna.4.

Peak, R. G., and B. N. Moe. 2010. *Population Trends of the Golden-cheeked Warbler on Fort Hood, Texas 1992–2010.* Endangered Species Monitoring and Management at Fort Hood, Texas: 2010 Annual Report. Fort Hood, TX: The Nature Conservancy, Fort Hood Project.

Pedersen, Ellen K., William E. Grant, and Michael T. Longnecker. 1996. "Effects of Red Imported Fire Ants on Newly-Hatched Northern Bobwhite." *Journal of Wildlife Management* 60 (1): 164–169.

Petty, Blake D., Roel R. Lopez, James C. Cathey, Shawn L. Locke, Markus J. Peterson, and Nova J. Silvy. 2005. "Effects of Feral Hog Control on Nest Fate of Eastern Wild Turkey in the Post Oak Savannah of Texas." In *Wild Turkey Management: Accomplishments, Strategies, and Opportunities. Proceedings of the Ninth National Wild Turkey Symposium,* ed. C. Alan Stewart and Valerie R. Frawley, 169–172. Grand Rapids: Michigan Department of Natural Resources. https://www.michigan.gov/documents/dnr/Ninth_National_Wild_Turkey_Symposium_Proceedings_450287_7.pdf.

Pietzi, Pamela J., and Diane A. Granfors. 2000. "White-tailed Deer (*Odocoileus virginianus*) Predation on Grassland Songbird Nestlings." *American Midland Naturalist* 144 (2): 419–422.

Point Reyes Bird Observatory (PRBO) Conservation Science. 2003. http://www.prbo.org/.

Poole, J. M., W. R. Carr, D. M. Price, and J. R. Singhurst. Forthcoming. *Rare Plants of Texas: A Field Guide.* College Station: Texas A&M University Press.

Poole, J. M., J. R. Singhurst, D. M. Price, and W. R. Carr. 2007. *A List of the Rare Plants of Texas.* Austin: Wildlife Diversity Program, Texas Parks and Wildlife Department, and the Nature Conservancy of Texas.

Porter, Michael. 2008. "Fall and Winter Duck Foods in South Central U.S." Samuel Roberts Noble Foundation. http://www.noble.org/ag/wildlife/duck-foods/.

Project Feeder Watch. 2008–2012. "About Birds and Bird Feeding." http://www.birds.cornell.edu/pfw/AboutBirdsandFeeding/abtbirds_index.html.

Raby, C. R., D. M. Alexis, A. Dickinson, and N. S. Clayton. 2007. "Planning for the Future by Western Scrub-Jays" (abstract). *Nature* 445 (7130). doi:10.1038/nature05575.

Randel, Charles J., Dustin A. Jones, Beau J. Willsey, Raymond Aguirre, Jody N. Schaap, Markus J. Peterson, and Nova J. Silvy. 2005. "Nesting Ecology of Rio Grande Wild Turkey in the Edwards Plateau of Texas." In *Wild Turkey Management: Accomplishments, Strategies, and Opportunities. Proceedings of the Ninth National Wild Turkey Symposium,* ed. C. Alan Stewart and Valerie R. Frawley, 237–244. Grand Rapids: Michigan Department of Natural Resources. https://www.michigan.gov/documents/dnr/Ninth_National_Wild_Turkey_Symposium_Proceedings_450287_7.pdf.

Raphael, Martin G., and Marshall White. 1984. "Use of Snags by Cavity-Nesting Birds in the Sierra Nevada." *Wildlife Monographs* 86:3–66.

Rasmussen, G. Allen, Guy R. McPherson, and Henry A. Wright. 1986. *Prescribed Burning Juniper Communities in Texas.* Texas Tech Management Note 10. Department of Range and Wildlife Management. Lubbock: Texas Tech University.

Rawinski, Thomas J. 2008. "Impacts of White-tailed Deer Overabundance in Forest Ecosystems: An Overview." US Department of Agriculture, Northeastern Area State and Private Forestry, Forest Service. www.na.fs.fed.us.

Regosin, Jonathan V. 1998. "Scissor-tailed Flycatcher (*Tyrannus forficatus*)." In *The Birds of North America Online,* ed. A. Poole. Ithaca, NY: Cornell Lab of Ornithology. http://bna.birds.cornell.edu/bna/species/342; doi:10.2173/bna.342.

Reynolds, Michael C., and Paul R. Krausman. 1998. "Effects of Winter Burning on Birds in Mesquite Grassland." *Wildlife Society Bulletin* 26 (4): 867–876.

Ringelman, James K. 1990. "Life History, Traits, and Management of the Gadwall." In *Waterfowl Management Handbook 13.1.2.* Fort Collins, CO: Colorado Division of Wildlife.

Riparian Road Guide. *Managing Roads to Enhance Riparian Areas.* 1994. Washington, DC: Terrene Institute. http://www.fs.fed.us/rm/boise/AWAE/labs/awae_flagstaff/Hot_Topics/ripthreatbib/reeder_riproadguide.pdf.

Rising, James D., and Pamela L. Williams. 1999. "Bullock's Oriole (*Icterus bullockii*)." In *The Birds of North America Online,* ed. A. Poole. Ithaca, NY: Cornell Lab of Ornithology. http://bna.birds.cornell.edu/bna/species/416; doi:10.2173/bna.416.

Robinson, W. Douglas. 2012. "Summer Tanager (*Piranga rubra*)." In *The Birds of North America Online,* ed. A. Poole. Ithaca, NY: Cornell Lab of Ornithology. http://bna.birds.cornell.edu/bna/species/248; doi:10.2173/bna.248.

Rodewald, Paul G., and Ross D. James. 1996. "Yellow-throated Vireo (*Vireo flavifrons*)." In *The Birds of North America Online,* ed. A. Poole. Ithaca, NY: Cornell Lab of Ornithology. http://bna.birds.cornell.edu/bna/species/247; doi:10.2173/bna.247.

Rohwer, Frank C., William P. Johnson, and Elizabeth R. Loos. 2002. "Blue-winged Teal (*Anas discors*)." In *The Birds of North America Online,* ed. A. Poole. Ithaca, NY: Cornell Lab of Ornithology. http://bna.birds.cornell.edu/bna/species/625; doi:10.2173/bna.625.

Rollins, Dale, and John P. Carroll. 2001. "Impacts of Predation on Northern Bobwhite and Scaled Quail." *Wildlife Society Bulletin* 29 (1): 39–51.

Rosenberg, K. V., R. W. Rohrbaugh Jr., S. E. Barker, J. D. Lowe, R. S. Hames, and A. A. Dhondt. 1999. "A Land Manager's Guide to Improving Habitat for Scarlet Tanagers and Other Forest-Interior Birds." http://www.birds.cornell.edu/conservation/tanager/tanager.pdf.

Runkle, James R. 2000. "Canopy Tree Turnover in Old-Growth Mesic Forests of Eastern North America." *Ecology* 81(2): 554–567.

Russell, F. L., and N. L. Fowler. 2004. "Effects of White-tailed Deer on the Population Dynamics of Acorns, Seedlings and Small Saplings of *Quercus buckleyi*." *Plant Ecology* 173:59–72.

Russell, F. Leland, David B. Zippin, and Norma L. Fowler. 2001. "Effects of White-tailed Deer (*Odocoileus virginianus*) on Plants, Plant Populations and Communities: A Review." *American Midland Naturalist* 146 (1): 1–26.

Russell, Robin E., Victoria A. Saab, and Jonathan G. Dudley. 2007. "Habitat-Suitability Models for Cavity-Nesting Birds in a Postfire Landscape." *Journal of Wildlife Management* 71 (8): 2600–2611.

Sallabanks, Rex, and Frances C. James. 1999. "American Robin (*Turdus migratorius*)." In *The Birds of North America Online,* ed. A. Poole. Ithaca, NY: Cornell Lab of Ornithology. http://bna.birds.cornell.edu/bna/species/462; doi:10.2173/bna.462.

Santiago, Melissa J., and Amanda D. Rodewald. *Dead Trees as Resources for Forest Wildlife.* Columbus: Ohio State University Extension W-18-04. ohioline.osu.edu/w-fact/0018.html.

Sauer, J. R., J. E. Hines, J. E. Fallon, K. L. Pardieck, D. J. Ziolkowski Jr., and W. A. Link. 2011. *The North American Breeding Bird Survey, Results and Analysis 1966–2010.* Version 12.07.2011. Laurel, MD: USGS Patuxent Wildlife Research Center.

Schaap, Jody N., Markus J. Peterson, Nova J. Silvy, Raymond Aguirre, and Humberto L. Perotto Baldivieso. 2005. "Spatial Distribution of Female Rio Grande Wild Turkeys during the Reproductive Season." In *Wild Turkey Management: Accomplishments, Strategies, and Opportunities. Proceedings of the Ninth National Wild Turkey Symposium,* ed. C. Alan Stewart and Valerie R. Frawley, 231–236. Grand Rapids: Michigan Department of Natural Resources. https://www.michigan.gov/documents/dnr/Ninth_National_Wild_Turkey_Symposium_Proceedings_450287_7.pdf.

Scharf, William C., and Josef Kren. 2010. "Orchard Oriole (*Icterus spurius*)." In *The Birds of North America Online,* ed. A. Poole. Ithaca, NY: Cornell Lab of Ornithology. http://bna.birds.cornell.edu/bna/species/255; doi:10.2173/bna.255.

Schwertner, T. W., H. A. Mathewson, J. A. Roberson, M. Small, and G. L. Waggerman. 2002. "White-winged Dove (*Zenaida asiatica*)." In *The Birds of North America Online,* ed. A. Poole. Ithaca, NY: Cornell Lab of Ornithology. http://bna.birds.cornell.edu/bna/species/710; doi:10.2173/bna.710.

Sedgwick, James A., and Fritz L. Knopf. 1987. "Breeding Bird Response to Cattle Grazing of a Cottonwood Bottomland." *Journal of Wildlife Management* 51 (1): 230–237.

Shackelford, C. E., N. R. Carrie, C. M. Riley, and D. K. Carrie. 2001. *Project Prairie Birds—a Citizen Science Project for Wintering Grassland Birds.* PWD BK @7000-485 (1/01). Austin: Texas Parks and Wildlife Department.

Shenandoah National Park. 2010. *Deer Exclosures.* Natural Resource Fact Sheet. National Park Service, US Department of the Interior. http://www.nps.gov/shen/naturescience/upload/SHEN_NR_112_Deer_Exclosures.pdf.

Sibley, David Allen. 2003. *Causes of Bird Mortality.* Sibley Guides. http://www.sibleyguides.com/conservation/causes-of-bird-mortality.

———. 2009. *The Sibley Guide to Trees.* New York: Alfred A. Knopf.

Singhurst, Jason R., Laura L. Hansen, Jeffrey N. Mink, Bill Armstrong, Donnie Frels Jr., and Walter C. Holmes. 2010. "The Vascular Flora of Kerr Wildlife Management Area, Kerr County, Texas." *Journal of the Botanical Research Institute of Texas* 4 (1): 497–521.

Smallwood, John A., and David M. Bird. 2002. "American Kestrel (*Falco sparverius*)." In *The Birds of North America Online,* ed. A. Poole. Ithaca, NY: Cornell Lab of Ornithology. http://bna.birds.cornell.edu/bna/species/602; doi:10.2173/bna.602.

Smith, Timothy A., Deanna L. Osmond, Christopher E. Moorman, Jon M. Stucky, and J. Wendell Gilliam. 2008. "Effect of Vegetation Management on Bird Habitat in Riparian Buffer Zones." *Southeastern Naturalist* 7 (2): 277–288.

"Snag (ecology)." 2012. Wikipedia. http://en.wikipedia.org/w/index.php?title=Snag_(ecology)&oldid=521391695.

Stout, Susan. 2004. "The Forest Nobody Knows." *Forest Science* 1 (Winter). USDA Forest Service Northeastern Research Station.

Stutzenbaker, Charles D. 2010. *Aquatic and Wetland Plants of the Western Gulf Coast.* College Station: Texas A&M University Press.

Swanson, David L., James L. Ingold, and Robert Galati. 2012. "Golden-crowned Kinglet (*Regulus satrapa*)." In *The Birds of North America Online,* ed. A. Poole. Ithaca, NY: Cornell Lab of Ornithology. http://bna.birds.cornell.edu/bna/species/301; doi:10.2173/bna.301.

Tarof, Scott, and Charles R. Brown. 2013. "Purple Martin (*Progne subis*)." In *The Birds of North America Online,* ed. A. Poole. Ithaca, NY: Cornell Lab of Ornithology. http://bna.birds.cornell.edu/bna/species/287; doi:10.2173/bna.287.

Tarvin, Keith A., and Glen E. Woolfenden. 1999. "Blue Jay (*Cyanocitta cristata*)." In *The Birds of North America Online,* ed. A. Poole. Ithaca, NY: Cornell Lab of Ornithology. http://bna.birds.cornell.edu/bna/species/469; doi:10.2173/bna.469.

Taylor, Richard B. 2013. *History and Distribution of Feral Hogs in Texas.* Texas Natural Wildlife. http://agrilife.org/texnatwildlife/feral-hogs/history-and-distribution-of-feral-hogs-in-texas/.

Taylor, Richard B., and Eric C. Hellgren. 1997. "Diet of Feral Hogs in the Western South Texas Plains." *Southwestern Naturalist* 42 (1): 33–39.

Taylor, Richard B., Eric C. Hellgren, Timothy M. Gabor, and Linda M. Ilse. 1998. "Reproduction of Feral Pigs in Southern Texas." *Journal of Mammalogy* 79 (4): 1325–1331.

Taylor, Rick. 2003. *The Feral Hog in Texas.* PWD BK W7000-195. Austin: Texas Parks and Wildlife Department. http://www.tpwd.state.tx.us/publications/pwdpubs/media/pwd_bk_w7000_0195.pdf.

Teer, J. G., J. W. Thomas, and E. A. Walker. 1965. "Ecology and Management of White-tailed Deer in the Llano Basin of Texas." *Wildlife Monographs* 15:1–62.

Terres, John K. 1980. *The Audubon Society Encyclopedia of North American Birds.* New York: Alfred A. Knopf.

Texas A&M AgriLife Extension. "Aquaplant—A Pond Manager Diagnostics Tool." Accessed December 27, 2015. http://aquaplant.tamu.edu/.

Texas Center for Policy Studies (TCPS). 2001. *Parkland and Open Space in the Texas Hill Country.* http://www.texascenter.org/publications/openspace.pdf.

Texas Department of Agriculture. 2014. "Noxious and Invasive Plant List." http://texreg.sos.state.tx.us/fids/200701978-1.html.

Texas Forest Service. 2014. "Best Management Practices: Stream Crossings." Forestry Management Sheet, pt. II, 29–68. Texas Forestry Association. http://texasforestservice.tamu.edu/uploadedFiles/Sustainable/bmp/Publications/BMP%20Manual_March2014-web.pdf.

Texas Health and Human Services Commission. 2004. *Growth & Poverty Projections, 2001–2010.* Austin: Texas State Data Center and Research Department. http://www.hhsc.state.tx.us/research/maps.html.

"Texas Invasives: Invasive Plant Database." Produced by a collaboration of government agencies and nonprofit organizations: U.S. Forest Service, United States Department of Agriculture Animal and Plant Health Inspection Service, Lady Bird Johnson Wildflower Center, Texas Parks & Wildlife, Texas A&M AgriLife Extension, U.S. Fish & Wildlife Service Sport Fish Restoration, Texas A&M Forest Service, The Texas State University System. Accessed December 27, 2015. www.texasinvasives.org/invasives_database/.

Texas Nature Conservancy. 2005. "What Are the Worst Weeds?" Global Invasive Species Team. http://www.invasive.org/gist/worst.html.

Texas Parks and Wildlife Department (TPWD). 1990. "White-tailed Deer." http://www.tpwd.state.tx.us/publications/nonpwdpubs/introducing_mammals/white_tailed_deer/.

———. 2006. "Rare and Threatened Species of Texas by County." http://tpwd.texas.gov/gis/rtest/.

———. 2008. "Managing Habitat for White-tailed Deer in the Hill Country of Texas." PWDRPW7000-0193. https://tpwd.texas.gov/publications/pwdpubs/media/pwd_rp_w7000_0193.pdf.

Thomas, Jack Ward, James G. Teer, and E. A. Walker. 1964. "Mobility and Home Range of White-tailed Deer on the Edwards Plateau in Texas." *Journal of Wildlife Management* 28 (3): 463–472.

Tilghman, Nancy G. 1989. "Impacts of White-tailed Deer on Forest Regeneration in Northwestern Pennsylvania." *Journal of Wildlife Management* 53 (3): 524–532.

Timmons, Jared B., Blake Alldredge, William E. Rogers, and James C. Cathey. 2012. *Feral Hogs Negatively Affect Native Plant Communities.* AgriLife Extension SP-467. College Station: Texas A&M University System. http://plumcreek.tamu.edu/media/8139/feral_hogs_negatively_affect_native_plant_communities.pdf.

Timmons, Jared, James C. Cathey, Nikki Dictson, and Mark McFarland. 2011a. *Feral Hogs and Water Quality in Plum Creek.* Texas AgriLife Extension Service SP-422. College Station: Texas A&M University System. http://plumcreek.tamu.edu/media/7031/feral-hogs-and-water-quality-in-plum-creek.pdf.

———. 2011b. *Feral Hog Laws and Regulations in Texas.* Texas AgriLife Extension Service SP-420. College Station: Texas A&M University System. http://feralhogs-tamu-edu.wpengine.netdna-cdn.com/files/2011/08/Feral-Hog-Laws-and-Regulations-in-Texas.pdf.

———. 2011c. *Feral Hog Transportation Regulations.* Texas AgriLife Extension Service SP-423. College Station: Texas A&M University System. http://plumcreek.tamu.edu/media/7037/feral-hog-transportation-regulations.pdf.

Timmons, Jared, James C. Cathey, Dale Rollins, Nikki Dictson, and Mark McFarland. 2011. *Feral Hogs Impact Ground-Nesting Birds.* Texas AgriLife Extension Service SP-419. College Station: Texas A&M University System. http://plumcreek.tamu.edu/media/7034/feral-hogs-impact-ground-nesting-birds.pdf.

Titman, Rodger D. 1999. "Red-breasted Merganser (*Mergus serrator*)." In *The Birds of North America Online,* ed. A. Poole. Ithaca, NY: Cornell Lab of Ornithology. http://bna.birds.cornell.edu/bna/species/443; doi:10.2173/bna.443.

Tolleson, D., D. Rollins, W. Pinchak, M. Ivy, and A. Hierman. 1993. *Impact of Feral Hogs on Ground-Nesting Gamebirds.* Texas Natural Wildlife. San Angelo, TX: District 7 AgriLife Research and Extension Center. http://agrilife.org/texnatwildlife/feral-hogs/impact-of-feral-hogs-on-ground-nesting-gamebirds/.

Traweek, Max S. 1985. *Statewide Census of Exotic Big Game Animals: Performance Report as Required by Federal Aid in Wildlife Restoration Act.* Federal Aid Project No. W-109-R-8. http://books.google.com/books/about/Statewide_Census_of_Exotic_Big_Game_Anim.html?id=MH0EHAAACAAJ.

Trocki, Carol L., and Peter W. C. Paton. 2005. *Developing a Conservation Strategy for Grassland Birds at Saratoga National Historical Park.* Appendices C and D. Natural Resources Report. NPS/NER/NRR-2005/004. Boston: US Department of the Interior, National Park Service Northeast Region. irmafiles.nps.gov/reference/holding/440218/SARA_grasslandbirds_NRR205004.pdf.

Tveten, John L. 1993. *The Birds of Texas.* Fredericksburg, TX: Shearer Publishing.

Tweit, Robert C. 2005. "Bullock's Oriole." In *The Texas Breeding Bird Atlas.* College Station: Texas A&M University System. http://txtbba.tamu.edu/species-accounts/bullocks-oriole/.

Ueckert, Darrell N. "Biology and Ecology of Redberry Juniper." 1997a. Texas Natural Resources Server. http://texnat.tamu.edu/library/symposia/juniper-ecology-and-management/biology-and-ecology-of-redberry-juniper/.

———. "Juniper Control and Management." 1997b. Texas Natural Resources Server. http://texnat.tamu.edu/library/symposia/juniper-ecology-and-management/juniper-control-and-management/.

Ueckert, Darrell, and Allan McGinty. 1999. *Brush Busters: How to Estimate Costs for Controlling Small Mesquite.* AgriLife Extension E-131. College Station: Texas A&M

System. http://texnat.tamu.edu/about/brush-busters/mesquite/how-to-estimate-cost-for-controlling-mesquite/.

University of Texas Plant Resources Center. 2006. "Flora of Texas TEX-LL Database." http://prc-symbiota.tacc.utexas.edu/index.php.

Urabek, Raymond L. 1989. "Evaluation of Predator Guards for Black-bellied Whistling Duck Nest-Boxes." *Wildlife Management.* Internet Center for Great Plains Wildlife Damage Control Workshop Proceedings. Lincoln: University of Nebraska. http://digitalcommons.unl.edu/cgi/viewcontent.cgi?article=1417&context=gpwdcwp.

USDA. 1994. "The Use and Management of Browse in the Edwards Plateau of Texas." Temple, TX: Natural Resources Conservation Service. http://kinney.agrilife.org/files/2011/08/usemgmtbrowseeptx_22.pdf.

———. 2010. "Assistance with Waterfowl Damage." Wildlife Services Factsheet. http://www.aphis.usda.gov/publications/wildlife_damage/content/printable_version/fs_waterfowl.pdf.

———. 2011. "Game from Farm to Table." Food Safety and Inspection Service Food Safety Information. http://www.fsis.usda.gov/PDF/Game_from_Farm_to_Table.pdf.

———. 2012. "PLANTS Database." Greensboro, NC: Natural Resource Conservation Service (NRCS), National Plant Data Team. http://plants.usda.gov.

US Fish and Wildlife Service. 2002. *Birds of Conservation Concern.* Arlington, VA: US Division of Migratory Bird Management. https://www.fws.gov/Midwest/wind/references/BCC2002.pdf.

Vega, Jorge H., and John H. Rappole. 1994a. "Composition and Phenology of an Avian Community in the Rio Grande Plain of Texas." *Wilson Bulletin* 106 (2): 366–380.

———. 1994b. "Effects of Scrub Mechanical Treatment on the Nongame Bird Community in the Rio Grande Plain of Texas." *Wildlife Society Bulletin* 22 (2): 165–171.

Vennesland, R. G. 2004. "Great Blue Heron, *Ardea herodias.*" Identified Wildlife Management Strategy Guidelines. http://www.env.gov.bc.ca/wld/frpa/iwms/documents/Birds/b_greatblueheron.pdf.

Vennesland, Ross G., and Robert W. Butler. 2011. "Great Blue Heron (*Ardea herodias*)." In *The Birds of North America Online,* ed. A. Poole. Ithaca, NY: Cornell Lab of Ornithology. http://bna.birds.cornell.edu/bna/species/025; doi:10.2173/bna.25.

Verbeek, N. A., and P. Hendricks. 1994. "American Pipit (*Anthus rubescens*)." In *The Birds of North America Online,* ed. A. Poole. Ithaca, NY: Cornell Lab of Ornithology. http://bna.birds.cornell.edu/bna/species/095; doi:10.2173/bna.95.

Vickery, Peter D. 1996. "Grasshopper Sparrow (*Ammodramus savannarum*)." In *The Birds of North America Online,* ed. A. Poole. Ithaca, NY: Cornell Lab of Ornithology. http://bna.birds.cornell.edu/bna/species/239; doi:10.2173/bna.239.

Watt, Doris J., and Ernest J. Willoughby. 1999. "Lesser Goldfinch (*Carduelis psaltria*)." In *The Birds of North America Online,* ed. A. Poole. Ithaca, NY: Cornell Lab of Ornithology. http://bna.birds.cornell.edu/bna/species/392; doi:10.2173/bna.392.

Watts, Bryan D. 2011. "Yellow-crowned Night-Heron (*Nyctanassa violacea*)." In *The Birds of North America Online,* ed. A. Poole. Ithaca, NY: Cornell Lab of Ornithology. http://bna.birds.cornell.edu/bna/species/161; doi:10.2173/bna.161.

Wauer, Roland H. 1999. *The American Robin.* Austin: University of Texas Press.

Webb, Stephen L., David G. Hewitt, and Mickey W. Hellickson. 2007. "Scale of Management for Mature Male White-tailed Deer as Influenced by Home Range and Movements." *Journal of Wildlife Management* 71 (5): 1507–1512.

Weeks, Harmon P., Jr. 1994. "Eastern Phoebe (*Sayornis phoebe*)." In *The Birds of North America Online,* ed. A. Poole. Ithaca, NY: Cornell Lab of Ornithology. http://bna.birds.cornell.edu/bna/species/094; doi:10.2173/bna.94.

Wheelwright, N. T., and J. D. Rising. 2008. "Savannah Sparrow (*Passerculus sandwichensis*)." In *The Birds of North America Online,* ed. A. Poole. Ithaca, NY: Cornell Lab of Ornithology. http://bna.birds.cornell.edu/bna/species/045; doi:10.2173/bna.45.

"Where the White-tail Roam." 2007. *New York Times,* December 2. Week in Review: Ideas and Trends. http://www.nytimes.com/2007/12/02/weekinreview/02deer.html?_r=0.

Wiens, John A., and John T. Rotenberry. 1981. "Habitat

Associations and Community Structure of Birds in Shrubsteppe Environments." *Ecological Monographs* 51 (1): 21–42.

Wilcox, Bradford P., and Yun Huang. 2010. "Woody Plant Encroachment Paradox: Rivers Rebound as Degraded Grasslands Convert to Woodlands." *Geophysical Research Letters* 37. doi:10.1029/2009GL041929, 2010.

Winter, Maiken, Douglas H. Johnson, and John Faaborg. 2000. "Evidence for Edge Effects on Multiple Levels in Tallgrass Prairie." *The Condor* 102 (2): 256–266.

Witmer, M. C., D. J. Mountjoy, and L. Elliot. 1997. "Cedar Waxwing (*Bombycilla cedrorum*)." In *The Birds of North America Online,* ed. A. Poole. Ithaca, NY: Cornell Lab of Ornithology. http://bna.birds.cornell.edu/bna/species/309; doi:10.2173/bna.309.

Wolf, Blair O. 1997. "Black Phoebe (*Sayornis nigricans*)." In *The Birds of North America Online,* ed. A. Poole. Ithaca, NY: Cornell Lab of Ornithology. http://bna.birds.cornell.edu/bna/species/268; doi:10.2173/bna.268.

Woodin, Marc C., M. K. Skoruppa, B. D. Pearce, A. J. Ruddy, and G. C. Hickman. 2010. *Grassland Birds Wintering at U.S. Navy Facilities in Southern Texas.* USGS Open File Report 2010-1115. U.S. Geological Survey, Reston, Virginia, in cooperation with Texas A&M University, Corpus Christi. http://pubs.usgs.gov/of/2010/1115/pdf/OFR2010-1115.pdf.

Wrede, Jan. 2005. *Trees, Shrubs, and Vines of the Texas Hill Country: A Field Guide.* College Station: Texas A&M University Press.

Yasukawa, Ken, and William A. Searcy. 1995. "Red-winged Blackbird (*Agelaius phoeniceus*)." In *The Birds of North America Online,* ed. A. Poole. Ithaca, NY: Cornell Lab of Ornithology. http://bna.birds.cornell.edu/bna/species/184; doi:10.2173/bna.184.

Yosef, Reuven. 1996. "Loggerhead Shrike (*Lanius ludovicianus*)." In *The Birds of North America Online,* ed. A. Poole. Ithaca, NY: Cornell Lab of Ornithology. http://bna.birds.cornell.edu/bna/species/231; doi:10.2173/bna.231.

Index

Note: Page numbers in *italic* indicate photos. Page numbers in **bold** indicate maps.

agarita
  deer browse evaluation, 431
  live oak savannah, 22
  moist and dry mixed wooded slopes, 12
  post oak savannah, 17
  scientific name, 451
  shin oak savannah, 20
agricultural tax valuation, 1
all year (residents) bird species
  backyards, 285
  canyons, springs, and seeps, 177
  dry mixed wooded slopes, 16
  grasslands, 81
  live oak savannah, 22
  moist mixed wooded slopes, 14
  post oak savannah, 18
  rivers and creeks, 129
  shin oak savannah, 20
  tanks, ponds, and lakes, 233
American beautyberry, 141, 451
American elm, 448, 450
American Goldfinch, *298, 334*
  backyards, 284, 285
  canyons, springs, and seeps, 177
  dry mixed wooded slopes, 16
  habitat problem summary, 332
  live oak savannah, 22
  post oak savannah, 18
  rivers and creeks, 129
  shin oak savannah, 20
  summary, 334
American Kestrel, *27, 53*
  habitat problem summary, 52, 111
  live oak savannah, 22
  native predator of other birds, 382–83
  native to Hill Country, 315
  natural disturbance in grassland and savannah, 23
  nest boxes, 43, 314–15, 317
  other birds as prey, 382–83
  post oak savannah, 18
  predatory impact, 378
  summary, 53
  wooded slopes and savannahs, 9

American Pipit, *100, 112*
  grasslands, 73, 81, 92
  summary, 112
American Redstart
  canyons, springs, and seeps, 177
  habitat problem summary, 52
  live oak savannah, 22
  moist mixed wooded slopes, 14
  need for diverse wooded slope and savannah habitat, 31
  post oak savannah, 18
  rivers and creeks, 129
  shin oak savannah, 20
American Robin, *304, 335*
  backyards, 284, 285
  cedar usage by, 412
  grasslands, 81
  plants beneficial to, 406
  redberry juniper usage, 428
  summary, 335
  tree needs, 284–85
American smoke tree, 16, 448, 450
American sycamore, 131, 135, 139, 141, 450
American Wigeon, *154, 160*
  competition from exotic waterfowl, 153
  habitat problem summary, 159
  riparian bird, 129
  rivers and creeks, 127, 129
  summary, 160
  tanks, ponds, and lakes, 233
  tank structure for, 244
annual sunflower, 453
antelope horns, 16, 453
ants
  native ants, protecting, 372–75
  red imported fire ant, 369–72
aparejo muhly, 180, 455
aquatic plants
  coon's tail, 256
  duckweed, 256
  filamentous algae, 256
  inadequate at tanks, ponds, and lakes, 253–56
  invasion at springs, 198–99

pondweeds, 256
southern water nymph, 256
stoneworts, 256
water milfoil, 256
water primrose, 256
aromatic sumac, 12, 430, 451
arrowhead, 256, 453
artificial feeding of birds and other animals. *See* feeders and feeding
Ashe juniper. *See* cedar
Ash-throated Flycatcher, *44, 54*
  habitat problem summary, 52
  live oak savannah, 22
  moist mixed wooded slopes, 14
  native to Hill Country, 315
  nest boxes, 43
  overgrazing problem, 28
  post oak savannah, 18
  shin oak savannah, 20
  summary, 54
  wooded slopes and savannahs, 9
aspect of hillside, characteristics of, 10
aspergillosis, 307
asters, white heath, 12
avian diseases at bird feeders, 305–9
avian pox, 307
axis deer, *36, 37, 199, 393*

Bacillus thuringiensis (BT), 309
backyards, *284–354, 286, 287, 289, 290, 291, 292, 295, 297, 298, 302, 304, 308, 310, 312, 316, 328*
  all year bird species, 285
  bird summaries, 334–54
  cat predator challenge, 366
  cedar management, 423
  common and popular birds, 288–91
  description, 284–88
  disease spread from high bird concentration at feeders, 305–9
  favorite bird species, 284
  ground cover, 286–88, 293–96
  habitat problem summary, 332–33
  inadequate nest sites for cavity-nesting birds, 309–18

479

backyards (*cont.*)
   introduction, 284
   invasive native urban pest birds, 326–29
   lack of native plant species diversity, 292–93
   lack of reliable and safe places for birds to drink and bathe, 301–5
   outlaw bird species, 284
   pesticide and herbicide effects, 309
   pitfalls to feeding birds, 296–301
   priority bird species, 284
   problems, 292–331
   Purple Martin colonies, 318–26
   under the radar bird species, 284
   spring and fall bird species, 285
   summary, 332–33
   summer bird species, 285
   tree canopy, 284–85
   understory, 285, 293–96
   window threat to birds, 329–31
   winter bird species, 285
balanced grassland management, 90–91
Balcones Escarpment Canyonlands, **8,** 10–11, 76–77, 131–32
bald cypress, 131, 135, 139, 141, 450
Bald Eagle, 383
ball cedar, 413
balsam gourd (Lindheimer's globeberry), 17, 141, 452
bare soil. *See* erosion and bare ground
Barn Owl, 315
Barn Swallow, *260, 266*
   backyards, 285
   habitat problem summary, 332
   summary, 266
   tanks, ponds, and lakes, 230, 233
bastard cabbage, 453
bearded sprangletop, 256, 455
beebalm, 453
beggarticks, 256, 453
Bell's Vireo, *176, 215*
   Balcones Canyonlands, 132
   canyons, springs, and seeps, 175, 177
   dependence on dense understory, 200
   habitat problem summary, 159, 213
   live oak savannah, 22
   parasitism rate, 358
   post oak savannah, 18
   riparian habitat, 141
   rivers and creeks, 129
   shin oak savannah, 20
   summary, 215

Belted Kingfisher, *257, 267*
   fish habitat for constructed tank, 248
   habitat problem summary, 159, 265
   need for clear water to hunt, 256
   rivers and creeks, 129
   summary, 267
   tanks, ponds, and lakes, 230, 233
   wetland vegetation, 233
bench seeps, 180
bermudagrass, 455
Bewick's Wren, *295, 338*
   backyards, 284, 285
   canyons, springs, and seeps, 177
   cavity-nester, 310
   deer impact on, 401
   dry mixed wooded slopes, 16
   grasslands, 81
   habitat problem summary, 332–33
   importance of native plants to, 293
   live oak savannah, 22
   moist mixed wooded slopes, 14
   native to Hill Country, 315
   need for vibrant understory, 285
   nest boxes, 43
   plants beneficial to, 406
   post oak savannah, 18
   riparian habitat understory, 135
   rivers and creeks, 129
   shin oak savannah, 20
   summary, 338
big bluestem, 17, 22, 455
big red sage, 453
bigtooth maple, 14, 448, 450
biological controls, 196, 373
birdbaths, 302–3, 305
Bird Crash Preventer, 331
bird diversity, cedar management effects on, 94–97
bird habitat. *See* habitat
birds of the Texas Hill Country
   master bird chart, 432–49
   use of cedar by, 411–13
bird strikes on windows, 329–31
bird summaries
   backyards, 334–54
   canyons, springs, and seeps, 215–29
   grasslands, 112–26
   purpose, 6–7
   rivers and creeks, 160–74
   tanks, ponds, and lakes, 266–83
   wooded slopes and savannahs, 53–72
bird-watching, protecting springs from, 206

bison, ecological disturbance role of, 23
Black-and-white Warbler, *194, 216*
   canyons, springs, and seeps, 132, 175, 177
   habitat problem summary, 213
   rivers and creeks, 129
   summary, 216
Black-bellied Whistling-Duck, *245, 268*
   habitat problem summary, 159, 265
   maintaining or adding snags, 40
   native to Hill Country, 315
   nest boxes, 317
   riparian bird, 129
   rivers and creeks, 129
   summary, 268
   tanks, ponds, and lakes, 230, 233
blackbirds
   Red-winged (*See* Red-winged Blackbird)
   Yellow-headed, 81
Black-capped Vireo, *39, 55*
   Brown-headed Cowbird threat to, 357, 359
   deer impact on, 401
   habitat problem summary, 52
   live oak savannah, 22
   parasitism rate, 358
   redberry juniper usage, 428
   shin oak savannah, 20
   summary, 55
   wooded slopes and savannahs, 9, 31
Black-chinned Hummingbird, *290, 336*
   backyards, 284, 285, 288
   canyons, springs, and seeps, 177, 190
   dry mixed wooded slopes, 16
   feeding recommendations, 288–89
   habitat problem summary, 52, 213
   indiscriminate cedar clearing, 193
   live oak savannah, 21, 22
   moist mixed wooded slopes, 14
   overgrazing problem, 28
   post oak savannah, 18
   rivers and creeks, 129
   summary, 336
Black-crested Titmouse, *312, 337*
   backyards, 284, 285, 332–33
   canyons, springs, and seeps, 177
   cavity-nester, 310
   dry mixed wooded slopes, 16
   extensive cedar as problematic, 37
   habitat problem summary, 52, 213
   indiscriminate cedar clearing, 193
   live oak savannah, 22

maintaining or creating snags, 40
moist mixed wooded slopes, 12, 14
native to Hill Country, 315
need for vibrant understory, 285
nest boxes, 43
plants beneficial to, 406
post oak savannah, 18
riparian bird, 131
rivers and creeks, 129
shin oak savannah, 20
summary, 337
black dalea, 16, 448, 451
blackjack oak, 17, 430, 450
black-oil sunflower seed, 299
Black Phoebe, *137, 161*
    habitat problem summary, 159
    riparian habitat management, 136, 154
    rivers and creeks, 127, 129
    summary, 161
black samson echinacea (purple coneflower), 22, 142, 453
Black-throated Green Warbler, 14, 129, 177
Black-throated Sparrow, *76, 113*
    grasslands, 73, 81
    habitat problem summary, 111
    live oak savannah, 22
    plateau prairie, 76
    shin oak savannah, 20
    summary, 113
black willow, 141, 450
blanco crabapple, 14, 406, 448, 450
blown-out dams, repairing, 259–61
bluebonnet, 453
Blue-gray Gnatcatcher, *45, 56*
    dry mixed wooded slopes, 16
    erosion effect on, 43
    habitat problem summary, 52
    moist mixed wooded slopes, 14
    parasitism rate, 358
    post oak savannah, 18
    shin oak savannah, 20
    summary, 56
    wooded slopes and savannahs, 9
Blue Grosbeak, *302, 339*
    backyards, 284, 285
    habitat problem summary, 332
    live oak savannah, 22
    need for grasses for seed, 287
    parasitism rate, 358
    plants beneficial to, 406
    rivers and creeks, 129
    summary, 339
Blue-headed Vireo

canyons, springs, and seeps, 177
live oak savannah, 22
moist mixed wooded slopes, 14
post oak savannah, 18
rivers and creeks, 129
shin oak savannah, 20
Blue Jay, *328, 340*
    habitat problem summary, 333
    pest bird, 326, 327–29
    summary, 340
blue mistflower, 142, 453
bluestems
    big, 17, 22
    cane/silver, 12, 22
    KR, 29–30, 93, 99–102, 104–6
    little, 14, 17, 22, 142
Blue-winged Teal, *247, 269*
    competition from exotic waterfowl, 153
    habitat problem summary, 265
    importance of riparian understory, 138
    summary, 269
    tanks, ponds, and lakes, 230, 233
    tank structure for, 244, 247
bobcat, 382, *383, 385*
bobwhite. *See* Northern Bobwhite
box canyon, *134. See also* Balcones Escarpment Canyonlands
box traps for feral hogs, 363, 365
bracted passionflower, 448, 452
Brazilian vervain, 453
breeding birds. *See* spring and summer (breeding birds) bird species
breeding territory. See *individual bird species by name at "summary"*
brickell-bush, 141, 451
bristlegrasses, 12, 455
broadleaf woodoats, 14, 142, 455
Brown Creeper, 14
brown-eyed Susan, 142, 453
Brown-headed Cowbird, *360, 387*
    canyons, springs, and seeps, 175
    cattle presence as creator of habitat, 28
    grasslands, 73
    parasite/predator of other birds, 356–59, 387
    predation impact, 375
    rivers and creeks, 127
    summary, 387
    tanks, ponds, and lakes, 230
    trapping, 359
    wooded slopes and savannahs, 9
Brown Thrasher, 24, 406

browse, as deer forage, 395. *See also* deer and exotic animal browsing
Brune, Gunnar M., 175
brush management
    fallacy of controlling water by, 414–15
    invasive woody plant control, 95–96
brush piles, 297
    adding to backyard habitats, 294
    quick fix for missing understory, 37
    tepee style, 296
BT (Bacillus thuringiensis), 309
buffalograss, 22, 455
Bufflehead, *157, 162*
    habitat problem summary, 159, 265
    riparian bird, 129
    rivers creeks, 127, 129
    summary, 162
    tanks, ponds, and lakes, 233
bulldozer for cedar control, 425
Bullock's Oriole, *133, 163*
    mesquite, 96
    moist mixed wooded slopes, 14
    plants beneficial to, 406
    post oak savannah, 18
    riparian bird, 131
    rivers and creeks, 127, 129
    summary, 163
bulrushes, 256, 455
bunchgrasses
    ecological succession, 81
    importance for healthy grasslands, 91, 94
    old field restoration, 106–7
    resting the land, 29–30
bundleflower, 14, 20, 22, 142, 453
buntings
    Indigo (*See* Indigo Bunting)
    Lark, 81
    Painted (*See* Painted Bunting)
bur-clover, 453
bush palmetto, 448
bush sunflower, 17, 22, 142, 453
Bushtit, 428
bushy bluestem, 455

Cactus Wren, *15, 57*
    backyards, 285
    canyons, springs, and seeps, 177
    dry mixed wooded slopes, 12
    live oak savannah, 22
    shin oak savannah, 20
    summary, 57
    wooded slopes and savannahs, 9

Cain, Alan, 393
Canada wildrye, prairie wildrye, 455
cane bluestem/silver bluestem, 12, 22, 455
canyon mock orange, 448, 451
canyons, 175–229, *176*, *178*, *179*, *182*, *194*, *201*, *202*, *203*
   all year bird species, 177
   bird summaries, 215–29
   canyons description, 175
   cedar management, 190–91, 193–94, 422
   deer and other browser control, 200
   exotic plant invasions, 196–99
   favorite bird species, 175
   feral hog problem, 181–82
   human use modifications, 204–5
   introduction, 175
   livestock control to prevent overgrazing, 185–86
   outlaw bird species, 175
   overbrowsing, 200–202
   priority bird species, 175
   problems, 181–214
   under the radar bird species, 175
   spring and fall bird species, 177
   spring and summer bird species, 177
   summary, 213–14
   typical tree canopy, 284
   water flow loss and poor water quality, 208–12
   winter bird species, 177
   See also springs and seeps
Canyon Towhee, *38*, *58*
   dry mixed wooded slopes, 16
   habitat problem summary, 52
   live oak savannah, 22
   moist mixed wooded slopes, 14
   need for diverse wooded slope and savannah habitat, 31
   redberry juniper usage, 428
   rivers and creeks, 129
   shin oak savannah, 20
   summary, 58
   wooded slopes and savannahs, 9
Canyon Wren, *179*, *217*
   canyons, springs, and seeps, 175, 177, 178
   dry mixed wooded slopes, 16
   rivers and creeks, 129
   summary, 217
cardinal. *See* Northern Cardinal
Carolina buckthorn
   benefits to birds, 406
   deer browse evaluation, 430
   live oak savannah, 22
   moist mixed wooded slopes, 14
   priority woody plant, 448
   scientific name, 450
   shin oak savannah, 20
Carolina Chickadee, *13*, *59*
   backyards, 285
   canyons, springs, and seeps, 177
   cavity-nester, 310
   dry mixed wooded slopes, 16
   habitat problem summary, 52, 213, 332–33
   indiscriminate cedar clearing, 193
   live oak savannah, 22
   maintaining or adding snags, 40
   moist mixed wooded slopes, 12, 14
   native to Hill Country, 315
   nest boxes, 43
   plants beneficial to, 406
   post oak savannah, 18
   rivers and creeks, 129
   summary, 59
   tree needs, 285
   wooded slopes and savannahs, 9
Carolina snailseed, 141, 430, 448, 452
Carolina Wren
   backyards, 285
   canyons, springs, and seeps, 177
   habitat problem summary, 332–33
   live oak savannah, 22
   moist mixed wooded slopes, 14
   native to Hill Country, 315
   need for vibrant understory, 285
   nest boxes, 43
   post oak savannah, 18
   rivers and creeks, 129
Cassin's Sparrow, *75*, *114*
   grasslands, 73, 81
   habitat problem summary, 111
   live oak savannah, 22
   old field, 80
   plateau prairie, 74
   post oak savannah, 18
   shin oak savannah, 20
   summary, 114
catclaw, Roemer's acacia, 430
catclaw acacia, 16, 22, 451
cats, domestic, 303, 365–69, 375, 390
cattle
   benefits to savannah, 26
   contribution to Brown-headed Cowbird invasion, 28
   impact on grasslands and savannahs, 92, 93
   preferred livestock for bird diversity, 186
   See also livestock
cavity-nesting birds
   grasslands lacking good nest sites, 110
   inadequate backyard nest sites, 309–18
   maintaining or adding snags, 39–43
cedar (Ashe juniper)
   birds' use of, 411–13
   contribution to erosion, 46–47
   deer browse evaluation, 431
   deer exclosures, 35, *36*, 37
   dry mixed wooded slopes, *41*
   effects of invasion in riparian habitat, 152
   invasion of rivers and creeks habitat, 151–53
   live oak savannah, 22
   moist and dry mixed wooded slopes, 12
   old-growth, importance of preserving, 415, 425
   other animals' use of, 413
   proliferation, 9
   scientific name, 450
   seeps, 180
   unsuitability for snag creation, 43
   See also cedar management
cedar elm
   deer browse evaluation, 430
   live oak savannah, 22
   post oak savannah, 17
   priority woody plant, 448
   riparian habitat, 131
   scientific name, 450
cedar management, 410–29
   avoiding erosion, 47
   avoiding goats for, 32
   backyards, 423
   canyons, 190–91, 193–94, 422
   clearing guidelines, 414–15, 417
   feral hogs' damage to canyon habitat, 182
   grasslands, 97, 411, 419–20
   history, 410–11
   indiscriminate clearing, 192–93
   introduction, 410
   limited bird diversity due to, 94–97
   planning, 417–23
   prescribed burns to control, 25
   removal methods, 424–27

rivers and creeks, 420–21
savannahs, 37–43, 419–20
selective clearing methods, 194–95
springs or seeps, 192, 195, 422–23
tanks, ponds, and lakes, 423
trail building with cedar, 416
water usage, 413–15
wooded slopes, 37–43, 419–20
cedar sage, 142, 453
cedar sedge, 14, 20, 22, 455
Cedar Waxwing, *201, 218*
backyards, 285
canyons, springs, and seeps, 175, 177
deer control to attract, 200
extensive cedar as problematic, 37
food sources at springs and seeps, 181
habitat problem summary, 52, 213, 214
indiscriminate cedar clearing, 195
moist mixed wooded slopes, 14
plants beneficial to, 406
redberry juniper usage, 428
rivers and creeks, 129
summary, 218
tree needs, 284–85
cenizo, 16, 430, 451
chaining for cedar control, 425
chemical controls
cedar management, 426–27
invasive exotic plant removal, 102, 196
KR bluestem, 101–2
mesquite, 97
redberry juniper, 429
Two-Step Method for RIFA control, 370–72
chickadee. *See* Carolina Chickadee
Chile pequin, 142, 453
Chimney Swift, *310, 341*
backyards, 284, 285
habitat problem summary, 332
summary, 341
wetland vegetation, 233
chinaberry, 102, 450
Chinese tallow, 102, 152, 450
chinkapin oak, 141, 430, 450
Chipping Sparrow, *292, 342*
all year bird species, 233
backyards, 284, 285
canyons, springs, and seeps, 177
dry mixed wooded slopes, 16
habitat problem summary, 332
live oak savannah, 22
moist mixed wooded slopes, 14
need for ground cover diversity, 287

post oak savannah, 18
rivers and creeks, 129
summary, 342
tanks, ponds, and lakes, 233
Clay-colored Sparrow
backyards, 285
canyons, springs, and seeps, 177
dry mixed wooded slopes, 16
grasslands, 81
live oak savannah, 22
moist mixed wooded slopes, 14
need for diverse wooded slope and savannah habitat, 31
post oak savannah, 18
rivers and creeks, 129
shin oak savannah, 20
clematis, 141, 452
climate change, rivers and creeks, 136
coast cockspur grass, 256
columbine, 12, 453
Comanche Springs, 209
common buttonbush, 141, 256, 451
common elderberry, 142, 453
common four o'clock, 453
Common Nighthawk, *363, 388*
backyards, 285
feral hog threat to, 362
grasslands, 81
live oak savannah, 22
post oak savannah, 18
shin oak savannah, 20
summary, 388
tanks, ponds, and lakes, 233
Common Poorwill, 22
Common Raven, 139, 159, 406
Common Yellowthroat, *197, 219*
all year bird species, 233
backyards, 285
canyons, springs, and seeps, 175, 177
dependence on spring habitat, 205
habitat problem summary, 159, 213, 214
riparian habitat, 135, 138, 141, 154
rivers and creeks, 129
spring and seep habitats, 180
spring habitat, 198
summary, 219
tanks, ponds, and lakes, 233
conservation gardening, 294, 405, 408
conservation hunting, 403
constructed tanks, ponds, and lakes, 230–83
all year bird species, 233
alternative water sources for birds, 263

artificial overfeeding of waterfowl, 264
bare soil around water, 253
bird summaries, 266–83
cedar management around, 423
description, 231–35
emergent plants and wet-dry cycle, 235
favorite bird species, 230
inadequate aquatic vegetation, 253–56
introduction, 230–31
management, 235–50
outlaw bird species, 230
poor design resulting in poor habitat, 250–53
priority bird species, 230
problems, 250–65
under the radar bird species, 230
reliable water and wetland vegetation, 233
shallow flats, 233–35
shoreline and wetland vegetation, 232–33
spring and fall bird species, 233
spring and summer bird species, 233
summary, 265
turbid or cloudy water, 256–59
unstable water level, 259–62
water-control structures, 235
winter bird species, 233
See also tank building
contraceptives, avoiding in deer management, 404
control of exotic, invasive, or predatory animals
Brown-headed Cowbird, 358–59, 387
domestic cats, 367–69, 390
Egyptian Goose, 271
European Starling, 324–25, 344
feral hogs, 185, 257, 362–63, 391
Great-tailed Grackle, 326–27, 346
House Sparrow, 300–301, 324–25, 348
Muscovy Duck, 276
coon's tail, 256, 456
Cooper's Hawk, 289, 378–79
coral honeysuckle, 17, 448, 452
corral traps for feral hogs, 363, *364,* 365
cover
cat, 390
feral hog, 391
See also brush management; ground cover; tree canopy; understory; *individual bird species by name at "summary"*

cowbird. *See* Brown-headed Cowbird
cowitch vine, 430, 452
coyote, 380–82
coyotillo, 16, 451
creek plum, 14, 448, 451
creeks and rivers. *See* riparian zone or habitat
Crested Caracara, 383–84, *384*, 389, *389*
critical area for habitat, and balancing predator effects, 355
cross-fencing for rotational grazing, 90
croton, 453
crow poison, 12, 453
cuckoo. *See* Yellow-billed Cuckoo
cultural methods, invasive exotic plant removal, 196
culverts, avoiding near stream banks, 143
curly mesquite, 16, 20, 22, 455

dam for constructed tank, 248–49, 423
damianita, 16, 453
Dark-eyed Junco, *241,* 270
   canyons, springs, and seeps, 177
   rivers and creeks, 129
   summary, 270
   tanks, ponds, and lakes, 230, 233
dayflower, 20, 453
deep ground cover, maintaining, 94
deer and exotic animal browsing management, 374, 393–409
   avoiding contraceptives, 404
   avoiding feeding, 402–3
   axis deer, *36, 37,* 199, 393
   backyards, 294
   browse evaluation, 430–31
   canyons, 200
   competition impact, 375
   conservation gardening, 405, 408
   effect on cedar overgrowth, 422
   exclosures, 35, *35–36, 37,* 405, 407, 409
   goat compared to deer browsing, 93
   grasslands, 97–98
   habitat recovery, 407
   harvesting (hunting), 34, 149, 395, 403, 404
   high fencing, 34, 408
   identifying browse lines, 397, *398,* 399
   indiscriminate cedar clearing, 193
   introduction, 393
   life cycle of deer, 393–96
   management issues for deer, 402–9
   overpopulation problem for deer, 396–99
   plant diversity, 373–74, 395
   plants eaten by deer, 293
   relationship to birds, 400–402
   restoration of rangeland, 86
   riparian habitat management, 138, 148–49, 202–3, 204
   savannahs, 33–37
   signs of good population management, 397–99
   understory damage by deer, 374–75, 401
   urban trap and removal, 403–4
   wooded slopes, 33–37
dense litter, and quality of grassland bird habitat, 98
desert sumac, 16, 431, 451
detritus, defined, 41
Dickcissel
   benefits of restored rangeland, 85
   grasslands, 81, 91
   habitat problem summary, 111
   parasitism rate, 358
diseases
   at feeders, 305–9
   feral hogs, 365
disking, old field restoration, 106
disturbance, ecological. *See* ecological disturbance
diversity of plant and animal life. *See* bird diversity; plant diversity
dock, 256, 453
domestic cat. *See* cats
domestic waterfowl, separating from wildlife, 264
doves. *See* Mourning Dove; White-winged Dove
Downy Woodpecker, 315–16, 317
dropseed, 17, 455
drought, 91, 211–12
Drummond's wild petunia, 142, 454
dry mixed wooded slopes, *15, 38, 41*
   all year bird species, 16
   characteristics, 12
   erosion control, 48–51
   spring and fall bird species, 16
   spring and summer bird species, 16
   winter bird species, 16
ducks. See *individual species by name*
duckweed, 256, 456
dwarf palmetto, 448
dwarf palmetto/bush palmetto, 451

early spring, summer, and fall bird species, 223
Eastern Bluebird, *109, 115*
   backyards, 285
   deer impact on, 401
   grasslands, 73, 81, 92
   habitat problem summary, 52, 111, 213, 332–33
   importance of insects, 108
   maintaining or creating snags, 40
   native to Hill Country, 315
   need for ground cover diversity, 286
   nest boxes, 313
   nest box preference, 311
   plants beneficial to, 406
   post oak savannah, 18
   summary, 115
eastern cottonwood, 131, 450
eastern gamagrass, 142, 455
Eastern Meadowlark, *80, 116*
   benefits of restored rangeland, 85
   grasslands, 73, 81, 92
   habitat problem summary, 111
   live oak savannah, 22
   old field, 80
   parasitism rate, 358
   plateau prairie, 76
   post oak savannah, 18
   shin oak savannah, 20
   summary, 116
Eastern Phoebe, *191, 220*
   all year bird species, 233
   backyards, 285
   canyons, springs, and seeps, 175, 177, 190
   habitat problem summary, 159, 213
   moist mixed wooded slopes, 14
   plants beneficial to, 406
   rivers and creeks, 129
   summary, 220
   tanks, ponds, and lakes, 233
Eastern Screech-Owl, *316, 343*
   backyards, 284
   habitat problem summary, 52, 333
   live oak savannah, 22
   maintaining or adding snags, 40
   native to Hill Country, 315
   nest boxes, 43
   summary, 343
Eastern Wood-Pewee, 14
ecological disturbance
   healthy grassland and savannah, 23, 82

imitating natural disturbances, 23–24
loss of plant diversity from lack of, 23–26
preventing water disturbance, 256–59
ecological equilibrium, working on, 376–77
ecological succession, 75, 80, 104–9, 401–2
ecological traps, 356
Edwards Plateau
    deer overpopulation in, 396
    wooded slope and savannah habitat, 11
Edwards Plateau Woodland, **8**
Egyptian Goose, 230, 263, 271, *271*
elbow bush (spring herald), 14, 20, 22, 141, 431, 448, 451
electric fencing for rotational grazing, 80
elephant ear, 454
emergent vegetation, *154, 234*, 235, *247, 255*, 256
Engelmann daisy (cutleaf daisy), 17, 22, 142, 454
Engelmann's evening-primrose, 454
erosion and bare ground
    around water features, 253
    cedar management, 97, 193
    flooding's role in preventing, 135
    grasslands, 47–48, 86–87, 94
    human overuse of stream banks, 142
    livestock damage around springs, 186
    riparian habitat loss due to, 138, 145, 147–51
    savannahs, 26–30, *31*, 43–48
    wooded slopes, 26–30, *31*, 43–51
escarpment black cherry, 14, 406, 431, 448, 450
European Starling, *325, 344*
    backyards, 284
    habitat problem summary, 333
    summary, 344
    threat to Purple Martin colony, 324–25
evaporation and water loss from tanks
    tank building, 239
evergreen savannah. *See* live oak savannah
evergreen sumac, 12, 22, 431, 450, 451
Eve's necklace, 22, 430, 448, 450
excavator for cedar control, 424–25, *426*
exclosures, deer, 35, *35–36*, 37, 405, 407, 409
exotic animal species
    axis deer, *36*, 37, 199, 393

fish in riparian habitat, 153
invasion of rivers and creeks habitat, 151–53
removing from property, 37
See also deer and exotic animal browsing management; feral hogs
exotic bird species
    discouraging from feeders, 300–301
    protecting Purple Martin colonies from, 324–25
    waterfowl competition, 153
exotic plant species
    canyons, 196
    grasslands, 99–104
    livestock's contribution to invasion by, 153, 196, 198
    old field succession problem, 80, 104–9
    riparian habitat, 151–53, 196–99
    See also KR (King Ranch) bluestem

fall and/or spring (migrants) bird species. *See* spring and/or fall (migrants) bird species
Fallon, Ann, 127
fall witchgrass, 12, 455
false indigo, 448, 451
favorite bird species
    American Goldfinch, 284, 334
    American Robin, 284, 335
    backyards, 284
    Barn Swallow, 230, 266
    Belted Kingfisher, 230, 267
    Black Phoebe, 127, 161
    Blue-winged Teal, 230, 269
    Bullock's Oriole, 127, 163
    canyons, springs, and seeps, 175
    Cedar Waxwing, 175, 218
    Dark-eyed Junco, 230, 270
    Eastern Bluebird, 73, 115
    Eastern Phoebe, 175, 220
    grasslands, 73
    Great Blue Heron, 127, 164
    Greater Roadrunner, 9, 61
    Green-winged Teal, 230, 273
    Indigo Bunting, 175, 223
    Northern Cardinal, 284, 350
    Northern Mockingbird, 284, 351
    Purple Martin, 284, 352
    Red-tailed Hawk, 9, 68
    Ring-necked Duck, 230, 279
    rivers and creeks, 127
    tanks, ponds, and lakes, 230

    Western Scrub-Jay, 9, 71
    Wood Duck, 127, 171
    wooded slopes and savannahs, 9
featured bird species
    backyards, 284
    canyons, springs, and seeps, 175
    grasslands, 73
    rivers and creeks, 127
    tanks, ponds, and lakes, 230
    wooded slopes and savannahs, 9
feeder halo, 300–301
feeders and feeding of birds
    avian diseases at bird feeders, 305–9
    avoiding overfeeding of waterfowl, 264
    hummingbird recommendations, 288–89
    pitfalls for backyard, 296–301
    protecting feeders, 300
feeding of deer, avoiding, 402–3
feral hogs, 360–65, *391*
    canyon damage from, 181–85
    control recommendations, 185, 257, 362–63
    habitat damage caused by, 362–63
    omnivorous appetite threat from, 361
    personal story, 364
    population and impact, 360
    processing and consuming meat, 365
    profile, 391
    riparian habitat damage, 148–49, 157, 181–85
    summary, 391
fern-covered seeps, 180
Field Sparrow, *107, 117*
    grasslands, 73, 81
    habitat problem summary, 111
    live oak savannah, 22
    old field, 80, 105, 107
    parasitism rate, 358
    plateau prairie, 76
    post oak savannah, 18
    predators of, 379
    shin oak savannah, 20
    summary, 117
fig, 451
filamentous algae, 256, 456
finches. *See* American Goldfinch; House Finch; Lesser Goldfinch
fire. *See* prescribed burns
firewheel, 454
fish habitat, tank building consideration, 248

flameleaf sumac (prairie flameleaf sumac)
   deer browse evaluation, 431
   live oak savannah, 22
   moist and dry mixed wooded slopes, 12
   scientific name, 450, 451
   shin oak savannah, 20
flatsedges, 256, 455
flood debris, and riparian habitat management, 145–46
floodplains and flooding, 132–36, 139–40, 141, 150–51
flotsam and jetsam, defined, 145
flycatchers. See Ash-throated Flycatcher; Scissor-tailed Flycatcher; Vermilion Flycatcher
food
   cat, 390
   fall and winter duck foods, 256
   feral hog, 391
   See also individual bird species by name at "summary"
forbs
   annual vs. perennial and rotational grazing methods, 88
   antelope horns, 16
   black samson echinacea, 22, 142
   blue mistflower, 142
   brown-eyed Susan, 142
   bundleflower, 14, 20, 22, 142
   bush sunflower, 17, 22, 142
   cedar sage, 142
   cedar sedge, 14, 20, 22
   Chile pequin, 142
   columbine, 12
   common elderberry, 142
   controlling invasive exotics, 103–4
   crow poison, 12
   damianita, 16
   dayflower, 20
   deer forage, 97–98, 395
   Drummond's wild petunia, 142
   Engelmann daisy, 17, 22, 142
   frostweed, 142
   gayfeather, 20, 22
   goldenrod, 142
   greenthread, 22
   hairyfruit chervil, 20
   healthy riparian habitat, 142
   importance for healthy wooded slope and savannah habitats, 21
   important insect habitat, 83
   live oak savannah, 22
   loss of due to mowing, 108
   Maximilian sunflower, 22, 142
   mealy blue sage, 22
   milkpea, 20
   moist mixed wooded slopes, 14
   penstemon, 20, 22
   pigeonberry, 142
   plateau goldeneye, 142
   prairie clover, 22
   prairie paintbrush, 16
   rock daisy, 14
   seeds, 256
   sensitive briar, 17
   sheep's preference, 93
   shin oak savannah, 20
   shoreline vegetation, 254
   snoutbean, 22
   straggler daisy, 17
   Texas bush-clover, 20
   Texas milkweed, 12
   Texas star, 17
   tropical sage, 142
   Turk's cap, 142
   western ironweed, 142
   white boneset, 142
   white heath aster, 12, 142
   wild petunia, 12, 17
   wood-sorrel, 14
   yarrow, 12
   zexmenia, 20, 22
   See also grasses and forbs
fox, 380
frostweed, 142, 454

Gadwall, 255, 272
   competition from exotic waterfowl, 153
   habitat problem summary, 159, 265
   riparian bird, 129, 138, 253, 254–55
   rivers and creeks, 129
   summary, 272
   tanks, ponds, and lakes, 230, 233
   tank structure for, 244, 247
gayfeather, 20, 22, 454
geese (Egyptian Goose), 230, 263, 271, 271
giant reed, 455
Gilbertson, Steve, 311
Gilwood nest box design, 311, 314
gnatcatcher. See Blue-gray Gnatcatcher
goals for land stewardship
   cedar management, 417–18
   deer management, 402, 403
   defined, 4
   grasslands, 81–85
   rivers and creeks, 136
   tanks, ponds, and lakes, 235–37
   wooded slopes and savannahs, 23
goats
   impact on grasslands and savannahs, 92–93
   overgrazing of canyons, 185
   overgrazing on wooded slopes, 28, 32–33
   problematic cedar removal method, 427
goldenball leadtree, 430, 451
Golden-cheeked Warbler, 11, 60
   Balcones Canyonlands, 132
   canyons, springs, and seeps, 177
   cedar usage by, 411, 422
   deer impact on, 401
   dry mixed wooded slopes, 16
   habitat problem summary, 52, 213
   indiscriminate cedar clearing, 37, 193–94
   live oak savannah, 22
   mixed wooded slopes, 9, 14
   parasitism rate, 358
   post oak savannah, 18
   rivers and creeks, 129
   shin oak savannah, 20
   summary, 60
   wooded slopes and savannahs, 9
Golden-crowned Kinglet, 178, 221
   canyon habitat, 175
   canyons, springs, and seeps, 175, 177
   moist mixed wooded slopes, 14
   summary, 221
Golden Eagle, 383
Golden-fronted Woodpecker, 286, 345
   backyards, 284, 285
   canyons, springs, and seeps, 177
   dry mixed wooded slopes, 16
   habitat problem summary, 52, 214, 332
   live oak savannah, 22
   maintaining or adding snags, 40
   moist mixed wooded slopes, 14
   native to Hill Country, 315
   nest boxes for, 315–16, 317
   overbrowsing in canyons, 200–201
   plants beneficial to, 406
   post oak savannah, 18
   predators of, 378
   rivers and creeks, 129

summary, 345
tree needs, 284
goldenrod, 142, 454
goldenweave, 142, 454
goldfinches. *See* American Goldfinch; Lesser Goldfinch
grapes, 14, 141, 256, 431, 452
grasses and forbs
    birds' need for seed from, 287
    bunchgrasses, 29–30, 81, 91, 94, 106–7
    deer forage, 395–96
    ground cover needs of birds, 286–88
    mowing, 82–85, 101, 106–8, 145
    woodland slopes and savannahs, 16–22, *18, 19,* 23–26
    See also forbs; grasslands
Grasshopper Sparrow, *77, 118*
    grasslands, 73, 81
    habitat problem summary, 111
    parasitism rate, 358
    prairie habitats, 76, 77
    shin oak savannah, 20
    summary, 118
grasslands, 73–126, *75, 86*
    all year bird species, 81
    bird summaries, 112–26
    cedar management, 97, 411, 419–20
    cedar or mesquite invasion effects, 94–97, 411
    deer overpopulation effects, 97–98
    dense litter effects, 98
    description, 74–80
    ecological disturbance benefits, 23, 82
    erosion from short-mowed grass, 47–48
    favorite bird species, 73
    habitat problem summary, 111
    introduction, 73
    invasion of exotic plant species, 99–104
    lack of good cavity-nesting sites, 110
    management goals, 81–82
    mowing basics, 82–85
    old field succession problems, 104–9
    outlaw bird species, 73
    priority bird species, 73
    problems, 85–110
    under the radar bird species, 73
    spring and fall bird species, 81
    spring and summer bird species, 81
    storm erosion from poor ground cover, 94
    summary, 111

winter bird species, 81
    See also overgrazed habitat
grass seeds, 256
Gray Catbird, 406
grazing
    balancing intensity of, 90–91
    KR bluestem control, 101
    sustainable grassland role, 82
    See also overgrazed habitat; rotational grazing
Great Blue Heron, *158, 164*
    all year bird species, 233
    American sycamore as nesting site, 139
    fish habitat for constructed tank, 248
    habitat problem summary, 159
    rivers and creeks, 127, 129
    summary, 164
    tank building for, 237
    tanks, ponds, and lakes, 233
Great Egret
    habitat problem summary, 159, 214, 265
    importance of riparian understory, 138
    tank building for, 237
    tanks, ponds, and lakes, 233
Greater Roadrunner, *19, 61*
    cedar clearing as problematic, 37
    cedar usage by, 411
    habitat problem summary, 52
    live oak savannah, 22
    predators of, 379
    shin oak savannah, 20
    summary, 61
    wooded slopes and savannahs, 9, 18, 23
Great Horned Owl, 139
Great-tailed Grackle, 326–27, *327,* 333, *346, 346*
greenbrier, 17, 22, 141, 431, 452
green condalia, 16, 22, 451
Green Heron, *155, 165*
    all year bird species, 233
    competition from exotic waterfowl, 153
    habitat problem summary, 213, 265
    riparian habitat management, 154
    rivers and creeks, 127, 129
    summary, 165
    tanks, ponds, and lakes, 233
Green Kingfisher, *130, 166*
    all year bird species, 233
    habitat problem summary, 159
    riparian bird, 127–28, 138

rivers and creeks, 127, 129
    summary, 166
    tanks, ponds, and lakes, 233
green sprangletop, 12, 22, 455
greenthread, 22, 454
Green-winged Teal, *234, 273*
    competition from exotic waterfowl, 153
    habitat problem summary, 159, 265
    importance of riparian understory, 138
    rivers and creeks, 129
    summary, 273
    tanks, ponds, and lakes, 230, 233
    tank structure for, 244
ground cover, *49*
    backyards, 286–88, 293–96
    balanced grassland management effects, 90–91
    erosion and soil loss during seasonal storms, 94
    erosion control role, 94
    riparian habitat management, 156
    runoff from impervious, 150
    streamside "tidy" problem, 144–47
groundwater
    avoiding pumping to fill tanks, 259
    conservation district, 209–10
    water table decline, 208–9
grubbing method for tree removal, 96, 429
guayacan (soapbush), 16, 430, 451
gum bumelia
    deer browse evaluation, 431
    healthy riparian habitat, 141
    live oak savannah, 22
    moist mixed wooded slopes, 14
    post oak savannah, 17
    scientific name, 450
    shin oak savannah, 20
guzzlers as alternative to tank building, 239, *241*

habitats
    cat, 390
    characteristics of good bird, 74
    ecological succession's role, 75
    feral hog, 391
    improvement to give birds edge over predators, 375–76
    mosaic pattern for healthy, 377–78
    problems, 213–14
    See also individual bird species by name at "summary"; specific habitat types

hackberries
   benefits to birds, 406
   deer browse evaluation, 430
   healthy riparian habitat, 141
   live oak savannah, 22
   moist mixed wooded slopes, 14
   post oak savannah, 17
   priority woody plant, 449
   riparian habitat, 131
   scientific name, 450
hairyfruit chervil, 20, 454
hairy grama, 16, 17, 20, 22, 455
Hall's panicum, 12, 455
Hardin, Garrett, 230
Harris's Sparrow, 81
harvesting of deer. *See* hunting
hawks. *See individual species by name*
hawthorn, 256, 406, 430, 448, 451
heavy grazing level, 92
herbaceous flowering plants (non-grasses). *See* forbs
herbicides and pesticides. *See* chemical controls; pesticides and herbicides
Hermit Thrush, 211, 222
   canyons, springs, and seeps, 175, 177
   dependence on dense understory, 138, 200
   food sources at springs and seeps, 181
   habitat problem summary, 213, 214
   indiscriminate cedar clearing, 193–94
   plants beneficial to, 406
   riparian bird, 130, 205, 208
   rivers and creeks, 129
   summary, 222
herons. *See individual species by name*
hexazinone chemical control for cedar, 427
high fencing as deer control, 34, 408
high-intensity/low-frequency (HILF) grazing method, 88
hillside seeps, 180
hog plum, 431, 451
hog wallowing damage, 181–85. *See also* feral hogs
honey mesquite, 22, 431, 450
horses, 92, 93
House Finch, 308, 347
   backyards, 284
   Brown-headed Cowbird threat to, 357
   habitat problem summary, 333
   mycoplasmosis, 306–7
   parasitism rate, 358

   post oak savannah, 18
   summary, 347
House Sparrow, 325, 348
   avoiding feeding, 300–301
   backyards, 284
   grasslands, 73
   summary, 348
   threat to Purple Martin colony, 324–25
huisache, 16, 430, 451
human use modifications
   canyons, 204–5
   controlling to keep water clear in ponds, 257–59
   riparian habitat management, 141–42, 152
   springs and seeps, 205–8
   See also backyards
hummingbirds. *See individual species by name*
hunting perches, 24
hunting to manage deer population, 34, 149, 395, 403, 404
Hutton's Vireo, 50, 62
   dry mixed wooded slopes, 16
   habitat problem summary, 52
   moist mixed wooded slopes, 14
   summary, 62
   wooded slopes and savannahs, 9
hydraulic tree shears, 424, 425
hydrilla, 456
hydro-axe for cedar control, 424

identification
   cat, 390
   feral hog, 391
   See also individual bird species by name at "summary"
Illinois bundleflower, 454
impervious vs. pervious cover, 150
Indigo Bunting, 182, 223
   canyon habitat, 181
   canyons, springs, and seeps, 175, 177
   habitat problem summary, 213
   indiscriminate cedar clearing, 193
   parasitism rate, 358
   summary, 223
insects
   critical bird food, 83
   deer overbrowsing and decline, 401
   livestock impact on springs, 186–87
   loss of due to mowing, 108
   loss of in heavily grazed areas, 92
   pesticide effects on birds, 309

   protecting native ants, 372–75
   RIFA invasion effects, 369–72
   water quality and availability, 154
invasive animals and plants. *See* exotic animal species; exotic plant species
invasive native urban pest birds, backyards, 326–29. *See also* exotic bird species
iTrack Wildlife app, 385–86

Jacobs Well, 209, 210
Jamaican sawgrass, 256, 455
Japanese honeysuckle, 453
jays. *See* Blue Jay; Western Scrub-Jay
johnsongrass, 102, 455
junipers. *See* cedar (Ashe juniper); redberry juniper

kidneywood, 430
Killdeer, 236, 274
   all year bird species, 233
   grasslands, 81, 92
   live oak savannah, 22
   plateau prairie, 76
   pond structure needs, 251
   rivers and creeks, 129
   shallow flats, 234
   shin oak savannah, 20
   summary, 274
   tanks, ponds, and lakes, 230, 233
kingfishers. *See* Belted Kingfisher; Green Kingfisher
kinglets. *See* Golden-crowned Kinglet; Ruby-crowned Kinglet
KR (King Ranch) bluestem, 29
   controlling, 99, 100–102
   livestock preferences, 93
   old field succession, 104–5, 106
   profile, 29–30
   scientific name, 456
Krebs, Charles, 355

lacey oak, 12, 431, 450
Ladder-backed Woodpecker, 17, 63
   dry mixed wooded slopes, 16
   habitat problem summary, 52
   live oak savannah, 22
   maintaining or adding snags, 40
   moist mixed wooded slopes, 14
   post oak savannah, 16, 18
   rivers and creeks, 129
   shin oak savannah, 20

summary, 63
wooded slopes and savannahs, 9
lakes, defined, 230-31, 232
land management, defined, 4
land stewardship for birds in Hill Country
ecological succession role, 75
monitoring to measure success, 4–5
preparation for being own adviser, 3
resting the land, 87, 90–91, 145, 186, 190–91
setting goals and objectives, 4
surveying property, 4
WTV, 1–3
See also *individual bird species by name at "summary"*
Langford, David K., 209–10
Lark Bunting, 81
Lark Sparrow, *108, 119*
grasslands, 73, 81, 92, 108
habitat problem summary, 111
live oak savannah, 22
mowing impact on, 107–8
old field, 80
parasitism rate, 358
plateau prairie, 76
post oak savannah, 18
shin oak savannah, 20
summary, 119
laurel, 16
leaks in tanks/ponds, repairing, 261–62
leasing land situation and rotational grazing, 90
Least Sandpiper, 233
leatherflower, 14, 452
Le Conte's Sparrow, *99, 120*
grasslands, 73, 81, 91
habitat problem summary, 111, 213
seeps as winter oases, 181
summary, 120
Leopold, Aldo, 73
Lesser Goldfinch, *78, 121*
backyards, 285
canyons, springs, and seeps, 177
grasslands, 73, 81
habitat problem summary, 111, 332
live oak savannah, 22
pocket prairie, 78
post oak savannah, 18
rivers and creeks, 129
shin oak savannah, 20
summary, 121
Lesser Scaup, *232, 275*
habitat problem summary, 159, 265

rivers and creeks, 129
summary, 275
tanks, ponds, and lakes, 230, 233
light grazing level, 90–91
ligustrum (privet), 102, 451
limbing up problem for cedar, 423
Lincoln's Sparrow, *297, 349*
backyards, 284, 285
dry mixed wooded slopes, 16
habitat problem summary, 332
live oak savannah, 22
need for ground cover diversity, 287
post oak savannah, 18
shin oak savannah, 20
summary, 349
Lindheimer rosettegrass, 16, 180, 456
Lindheimer's maiden fern, 180, 456
Lindheimer's morning glory, 14, 448, 452
Lindheimer's muhly, 180, 456
Lindheimer's silktassel, 431, 451
little bluestem
healthy riparian habitat, 142
live oak savannah, 22
moist mixed wooded slopes, 14
post oak savannah, 17
scientific name, 456
shin oak savannah, 20
live oak (plateau live oak), 20, 431, 450
live oak savannah, *21, 27, 30–31, 32, 44, 45, 51, 358*
all year bird species, 22
characteristics, 9
description, 20–22
prescribed burns on, 25–26
spring and fall bird species, 22
spring and summer bird species, 22
winter bird species, 22
livestock
Brown-headed Cowbird, 357
cattle, 26, 28, 92, 93, 186
contribution to exotic plant invasion, 153, 196, 198
controlling waste to keep water clear in ponds, 187, 256, 258
goats, 28, 32–33, 92–93, 185, 427
plants eaten by, 293
sheep, 92, 93, 199
See also overgrazed habitat
livestock troughs
alternatives to tank building, 239
avoiding overflow from, 188, 211
Llano Uplift, **8**
Loggerhead Shrike, *24, 64*

benefits of restored rangeland, 85
fire as tool in small-mammal habitat, 98
grasslands, 81
habitat problem summary, 52, 111
live oak savannah, 22
post oak savannah, 18
rivers and creeks, 129
shin oak savannah, 20
story of, 24
summary, 64
wooded slopes and savannahs, 9, 23
lotebush, 431, 451
Louisiana Waterthrush, *139, 167*
canyons, springs, and seeps, 177
habitat problem summary, 159, 213
importance of riparian understory, 135, 138
parasitism rate, 358
rivers and creeks, 127, 129
summary, 167

MacGillivray's Warbler, 14, 52, 129, 177
malta star-thistle, 454
mast, as deer forage, 395
Maximilian sunflower, 142, 454
meadowlarks. *See* Eastern Meadowlark; Western Meadowlark
mealy blue sage, 22, 454
mechanical controls
cedar management, 424–25
invasive exotic plant removal, 196, 198
invasive plant species in grasslands, 102
mesquite, 96, 97
redberry juniper, 429
"Merrill" rotational grazing method, 88
mesquite
dry mixed wooded slopes, 16
honey mesquite, 22, 431, 450
management, 94–97
Mexican buckeye, 14, 141, 431, 451
Mexican hat, 454
Mexican pinyon, 16, 450
Mexican plum, 448, 451
Michalec, Milan, 209
migrants. *See* spring and/or fall (migrants) bird species
milkpea, 20, 454
milkweed vine, 14, 453
milo (grain sorghum), 456
mint, 454

"miracle glass" to prevent window strikes, 330–31
Mississippi Kite, 81
mistletoe, 454
mixed wooded slopes, *11, 13, 15, 38, 41, 412*
    avoiding prescribed burn or grazing, 26
    cedar management, 419
    characteristics, 11–16
    dry mixed wooded slopes, 12, 48–51
    moist mixed wooded slopes, 12
    most common type, 10
    overgrazing problem, 28
    overview, 9
moderate grazing level, 91–92
moist mixed wooded slopes, *13*
    all year bird species, 14
    characteristics, 12
    spring and fall bird species, 14
    spring and summer bird species, 14
    winter bird species, 14
morning glory, 14
mosaic pattern for healthy habitat, 377–78
mountain lion, 382
mountain mahogany, 16, 430, 451
Mourning Dove, *79, 122*
    grasslands, 73, 81
    habitat problem summary, 111
    need for grasses for seed, 287
    old field, 79, 80
    post oak savannah, 18
    shin oak savannah, 20
    summary, 122
Mourning Warbler
    live oak savannah, 22
    moist mixed wooded slopes, 14
    riparian bird, 131
    rivers and creeks, 129
    understory loss in wooded slopes and savannah, 37
mowing, 82–85, 101, 106–8, 145
mudflats, 233–35, *236*
mulch and slash, in reducing soil loss in grasslands, 86–87
multiple pasture rotational grazing, 88
Muscovy Duck, 230, *263*, 276, *276*
musk thistle, 454
mustang grape, 22, 453
mycoplasmosis (House Finch disease), 306–7

Nashville Warbler, *40*, 65
    canyons, springs, and seeps, 177

extensive cedar as problematic, 37
habitat problem summary, 52
live oak savannah, 22
moist mixed wooded slopes, 14
post oak savannah, 18
rivers and creeks, 129
summary, 65
wooded slopes and savannahs, 9
native ants, protecting, 372–75
native plants
    backyard habitats, 292–93
    conservation gardening to increase, 405–6
    damaged and missing in rivers and creeks habitat, 137–44
    deer pressure on, 393, 397
    mowing guidelines for grasses, 83
    recovering from deer overbrowsing, 408–9
    value for birds, 293–94
native urban pest birds, 326–29
nest boxes, *44, 109, 312*
    adding to backyard habitats, 311–18
    cavity-nesting birds, 43, 110, 310
    choosing for Hill Country, 311
    information chart, 315
    monitoring, 316–18, 378
    mounting recommendations, 311–12
    protecting from predators, 318, 322–23
    Purple Martin colonies, 319–26
nests, mowing damage to, 107–8. See also *individual bird species by name at "summary"*
netleaf hackberry, 431, 450
nolina, 22, 451
Norris, Chad, 177, 183, 198
Northern Bobwhite, *32*, 66
    cattle grazing impact on ground habitat, 93
    feral hog threat to, 361
    grasslands, 81, 91, 92, 100–101
    habitat problem summary, 52, 111
    live oak savannah, 22
    mesquite, 96
    mowing impact on, 107–8
    old field, 80
    overgrazing problem, 28
    plants beneficial to, 406
    rivers and creeks, 129
    shin oak savannah, 20
    summary, 66
    wooded slopes and savannahs, 9
Northern Cardinal, *289, 350*

    adaptability to suburban backyards, 288
    backyards, 284, 285
    diet, 288
    habitat problem summary, 213
    moist mixed wooded slopes, 14
    parasitism rate, 358
    plants beneficial to, 406
    summary, 350
Northern Flicker
    habitat problem summary, 52
    live oak savannah, 22
    plants beneficial to, 406
    post oak savannah, 18
    shin oak savannah, 20
Northern Goshawk, 289
Northern Harrier, 81, 98, 111
Northern Mockingbird, *291, 351*
    adaptibility to suburban backyards, 288
    backyards, 284, 285
    diet, 290–91
    food sources at springs and seeps, 181
    Loggerhead Shrike competitor, 24
    plants beneficial to, 406
    redberry juniper usage, 428
    summary, 351
Northern Parula, *151, 168*
    habitat problem summary, 159
    rivers and creeks, 127, 129
    summary, 168
Northern Pintail
    competition from exotic waterfowl, 153
    habitat problem summary, 265
    riparian bird, 129
    tanks, ponds, and lakes, 233
    tank structure for, 244
Northern Shoveler, 244

oak/juniper woodlands. *See* mixed wooded slopes
oak wilt and snag-creation considerations, 42
oats, 456
objectives for land stewardship, defined, 4
old field, 79–80, 104–9
old-growth cedar, importance of preserving, 415, 425
old man's beard, 430, 453
Orange-crowned Warbler, *203, 224*
    backyards, 285
    canyons, springs, and seeps, 175, 177

dependence on dense understory, 37, 201-2, 285
dry mixed wooded slopes, 16
extensive cedar as problematic, 37
grasslands, 81
habitat problem summary, 214, 332
live oak savannah, 22
moist mixed wooded slopes, 14
post oak savannah, 18
rivers and creeks, 129
summary, 224
Orchard Oriole, *358*, *392*
    backyards, 285
    Brown-headed Cowbird threat to, 357
    canyons, springs, and seeps, 177
    live oak savannah, 22
    parasitism rate, 358
    post oak savannah, 18
    rivers and creeks, 129
    shin oak savannah, 20
    summary, 392
    tanks, ponds, and lakes, 233
orioles. *See* Bullock's Oriole; Orchard Oriole; Scott's Oriole
outlaw bird species
    backyards, 284
    canyons, springs, and seeps, 175
    Egyptian Goose, 230, *263*, 271, *271*
    European Starling, 284, 324-25, *325*, 333, 344, *344*
    grasslands, 73
    Muscovy Duck, 230, 276, *276*
    rivers and creeks, 127
    tanks, ponds, and lakes, 230
    wooded slopes and savannahs, 9
    See also Brown-headed Cowbird; House Sparrow
overgrazed habitat
    canyons, 185-86
    grasslands, 85-94
    RIFA proliferation, 370
    riparian habitat management, 143-44, 145, 148-49, 185-90
    rotational grazing, 88
    savannahs, 26-27, 28, *31*, 31-37, 43-46
    soil erosion, 43-47, 94
    unhealthy habitat, 22
    wooded slopes, 26-37
    See also rotational grazing
owls. *See individual species by name*

Painted Bunting, *21*, 67
    backyards, 285

Brown-headed Cowbird threat to, 357
canyons, springs, and seeps, 177, 186
deer impact on, 401
grasslands, 81
habitat problem summary, 52, 213, 332
live oak savannah, 22
loss of understory, 93
need for ground cover diversity, 286-87, 288
parasitism rate, 358
post oak savannah, 18
rivers and creeks, 129
shin oak savannah, 20
summary, 67
tanks, ponds, and lakes, 233
wooded slopes and savannahs, 9, 21, 23
pale buckeye, 451
panicgrasses, 256, 456
parasitic birds. *See* Brown-headed Cowbird
paspalum, 256, 456
passerine, defined, 311
passionflower vines, 14, 406, 453
pastureland vs. rangeland, 73
pearl milkweed, 20, 453
pecan, 131, 141, 450
Pectoral Sandpiper, 233
pedicles, defined, 47
penstemon, 20, 22, 454
pervious vs. impervious cover, 150
pesticides and herbicides
    backyards, 309
    dangers of using water sources, 198
    exotic aquatic plants, 199
    negative effects on birds, 24, 309
    redberry juniper, 429
    Two-Step Method for RIFA control, 370-72
Peterson, Roger Tory, 9
pH, soil, and post oak savannah, 17
phoebes. *See* Black Phoebe; Eastern Phoebe
phorid flies for RIFA control, 373
Pied-billed Grebe, 233, 265
pigeonberry, 142, 454
pigweeds, 256, 454
Pine Siskin, 14, 51, 285, 332
pink evening-primrose, 454
pink mimosa, 12, 431, 448, 452
pink smartweed, 256, 454
pipits. *See* American Pipit; Sprague's Pipit
plains, 73. *See also* grasslands

plains bristlegrass, 12, 456
plains lovegrass, 22, 456
plant diversity
    absolute importance for bird populations, 292-93
    deer impact on, 373-74
    grassland management, 95-96
    importance for deer habitat, 395
    importance of, 232-33
    riparian habitat, 131
    savannahs, 23-26, 31-37
    wooded slopes, 23-26, 31-37
plants
    dry mixed wooded slopes only, 16
    live oak savannah, 22
    master list in appendix, 450-57
    mixed wooded slopes, both types, 12
    moist mixed wooded slopes only, 14
    post oak savannah, 17
    riparian habitat, 141-42
    shin oak savannah, 20
plateau goldeneye, 142, 454
plateau live oak, 12, 22, 141
plateau milkvine, 20, 453
plateau prairie, 74-76, 363
pocket prairie, 78, 84
poison ivy, 12, 431, 453
ponds, 230-31, *231*, *234*, *236*, *245*, *247*, *249*, *252*, *255*, *260*, *263*. *See also* constructed tanks, ponds, and lakes
pondweeds, 256, 456
possumhaw, 14, 141, 431, 448, 452
post oak, 16, 17, 431, 450
post oak savannah, *17*
    all year bird species, 18
    cedar management, 419-20, *421*
    characteristics, 9
    description, 16-18
    least common in Hill Country, 11
    prescribed burns on, 25
    spring and fall bird species, 18
    spring and summer bird species, 18
    winter bird species, 18
poverty bush, 14, 431, 452
prairie, 73, *95*, *100*, *104*. *See also* grasslands
prairie clover, 22
prairie paintbrush, 16, 454
predator guards for nest boxes, 318
predators and other "dangerous" adversaries, 355-92
    American Kestrel, 382-83
    balancing predator effects, 355-56, 376
    Bald Eagle, 383

predators (*cont.*)
  bobcats, 382, *383, 385*
  cats, 303, 365–69, 375, 390
  cedar as shelter from, 413
  coyotes, 380–82
  Crested Caracara, 383–84
  decrease in deer's natural predators, 397
  foxes, 380
  Golden Eagle, 383
  introduction, 355–56
  mountain lions, 382
  native ants, protecting, 372–75
  native predator management, 375–78
  native predators of selected birds, 378–79
  raccoons, 299, 379–80
  red imported fire ant (RIFA), 369–72
  seasonal fluctuations for predators, 385
  skunk, 380
  snakes, 379
  tools to indicate presence of predators, 385–86
  Western Scrub-Jay, 384–85
  See also Brown-headed Cowbird; feral hogs
prescribed burns, *25, 33*
  cedar management, 425–26
  contribution to ecological disturbance, 23, 25
  dense litter control role, 98
  effect on mesquite vs. cedar, 96
  grassland restoration, 82, 87
  importance in dense litter control, 98
  KR bluestem control, 101
  redberry juniper, 428, 429
  savannah management, 18, 25–26
pricklypear, 22, 431, 452
primroses, 454
priority bird species
  American Kestrel, 9, 53
  backyards, 284
  Bell's Vireo, 175, 215
  Bewick's Wren, 284, 338
  Black-capped Vireo, 9, 55
  Black-chinned Hummingbird, 284, 336
  Black-crested Titmouse, 284, 337
  Black-throated Sparrow, 73, 113
  Cactus Wren, 9, 57
  canyons, springs, and seeps, 175
  Canyon Towhee, 9, 58
  Canyon Wren, 175, 217
  Carolina Chickadee, 9, 59
  Cassin's Sparrow, 73, 114
  Chimney Swift, 284, 341
  Chipping Sparrow, 284, 342
  Common Nighthawk, 388
  Common Yellowthroat, 175, 219
  Eastern Meadowlark, 73, 116
  Field Sparrow, 73, 117
  Golden-cheeked Warbler, 9, 60
  Golden-fronted Woodpecker, 284, 345
  Grasshopper Sparrow, 73, 118
  grasslands, 73
  Green Kingfisher, 127, 166
  Killdeer, 230, 274
  Ladder-Backed Woodpecker, 9, 63
  Lark Sparrow, 73, 119
  Le Conte's Sparrow, 73, 120
  Lesser Goldfinch, 73, 121
  Lesser Scaup, 230, 275
  Loggerhead Shrike, 9, 64
  Louisiana Waterthrush, 127, 167
  Mourning Dove, 73, 122
  Northern Bobwhite, 9, 66
  Orange-crowned Warbler, 175, 224
  Orchard Oriole, 392
  Painted Bunting, 9, 67
  Red-shouldered Hawk, 127, 169
  Red-winged Blackbird, 230, 278
  Rio Grande Turkey, 73, 123
  rivers and creeks, 127
  Rufous-crowned Sparrow, 69
  Scissor-tailed Flycatcher, 73, 125
  Sedge Wren, 175, 225
  Sora, 230, 280
  Spotted Sandpiper, 230, 281
  Summer Tanager, 127, 170
  tanks, ponds, and lakes, 230
  Vermilion Flycatcher, 230, 282
  wooded slopes and savannahs, 9
  Yellow-billed Cuckoo, 127, 174
  Yellow-crowned Night-Heron, 230, 283
  Yellow-throated Vireo, 175, 229
priority plant species
  benefits for birds, 406
  blanco crabapple, 14, 406, 448, 450
  Carolina buckthorn, 14, 20, 22, 406, 430, 448, 450
  conservation gardening, 294, 405–6
  escarpment black cherry, 14, 406, 431, 448, 450
  hawthorn, 256, 406, 430, 448, 451
  passionflower vines, 14, 406, 453
  red mulberry, 141, 406, 449, 450
  rusty blackhaw, 14, 22, 406, 430, 448, 452
  Texas mulberry, 406, 430, 448, 451, 452
  Virginia creeper, 14, 141, 406, 431, 448, 453
  See also hackberries
Pronone Power Pellet chemical control for cedar, 426–27
Proposition 11, 1
purple cone-flower, 142, 453, 454
Purple Martin, *319, 352*
  backyard colonies, 284, 285, 318–26
  habitat problem summary, 332–33
  house cleaning, 325
  summary, 352
  tanks, ponds, and lakes, 233
  wetland vegetation, 233
purple prairie clover, 454
purpletop, 14, 456
pyracantha, 102, 452

raccoon, 299, 379–80
ragweed, 454
rain lily, 454
rangeland, 73. *See also* grasslands
reaches, stream, 127
redberry juniper (pinchot juniper), 427–29
  vs. Ashe juniper (cedar), 410
  birds' use of, 428
  mechanical tool effects, 424, 425
  removal methods, 428–29
  scientific name, 450
Red-breasted Merganser, *249, 277*
  fish habitat for constructed tank, 248
  habitat problem summary, 265
  pond structure needs, 251
  summary, 277
  tanks, ponds, and lakes, 230, 233
Red-eyed Vireo, 18, 358
red grama, 16, 456
Redhead, 233
red imported fire ant (RIFA), 369–72
Redland ecological sites, 17
red mulberry, 141, 406, 449, 450
redroot, 448, 452
Red-shouldered Hawk, *140,* 169
  bald cypress as nesting site, 139
  fire as tool in small-mammal habitat, 98
  grasslands, 81
  habitat problem summary, 111, 159

natural disturbance in grassland and savannah, 23
rivers and creeks, 127, 129
river valley prairie, 77
summary, 169
tanks, ponds, and lakes, 233
Red-tailed Hawk, *27, 68*
    deer impact on, 401
    fire as tool in small-mammal habitat, 98
    natural disturbance in grassland and savannah, 23
    predatory impact, 379
    summary, 68
    wooded slopes and savannahs, 9
Red-winged Blackbird, *254, 278*
    all year bird species, 233
    canyons, springs, and seeps, 177
    habitat problem summary, 265
    need for grasses for seed, 287
    parasitism rate, 358
    rivers and creeks, 129
    summary, 278
    tanks, ponds, and lakes, 230, 233
    wetland vegetation, 233, 253
reseeding of grasslands, 87
residents. *See* all year (residents) bird species
resting the land, 87, 90–91, 145, 186, 190–91. *See also* rotational grazing
rice cutgrass, 256, 456
Ring-necked Duck, *234, 279*
    habitat problem summary, 265
    rivers and creeks, 129
    summary, 279
    tanks, ponds, and lakes, 230, 233
Rio Grande Turkey, *84, 123, 362*
    benefits of restored rangeland, 85
    canyon management, 186
    canyons, springs, and seeps, 177
    cattle grazing impact on ground habitat, 93
    dependence on seeps, 206
    feral hog threat to, 362
    grasslands, 73, 81, 91, 100–101
    habitat problem summary, 111, 213, 214
    habitat requirements, 84
    live oak savannah, 22
    mowing impact on, 107–8
    plants beneficial to, 406
    post oak savannah, 18
    rivers and creeks, 129

shin oak savannah, 20
summary, 123
riparian zone or habitat, 127–74, *128, 130, 133–34, 137, 139, 140, 144, 146, 151, 154–55, 157–58*
    all year bird species, 129
    Balcones Escarpment Canyonlands, 131–32
    bird summaries, 160–74
    cedar management, 420–21
    clearing problem, 138–41
    climate change, 136
    defined, 127
    description, 127–36
    erosion and bare ground, 147–51
    favorite bird species, 127
    floodplains, 132–36
    habitat problem summary at, 159
    introduction, 127
    invasive exotic plants and animals, 151–53
    management, 136
    native plants damaged or missing, 137–44
    outlaw bird species, 127
    priority bird species, 127
    problems, 136–58
    under the radar bird species, 127
    rivers and creeks, 130–31
    roads and habitat management, 142–43
    spring and fall bird species, 129
    spring and summer bird species, 129
    streamside "tidy" problem, 144–47
    summary, 159
    typical tree canopy, 284
    water quality and quantity issue, 154–58
    winter bird species, 129
rivers and creeks. *See* riparian zone or habitat
river valley prairie, 76–77, *99, 109*
roadrunners. *See* Greater Roadrunner
rock daisy, 14, 454
Roemer's acacia, 452
rose pavonia, 430, 452
rotational grazing
    canyons, 186
    contribution to ecological disturbance, 23, 25, 26
    example, 89
    grassland restoration, 88–90
    near springs, 188

restoring overgrazed pasture, 28
types, 88
roughleaf dogwood, 141, 448, 452
Ruby-crowned Kinglet
    backyards, 285
    habitat problem summary, 213, 332
    live oak savannah, 22
    moist mixed wooded slopes, 14
    need for vibrant understory, 285
    post oak savannah, 18
    preference for low areas, 51
    shin oak savannah, 20
Ruby-throated Hummingbird
    backyards, 285
    canyons, springs, and seeps, 177
    live oak savannah, 22
    moist mixed wooded slopes, 14
    post oak savannah, 18
    rivers and creeks, 129
Rufous-crowned Sparrow, *49, 69*
    canyons, springs, and seeps, 177
    erosion effect on, 43
    grasslands, 81, 92, 100–101
    habitat problem summary, 52, 111
    live oak savannah, 22
    mowing impact on, 107–8
    need for diverse wooded slope and savannah habitat, 31
    plateau prairie, 76
    redberry juniper usage, 428
    shin oak savannah, 20
    summary, 69
    wooded slopes and savannahs, 9
Rufous Hummingbird, 285
"rule of capture" law for groundwater pumping, 209
running or dripping water, attractiveness to birds, 303–5
runoff, water, 150, 156
rushes, 456
rusty blackhaw
    benefits to birds, 406
    deer browse evaluation, 430
    live oak savannah, 22
    moist mixed wooded slopes, 14
    priority woody plant, 448
    scientific name, 452

Sage Thrasher, 428
salmonellosis, 306
salt cedar, 452
Sandhill Crane, 233
sandpiper. *See* Spotted Sandpiper

savannahs, *384, 396*
   bird summaries, 53-72
   cedar management, 37-43, 419-20
   deer management, 33-37
   deer preferred habitat, 395
   description, 9, 16-22
   dryness/lack of water, 48-51
   ecological disturbance role, 23, 82
   erosion and bare ground, 26-30, *31*, 43-48
   grasses and forbs, 16-22, 18, *19*, 23-26
   livestock impact on, 26, 92-93
   near riparian habitat, 140-41
   overgrazed habitat, 26-27, 28, *31*, 43-46
   plant diversity, 23-26, 31-37
   prescribed burns on, 18, 25-26
   problems, 23-51
   See also live oak savannah; post oak savannah; shin oak savannah
Savannah Sparrow, *104, 124*
   deer population control, 98
   grasslands, 73, 81, 92
   habitat problem summary, 111
   seeps as winter oases, 181
   summary, 124
Scissor-tailed Flycatcher, *95, 125*
   grasslands, 73, 81, 92
   habitat problem summary, 111
   importance of insects, 108
   live oak savannah, 22
   post oak savannah, 18
   predators of, 378-79
   rivers and creeks, 129
   summary, 125
Scott's Oriole, *51, 70*
   habitat problem summary, 52
   live oak savannah, 22
   preference for higher areas, 51
   shin oak savannah, 20
   summary, 70
   wooded slopes and savannahs, 9
screwworm fly, as deer population control element, 396
SDG (short duration grazing) method, 88
seasonal occurrences of bird species
   backyards, 285
   canyons, springs, and seeps, 177
   dry mixed wooded slopes, 16
   grasslands, 81
   live oak savannah, 22
   moist mixed wooded slopes, 14
   post oak savannah, 18

rivers and creeks, 129
   shin oak savannah, 20
   tanks, ponds, and lakes, 233
Sedge Wren, *189, 225*
   canyons, springs, and seeps, 175, 177
   dependence on wetland habitat, 205
   grasslands, 91
   habitat problem summary, 213, 214, 265
   rivers and creeks, 129
   seeps as winter oases, 181
   summary, 225
   tanks, ponds, and lakes, 233
seeds for feeders, selecting proper, 299
seep muhly, 16, 180, 456
seeps, *184, 207, 211*
   bench seeps, 180
   cedar invasion, 192
   description, 178-81
   feral hog damage to, 183-84
   fern-covered seeps, 180
   hillside seeps, 180
   human use damage, 206-8
   livestock control to prevent overgrazing, 188-90
   loss of natural flow and water quality issues, 208
   overbrowsing at, 203-4
   sources for tank building, 240
   See also springs and seeps
sensitive briar, 17, 454
shallow flats, constructed tanks, 233-35
Sharp-shinned Hawk, 378
shears for cedar control, 424
sheep, 92, 93, 199
sheeting, rainwater, 87, 94
shin oak, 16, 20, 431, 449, 450
shin oak savannah, *19, 30, 33, 39, 49, 50*
   all year bird species, 20
   cedar management 420
   characteristics, 9
   description, 18-20
   prescribed burns on, 25-26
   shallow soil, 11
   spring and fall bird species, 20
   spring and summer bird species, 20
   winter bird species, 20
shorelines. *See* wetland vegetation
short duration grazing (SDG) method, 88
short grasses vs. bunchgrasses, as erosion control, 94
shrike. *See* Loggerhead Shrike
sick or dying birds, proper care of, 307-8

sideoats grama, 16, 20, 22, 456
silktassel, 22
silting in of springs, 188, 239-40
skunk, 380
skunkbush sumac, 254, 452
slash in grasslands, 86-87
slim tridens, 20, 456
slippery elm, 430, 449, 450
slope, characteristics of, 10
small-mammal habitat
   fire as tool to improve, 98
   live oak savannah, 27
snags, 39-43, *42*, 110, *144*, 311
snares, avoiding for catching feral hogs, 365
snoutbean, 22, 454
snowbell, 14
soapberry, 449, 450
soil
   alluvial soil in river valleys, 77
   impact of moisture changes, 49
   live oak savannah, 20
   patience needed in restoring, 48
   post oak savannah, 17
   restoration of rangeland, 86-87
   shin oak savannah, 11, 18
   on wooded slopes, 9
   See also erosion and bare ground
Song Sparrow, *184, 226*
   canyons, springs, and seeps, 175, 177
   grasslands, 81
   habitat problem summary, 111, 213
   summary, 226
Sora, *251, 280*
   canyons, springs, and seeps, 177
   dependence on spring habitat, 205
   habitat problem summary, 159, 214, 265
   need for wetland vegetation for shelter, 253
   rivers and creeks, 129
   safety factor in tank building, 249-50
   summary, 280
   tanks, ponds, and lakes, 230, 233
southern maidenhair fern, 180, 456
southern water nymph (southern naiad), 256, 456
southwest bernardia, 430, 452
southwestern bristlegrass, 22, 456
Spanish oak, 12, 430, 449, 450
sparrows. *See individual species by name*
spicebush, 448, 452
spiderwot, 454

spikerushes (spike sedges), 256, 456
spot-grazing and damage to seeps, 188
Spotted Sandpiper, 238, 281
    all year bird species, 233
    pond structure needs, 251
    shallow flats, 234
    summary, 281
    tank building for, 237
    tanks, ponds, and lakes, 230, 233
Spotted Towhee, 287, 353
    backyards, 284, 285
    canyons, springs, and seeps, 177
    habitat problem summary, 52, 213, 332
    live oak savannah, 22
    loss of understory, 93
    moist mixed wooded slopes, 14
    overbrowsing problem, 32–33
    post oak savannah, 18
    redberry juniper usage, 428
    riparian habitat, 130, 138
    shin oak savannah, 20
    summary, 353
Sprague's Pipit, 92
spring and/or fall (migrants) bird species
    backyards, 285
    canyons, springs, and seeps, 177
    grasslands, 81
    live oak savannah, 22
    moist mixed wooded slopes, 14
    post oak savannah, 18
    rivers and creeks, 129
    shin oak savannah, 20
    tanks, ponds, and lakes, 233
spring and summer (breeding birds) bird species
    backyards, 285
    canyons, springs, and seeps, 177
    grasslands, 81
    live oak savannah, 22
    moist mixed wooded slopes, 14
    rivers and creeks, 129
    shin oak savannah, 20
    tanks, ponds, and lakes, 233
springs and seeps, 175–229, 179, 191, 197, 205
    all year bird species, 177
    bird summaries, 215–29
    birds' use of, 180
    cedar in, 190–95, 422–23
    description, 176–78
    exotic plant invasion at, 197–99
    favorite bird species, 175

feral hog damage to, 182–83
groundwater table decline's effect on, 208–9
human use damage, 205
indiscriminate cedar clearing, 195
introduction, 175
livestock control, 186–88
loss of natural flow and water quality issues, 208
outlaw bird species, 175
overbrowsing, 202–3, 204
priority bird species, 175
problems, 181–214
under the radar bird species, 175
restoring from human modification, 206
seeps description, 178–81
silting in of springs, 188, 239–40
sources for tank building, 239–40
spring and fall bird species, 177
spring and summer bird species, 177
springs description, 176–78
summary, 213–14
winter bird species, 177
See also seeps
starlings. See European Starling
stewardship. See land stewardship for birds in Hill Country
stoneworts, 256, 457
straggler daisy, 17, 454
stream habitat
    access management, 141–42
    ecosystem of, 132
    tank building hydraulic changes to, 238–39
suet as bird food, 299–300
sugar hackberry, 451
Summer Tanager, 128, 170
    backyards, 285
    canyons, springs, and seeps, 177
    dry mixed wooded slopes, 16
    habitat problem summary, 52, 213
    indiscriminate cedar clearing, 193–94
    moist mixed wooded slopes, 12, 14
    parasitism rate, 358
    plants beneficial to, 406
    predators of, 378
    rivers and creeks, 127, 129
    summary, 170
    wooded slopes and savannahs, 23, 37
sunflowers, 454
supplemental feeding of deer, avoiding, 402–3

surveying property, land stewardship guidelines, 4
Swainson's Hawk
    fire as tool in small-mammal habitat, 98
    grasslands, 81
    habitat problem summary, 52, 111
    natural disturbance in grassland and savannah, 23
Swainson's Thrush
    canyons, springs, and seeps, 177
    dry mixed wooded slopes, 16
    live oak savannah, 22
    moist mixed wooded slopes, 14
    rivers and creeks, 129
    shin oak savannah, 20
swallow. See Barn Swallow
swift. See Chimney Swift
switchgrass, 14, 142, 456
sycamore-leaf snowbell, 14, 448, 452

tall dropseed, 22, 456
tall goldenrod, 454
tall grama, 22, 456
tank building, 238
    alternatives to, 239–50
    avoiding disturbing spring and seeps, 206, 210
    construction of, 244
    dams, 248–49, 423
    effects on riparian system, 238–39, 256–62
    evaporation and water loss, 210, 239
    fish habitat, 248
    floating cover for bird access, 210
    guzzlers as alternative to, 239, 241
    inability to build a tank, 262–63
    limitations, 237
    location consideration, 242
    management, 235–36
    permits, 240
    protecting riparian habitat vegetation, 142
    repairing leaks, 261–62
    reshaping shoreline, 247–48, 250–53
    safety for birds, 249–50
    seasonal consideration, 245
    shallow flats, 233–35
    size consideration, 240–42
    stream habitat and hydraulics changes, 238–39
    structure, 244, 246, 250
    water-level control, 248–49

tank building (cont.)
    watershed condition, 243–44
    See also constructed tanks, ponds, and lakes
tanks, defined, 230–31
tasajillo, 452
Taylor, Richard, 360
teals. *See* Blue-winged Teal; Green-winged Teal
Tennessee Warbler, 14, 18, 22, 177
Texas almond, 16, 452
Texas ash, 451
Texas bush-clover, 20, 454
Texas cupgrass, 12, 456
Texas Hill Country, book's span of coverage, 7, **8**
Texas kidneywood, 22, 448, 452
Texas lantana, 12, 452
Texas madrone, 16, 430, 449, 451
Texas milkweed, 12, 454
Texas mountain laurel, 16, 452
Texas mulberry, 406, 430, 448, 451, 452
Texas Parks and Wildlife 24-Hour Communication Center, 308
Texas Parks and Wildlife Department (TPWD), 2, 402
Texas persimmon
    deer browse evaluation, 431
    live oak savannah, 22
    moist and dry mixed wooded slopes, 12
    post oak savannah, 17
    scientific name, 452
    shin oak savannah, 20
Texas pricklypear, 452
Texas redbud
    deer browse evaluation, 431
    live oak savannah, 22
    moist and dry mixed wooded slopes, 12
    priority woody plant, 449
    scientific name, 451
    shin oak savannah, 20
Texas sotol, 16, 452
Texas star, 17, 455
Texas thistle, 455
Texas walnut, 431, 451
Texas Wildscapes Certification program, 284
Texas wintergrass, 14, 22, 456
thin-leaf brookweed, 180, 455
thistle seed as bird food, 299
thorn shrub, 15

threeawns, 20, 22, 93, 456
thrushes. *See* Hermit Thrush; Louisiana Waterthrush; Swainson's Thrush
tickle-tongue, 16, 431, 452
Tordon 22K, redberry juniper, 429
towhees. *See* Canyon Towhee; Spotted Towhee
TPWD (Texas Parks and Wildlife Department), 2, 402
track identification, 385–86
trail building, 142–43, 416
trail cameras, 386
Trap/Neuter/Release (TNR) programs for cats, 367
tree canopy
    backyards, 284–85
    deer impact on, 401–2
    flooding's role in nourishing, 135
    riparian habitat, 127, 139, *147*, 152
trichomoniasis, 307
Trinity-Edwards Aquifer, 209
tropical sage, 142, 455
trumpet creeper, 141, 448, 453
turk's cap, 455
Turk's cap, 142
twist-leaf yucca, 12, 20, 452
Two-Step Method for RIFA control, 370–71

understory
    backyards, 285, 293–96
    browsing damage to, 31–37, 93, 374–75, 401
    canyons, *176*, 200, 201–2
    cedar's threat to, 46
    flood benefits, 135
    riparian habitat, 131, 135, 138–41
    wooded slopes and savannah, 28, 31–37
under the radar bird species
    American Pipit, 73, 112
    American Wigeon, 127, 160
    Ash-throated Flycatcher, 9, 54
    backyards, 284
    Black-and-White Warbler, 175, 216
    Black-bellied Whistling-Duck, 230, 268
    Blue-Gray Gnatcatcher, 9, 56
    Blue Grosbeak, 284, 339
    Bufflehead, 127, 162
    canyons, springs, and seeps, 175
    Crested Caracara, 389
    Domestic Cat, 390
    Eastern Screech-Owl, 284, 343

    Gadwall, 230, 272
    Golden-crowned Kinglet, 175, 221
    grasslands, 73
    Green Heron, 127, 165
    Hermit Thrush, 175, 222
    House Finch, 284, 347
    Hutton's Vireo, 9, 62
    Lincoln's Sparrow, 284, 349
    Nashville Warbler, 9, 65
    Northern Parula, 127, 168
    Red-breasted Merganser, 230, 277
    rivers and creeks, 127
    Savannah Sparrow, 73, 124
    Scott's Oriole, 9, 70
    Song Sparrow, 175, 226
    Spotted Towhee, 284, 353
    tanks, ponds, and lakes, 230
    Vesper Sparrow, 73, 126
    White-crowned Sparrow, 9, 72
    White-eyed Vireo, 175, 227
    Wilson's Snipe, 175, 228
    wooded slopes and savannahs, 9
    Yellow-rumped Warbler, 127, 173
    Yellow Warbler, 127, 172
Upland Sandpiper, 81
urban trap and removal, deer management, 403–4
USGS National Wildlife Health Center, 307

Velpar chemical control for cedar, 426–27
Vermilion Flycatcher, *231*, 282
    all year bird species, 233
    backyards, 285
    canyons, springs, and seeps, 177
    grasslands, 81
    habitat problem summary, 265, 332
    live oak savannah, 22
    mesquite, 96
    post oak savannah, 18
    rivers and creeks, 129
    shin oak savannah, 20
    summary, 282
    tanks, ponds, and lakes, 230, 233
Vesper Sparrow, *86*, *126*
    benefits of restored rangeland, 85
    grasslands, 73, 81, 92
    habitat problem summary, 111
    post oak savannah, 18
    shin oak savannah, 20
    summary, 126
vinca, 453
vine mesquite, 22, 456

vireos. See *individual species by name*
Virginia creeper
    benefits to birds, 406
    deer browse evaluation, 431
    healthy riparian habitat, 141
    moist mixed wooded slopes, 14
    priority woody plant, 448
    scientific name, 453

walnuts, 14, 451
warblers. See *individual species by name*
wasps and ants in nest boxes, avoiding, 318
water and water ways
    backyard availability to birds, 301–5
    cedar management, 413–15
    cedar's use of, 413–14
    controlling levels in ponds, 235
    evaporation loss in tanks, 239
    feral hogs' damage to water quality, 183
    livestock damage to spring water quality, 187, 256, 258
    making sources welcoming to birds, 210
    pocket prairie, 78
    quality for constructed tanks, 233, 249–50
    river and creek quantity and quality problems, 154–58
    sheeting by rainwater, 87, 94
    water flow loss and poor water quality, 208–12
    wooded slopes and savannahs, 48–51
    See also canyons; constructed tanks, ponds, and lakes; riparian zone or habitat
watercress, 457
water lettuce, 457
water milfoil, 256, 457
water pennywort, 455
water primrose, 256, 457
watershed management, 156–58, 192, 241–44, 259
weather, Purple Martin vulnerability to, 325–26
Web Soil Survey, 242
western ironweed, 142, 455
Western Kingbird, 18
Western Meadowlark, 76, 80, 85, 92
western ragweed, 455
Western Screech-Owl, 315
Western Scrub-Jay, *41, 71, 420*

cedar usage by, 37, 412
dry mixed wooded slopes, 16
habitat problem summary, 52, 213, 214, 332
indiscriminate cedar clearing, 195
live oak savannah, 22
moist mixed wooded slopes, 14
overbrowsing in canyons, 200–201
plants beneficial to, 406
post oak savannah, 18
redberry juniper usage, 428
summary, 71
threat to other birds, 384–85
wooded slopes and savannahs, 9
western soapberry, 12, 141, 431, 452
wet and dry cycle, 212, 235
wetland vegetation
    birds' dependence on, 205, 233 253
    reshaping shoreline, 247–48, 250–53
    restoration methods, 256
    shorelines for constructed tanks, 232–33, 253
wet meadow. *See* pocket prairie
white boneset, 142, 455
whitebrush, 16, 431
whitebrush, common beebush, 452
White-crowned Sparrow, *33, 72*
    backyards, 285
    canyons, springs, and seeps, 177
    dry mixed wooded slopes, 16
    grasslands, 81
    habitat problem summary, 52, 111, 332
    live oak savannah, 22
    overbrowsing problem, 32–33
    overgrazing problem, 28
    post oak savannah, 18
    shin oak savannah, 20
    summary, 72
    wooded slopes and savannahs, 9
White-eyed Vireo, *207, 227*
    backyards, 285
    canyons, springs, and seeps, 175, 177
    dependence on seeps, 206
    habitat problem summary, 214, 332
    moist mixed wooded slopes, 14
    need for vibrant understory, 285
    parasitism rate, 358
    riparian bird, 129
    summary, 227
white heath aster, 12, 142, 455
white honeysuckle, 22, 430, 448, 453
white millet as bird food, 299
white-tailed deer, 34, 199, 393, 395. *See*

also deer and exotic animal browsing management
White-throated Sparrow, 18
White-winged Dove, *330, 354*
    backyards, 285
    habitat problem summary, 333
    pest birds in backyards, 329
    summary, 354
wildlife management, and Texas WTV, 1–2
wildlife rehabilitator, contacting, 331
Wildlife Tax Valuation (WTV), 1–2
wild millet (barnyard grass, coast cockspur grass), 256, 456
wild onion, 455
wild petunia, 12, 17, 455
wildrye, 22, 142, 456
wild turkey, 84, 401–2. *See also* Rio Grande Turkey
Wilson's Phalarope, 233
Wilson's Snipe, *205, 228*
    canyons, springs, and seeps, 175, 177
    competition from exotic waterfowl, 153
    dependence on wetland habitat, 205, 253
    habitat problem summary, 213, 214, 265
    importance of riparian understory, 138
    safety factor in tank building, 249–50
    shallow flats, 234
    summary, 228
    tanks, ponds, and lakes, 233
Wilson's Warbler
    canyons, springs, and seeps, 177
    dry mixed wooded slopes, 16
    live oak savannah, 22
    moist mixed wooded slopes, 14
    post oak savannah, 18
    rivers and creeks, 129
    shin oak savannah, 20
    wooded slopes and savannahs, 37
window threat to birds, backyards, 329–31
winter bird species
    backyards, 285
    canyons, springs, and seeps, 177
    grasslands, 81
    live oak savannah, 22
    moist mixed wooded slopes, 14
    post oak savannah, 18
    rivers and creeks, 129
    shin oak savannah, 20
    tanks, ponds, and lakes, 233
witch hazel, 448, 452

Wood Duck, *144, 171*
    dependence on riparian understory, 138
    habitat problem summary, 159
    native to Hill Country, 315
    nest boxes, 43, 314, 317
    rivers and creeks, 127, 129
    summary, 171
wooded slopes and savannahs, 9–72
    bird summaries, 53–72
    cedar imbalances, 37–43
    dryness/lack of water, 48–51
    favorite bird species, 9
    introduction, 9–11
    loss of plant diversity from lack of natural disturbances, 23–26
    management goals, 23
    outlaw bird species, 9
    overbrowsed habitat, 31–37
    overgrazed habitat, 26–30
    priority bird species, 9
    problems, 23–51
    under the radar bird species, 9
    rapid loss of topsoil, 43–48
    summary, 52
    See also mixed wooded slopes; savannahs
woodlands
    browser destruction, 200
    deer overpopulation signs, 397
    riparian habitat, 139–40
woodpeckers. See *individual species by name*

wood-sorrel, 14, 455
woody plants
    assuring regeneration, 31–37
    brush management, 95–96
    overbrowsed habitat problem, 31
    plant majority shin oak savannah, 18
    rarity at seeps, 180
    seeds or fruits, 256
wrens. See *individual species by name*

yarrow, 12, 455
Yellow-bellied Sapsucker, 406
Yellow-billed Cuckoo, *146, 174*
    canyons, springs, and seeps, 177
    habitat problem summary, 159
    live oak savannah, 22
    plants beneficial to, 406
    rivers and creeks, 127, 129
    shin oak savannah, 20
    summary, 174
Yellow-breasted Chat, 22, 37
Yellow-crowned Night-Heron, *252, 283*
    backyards, 285
    canyons, springs, and seeps, 177
    habitat problem summary, 265
    need for wetland vegetation for shelter, 253
    pond structure needs, 251
    rivers and creeks, 129
    summary, 283
    tank building for, 237, 249–50
    tanks, ponds, and lakes, 230, 233
Yellow-headed Blackbird, 81

yellow indiangrass, 17, 22, 142, 456
Yellow-rumped Warbler, *146, 173*
    backyards, 285
    canyons, springs, and seeps, 177
    habitat problem summary, 159, 265
    moist mixed wooded slopes, 14
    need for vibrant understory, 285
    riparian habitat, 146
    rivers and creeks, 127, 129
    summary, 173
    tanks, ponds, and lakes, 233
Yellow-throated Vireo, *202, 229*
    Balcones Canyonlands, 132
    canyons, springs, and seeps, 175, 177
    habitat problem summary, 214
    overbrowsing in canyons, 200
    rivers and creeks, 129
    summary, 229
Yellow Warbler, *134, 172*
    Balcones Canyonlands, 132
    canyons, springs, and seeps, 177
    habitat problem summary, 159
    live oak savannah, 22
    moist mixed wooded slopes, 14
    riparian bird, 131
    rivers and creeks, 127, 129
    summary, 172
yucca, 431, 452

zexmenia (wedelia), 20, 22, 455